Advances in

ORGANOMETALLIC CHEMISTRY

VOLUME 23

CONTRIBUTORS TO THIS VOLUME

Thomas A. Blinka

Neil G. Connelly

Philip E. Garrou

William E. Geiger

Bradley J. Helmer

Colin P. Horwitz

Duward F. Shriver

Robert West

Nils Wiberg

Advances in Organometallic Chemistry

EDITED BY

F. G. A. STONE

DEPARTMENT OF INORGANIC CHEMISTRY
THE UNIVERSITY
BRISTOL, ENGLAND

ROBERT WEST

DEPARTMENT OF CHEMISTRY
UNIVERSITY OF WISCONSIN
MADISON, WISCONSIN

VOLUME 23

1984

ACADEMIC PRESS, INC.

(Harcourt Brace Jovanovich, Publishers)
Orlando   San Diego   New York   London
Toronto   Montreal   Sydney   Tokyo

ACADEMIC PRESS, INC.
Orlando, Florida 32887

United Kingdom Edition published by
ACADEMIC PRESS, INC. (LONDON) LTD.
24/28 Oval Road, London NW1 7DX

LIBRARY OF CONGRESS CATALOG CARD NUMBER: 64-16030

ISBN 0-12-031123-2

PRINTED IN THE UNITED STATES OF AMERICA

84 85 86 87   9 8 7 6 5 4 3 2 1

Contents

The Electron-Transfer Reactions of Mononuclear Organotransition Metal Complexes

NEIL G. CONNELLY and WILLIAM E. GEIGER

Redistribution Reactions of Transition Metal Organometallic Complexes

PHILIP E. GARROU

Silyl, Germyl, and Stannyl Derivatives of Azenes, N_nH_n: Part I. Derivatives of Diazene, N_2H_2

NILS WIBERG

Polarization Transfer NMR Spectroscopy for Silicon-29: The INEPT and DEPT Techniques

THOMAS A. BLINKA, BRADLEY J. HELMER, and ROBERT WEST

C- and O-Bonded Metal Carbonyls: Formation, Structures, and Reactions

COLIN P. HORWITZ and DUWARD F. SHRIVER

Contributors

Numbers in parentheses indicate the pages on which the authors' contributions begin.

THOMAS A. BLINKA (193), *Department of Chemistry, University of Wisconsin, Madison, Wisconsin 53706*

NEIL G. CONNELLY (1), *Department of Inorganic Chemistry, University of Bristol, Bristol BS8 1TS, England*

PHILIP E. GARROU (95), *Dow Chemical USA, Central Research—New England Laboratory, Wayland, Massachusetts 01778*

WILLIAM E. GEIGER (1), *Department of Chemistry, University of Vermont, Burlington, Vermont 05405*

BRADLEY J. HELMER (193), *Department of Chemistry, University of Wisconsin, Madison, Wisconsin 53706*

COLIN P. HORWITZ (219), *Department of Chemistry, Northwestern University, Evanston, Illinois 60201*

DUWARD F. SHRIVER (219), *Department of Chemistry, Northwestern University, Evanston, Illinois 60201*

ROBERT WEST (193), *Department of Chemistry, University of Wisconsin, Madison, Wisconsin 53706*

NILS WIBERG (131), *Institut für Anorganische Chemie, University of Munich, 8000 Munich 2, Federal Republic of Germany*

ADVANCES IN ORGANOMETALLIC CHEMISTRY, VOL. 23

The Electron-Transfer Reactions of Mononuclear Organotransition Metal Complexes

NEIL G. CONNELLY

Department of Inorganic Chemistry
University of Bristol
Bristol, England

and

WILLIAM E. GEIGER

Department of Chemistry
University of Vermont
Burlington, Vermont

I

INTRODUCTION

The electrochemistry of large numbers of organic species is well established, and the occurrence of multiple oxidation states is commonplace for transition metal compounds. By comparison, the redox reactions of organotransition metal complexes are relatively unexplored.

In the 1960s, R. E. Dessy and co-workers (ref. *1* and references therein) demonstrated that many different classes of organometallic complex were electroactive, though these studies focused largely on reductive processes. Since then, several reviews have appeared largely cataloging electrochemical reactions and rarely reporting on their application (*2–5*).

Recent developments, however, have suggested a more significant role for organometallic electrochemistry. X-Ray structural studies on redox-related couples have provided detailed descriptions of the bonding in polynuclear species (ref. *6* and references therein), and the intermediacy of one-electron transfer in fundamental processes such as oxidative addition has been demonstrated (*7, 8*). Redox-induced isomerization (*9–11*), redox-catalyzed substitution (ref. *12*; ref. *13* and references therein), and photoelectrochemical reactions (*14, 15*) can provide routes to otherwise inaccessible species, and it hardly need be stated that diamagnetic precursors can provide radical products of obvious synthetic utility. Regrettably, we are unable to include a discussion of the electrosynthesis of organometallic compounds from simple substrates (*16, 17*), but recent results (*18–20*) suggest this method as a rival to those based on metal atom vapors. The electrogeneration of homogeneous catalysts is particularly attractive (*21–24*).

It is our belief that a full and detailed understanding of the electron-transfer properties of organometallic complexes can be achieved only by a combination of chemical and electrochemical studies; the use of one alone can lead to erroneous conclusions. Because we have insufficient space to provide a discussion of the theory and practice of elementary electrochemical techniques we refer the reader to several excellent treatments which also include an explanation of commonly used terminology (*25–30*). The synthetic chemist should not be deterred from routinely using techniques such as cyclic voltametry (CV),[1] voltametry at rotating metal disk electrodes, or controlled potential electrolysis (CPE), coulometry, and chronoamperometry. The proper employment of such techniques, for which instrumentation is readily available, should prove sufficient for all but the most detailed studies.

---

[1] Abbreviations: acac, acetylacetonate; bd, butadiene; bipy, bipyridyl; chd, cyclohexadiene; chpt, cycloheptatriene; cod, cyclooctadiene; cot, cyclooctatetraene; Cp, η^5-cyclopentadienyl; CPE, controlled-potential electrolysis; CV, cyclic voltametry; Cy, cyclohexyl; das, *o*-phenylene-

We also note that the use of chemical reagents commonly regarded as one-electron oxidants or reductants is not without complication. For example, species such as Ag^+, NO^+, and arenediazonium ions are known as "noninnocent" oxidants because of their unpredictable tendency to act as electrophilic ligands. When coupled with electrochemical methods, however, synthetic studies become less haphazard.

This article is confined largely to the more recent studies of the electron-transfer properties of mononuclear organotransition metal complexes; the earlier work is covered in more detail elsewhere (1–5). Simple binuclear species such as $[Mn_2(CO)_{10}]$ and $[\{Cr(CO)_3Cp\}_2]$ are included due to their close relationship with the anions $[Mn(CO)_5]^-$ and $[Cr(CO)_3Cp]^-$, but other bi- and polynuclear compounds will be described at a later date (31).

The metal carbonyls and their derivatives are discussed in Section II, followed by the hydrocarbon–metal complexes, classified by the bonding capacity of the organic ligand, in Sections III–IX. When two such hydrocarbons are present the compound is included in the section dealing with the ligand of higher hapticity. Sandwich complexes, such as the metallocenes and $[M(\eta\text{-arene})_2]$, are described in Section X as a special, related group of molecules.

Sections II–X are not exhaustive. For example, they contain no reference to metallocarboranes or to macrocyclic species containing metal—carbon bonds, such as the vitamin B_{12} models (32). However, we have attempted to include material which relates electrochemistry to other fields. For example, we note the formation of electron-transfer products by ^{60}Co γ irradiation, negative ion mass spectrometry, and photoelectron spectroscopy. We also describe pairs of complexes apparently related by one-electron transfer but which have not been studied electrochemically. Clearly, such studies would provide confirmatory evidence for the proposed formulations.

Where appropriate, redox potentials have been included either in the text or in tables. Unless stated otherwise those potentials are relative to the saturated calomel electrode (sce).

---

bis(dimethylarsine); dmbd, 2,3-dimethylbutadiene; dme, 1,2-dimethoxyethane; dmf, N,N'-dimethylformamide; dmop, 2,6-dimethoxyphenyl; dmpe, $Me_2PCH_2CH_2PMe_2$; dpae, Ph_2-$AsCH_2CH_2AsPh_2$; dppe, $Ph_2PCH_2CH_2PPh_2$; dppm, $Ph_2PCH_2PPh_2$; dq, η^4-duroquinone; ECE, electrochemical–chemical–electrochemical; Fc, ferrocenyl; imid, $\overline{CN(Me)CH}{=}CHNMe$; mes, mesityl; mnt, $S_2C_2(CN)_2$; nbd, norbornadienes; O–O, o-quinone; o-phen, o-phenanthroline; pfdt, $S_2C_2(CF_3)_2$; pq, phenanthrenequinone; pyr, pyridine; pz, pyrazolyl; sce, saturated calomel electrode; tcbdt, tetrachlorobenzenedithiolate; tcnb, 1,2,4,5-tetracyanobenzene; tcne, tetracyanoethylene; tdt, toluene-3,4-dithiolate; thf, tetrahydrofuran; tnb, 1,3,5-trinitrobenzene.

II

CARBONYL COMPLEXES AND THEIR DERIVATIVES

A. *General Studies*

Many of the electrochemical studies of metal carbonyl derivatives, and indeed of other species, have sought to correlate redox potentials with other physical or spectroscopic properties. These studies will be described briefly where appropriate but the reader may wish to refer to a review (*33*) of such correlations in coordination chemistry. A more theoretical basis has also been provided for the often noted relationship between E^0 and infrared carbonyl, nitrosyl, or isocyanide stretching frequencies (*34*).

Recently (*35, 36*), attempts have been made to predict the potential of oxidation, $E_{1/2}$, of a complex M_sL on the basis of a ligand parameter, ρ_L [Eq. (1)]. E_s is the potential when the metal site M_s is occupied by CO, and is

$$E_{1/2}[M_sL] = E_s + \beta\rho_L \tag{1}$$

said to reflect the "electron-richness" of that site; β is a measure of the site polarizability. The ligand parameter, ρ_L, is defined by Eq. (2) and may be determined for a wide range of ligands, L, due to the abundance of complexes

$$\rho_L = E_{1/2}[Cr(CO)_5L] - E_{1/2}[Cr(CO)_6] \tag{2}$$

[$Cr(CO)_5L$]. Values of ρ vary from 1.46 V for nitride, a strong π acceptor, to -1.55 V for hydroxide, a good donor.

The use of Eq. (1) is exemplified by the chemistry of $[MoL(NO)(dppe)_2]^Z$ where M_s is $Mo(NO)(dppe)_2$ (*36*). For L = Cl^-, SCN^- ($Z = 0$), and NCR (R = Me or Ph, $Z = 1$), a plot of $E_{1/2}$ versus ρ_L is linear, but $[Mo(CO)(NO)(dppe)_2]^+$ is irreversibly oxidized at a potential ~ 0.3 V more positive than expected. ^{31}P NMR spectroscopy subsequently revealed the *cis*-nitrosylcarbonyl geometry, contrasting with the trans arrangement for the halide and nitrile complexes. Interestingly, the product of the oxidation of *cis*-$[Mo(CO)(NO)(dppe)_2]^+$ is a dication which is reversibly reduced at the potential predicted for the trans isomer. The redox-induced isomerization mimics that of isoelectronic *cis*-$[Mo(CO)_2(dppe)_2]$ (Section II,C,6). Other examples of the predictive power of Eq. (1) are to be found in Sections II,C,4 and II,D.

B. *Vanadium, Niobium, and Tantalum*

In CH_2Cl_2, $[V(CO)_6]$ and $[V(CO)_6]^-$ are interconvertible by reversible one-electron transfer ($E^0 = 0.23$ V, versus Ag/AgCl). The CV of $[V(CO)_6]^-$

in acetone also reveals reversible formation of $[V(CO)_6]$, but the latter is stable for only a few seconds. The neutral carbonyl cannot be studied directly; it rapidly disproportionates to $[V(CO)_6]^-$ and V(II) salts (37). The electrochemistry of the Lewis base derivatives of $[V(CO)_6]^Z$ ($Z = 0$ or -1) has not been studied, but redox-induced isomerizations of the kind described in Sections II,C,6 and II,D are to be expected.

The seven-coordinate complexes $[MX(CO)_2(dmpe)_2]$ (M = Nb, X = Cl; M = Ta, X = H, Cl, Me, etc.) (38, 39) are reversibly oxidized in acetone. Neither the radical cations nor the products of a second, irreversible electron-transfer step have been further characterized.

C. Chromium, Molybdenum, and Tungsten

Many derivatives of the Group VI metal carbonyls, $[M(CO)_6]$ (M = Cr, Mo, or W) undergo reversible one-electron oxidation or reduction, and stable, sometimes isolable, metal- or ligand-based radicals have been fully characterized. The redox potentials of representative complexes are listed in Table I.

1. The Hexacarbonyls

Despite the very positive potential for the reversible oxidation of $[Cr(CO)_6]$ (40–43) [e.g., $E^0 = 1.09$ V in CF_3CO_2H (42)], the radical cation can be generated by CPE and characterized by ESR spectroscopy (42). A second oxidation, like the first for $[M(CO)_6]$ (M = Mo or W) in MeCN (40), is totally irreversible (41, 43).

The reduction of $[M(CO)_6]$ (M = Cr, Mo, or W) is more complex (40, 44, 45). Irreversible one-electron transfer is probably followed by carbonyl loss from $[M(CO)_6]^-$ and dimerization of $[M(CO)_5]^-$ to $[M_2(CO)_{10}]^{2-}$ (40, 44).

2. Acyls, Carbenes, and Carbynes

At $-78°C$, the one-electron oxidation of the acyl complexes $[M(COR)(CO)_5]^-$ (M = Cr or W, R = alkyl or aryl) is quasi-reversible; the lifetime of $[Cr(COPh)(CO)_5]$ is ~ 20 msec. The ESR spectrum of the chromium complex is similar to that of the benzaldehyde anion, shows no metal hyperfine coupling, and suggests the unpaired electron to be largely based on the acyl ligand.

Electrolytic oxidation of $[NEt_4][Cr(COPh)(CO)_5]$ gives the hydroxycarbene complex $[Cr\{C(OH)Ph\}(CO)_5]$ via hydrogen atom transfer from $[NEt_4]^+$ to $[Cr(COPh)(CO)_5]$. The reduction of coordinated acyls to

TABLE I

THE REDOX PROPERTIES OF $[M(CO)_6]$ (M = Cr, Mo, OR W) DERIVATIVES

Complex	Process	E^0 (V)	Reference[a] electrode	Solvent	Base electrolyte	Reference
$[W(CO)_5\{C(ONMe_4)Fc\}]$	$0 \rightleftharpoons 1$	-0.04		CH_2Cl_2	$[NEt_4][ClO_4]$	50
$[Cr(CO)_5\{C(OEt)Ph\}]$	$0 \rightleftharpoons 1$	0.90		CH_2Cl_2	$[NEt_4][ClO_4]$	50
$[CrCl(CO)_5]^-$	$-1 \rightleftharpoons 0$	0.63	Ag/AgCl	Acetone	$[NBu_4][ClO_4]$	116
	$0 \rightleftharpoons 1$[b]	1.13				
$[MoCl(CO)_5]^-$	$-1 \rightleftharpoons 0$	0.83	Ag/AgCl	Acetone	$[NBu_4][ClO_4]$	116
	$0 \rightleftharpoons 1$[b]	1.32				
$[Cr(CN)(CO)_5]^-$	$-1 \rightleftharpoons 0$	0.50	Ag/AgCl	CH_2Cl_2	$[NEt_4][ClO_4]$	50
$[Cr(CO)_5(NH_3)]$	$0 \rightleftharpoons 1$	0.71	Ag/AgCl	CH_2Cl_2	$[NEt_4][ClO_4]$	50
$[Cr(CO)_5\{CN(t\text{-}Bu)\}]$	$0 \rightleftharpoons 1$	1.10	Ag/AgCl	CH_2Cl_2	$[NEt_4][ClO_4]$	50
$[Cr(CO)_5(PMe_3)]$	$0 \rightleftharpoons 1$	1.18	Ag/AgCl	CH_2Cl_2	$[NEt_4][ClO_4]$	50
$[Cr(CO)_5\{P(OPh)_3\}]$	$0 \rightleftharpoons 1$	1.32	Ag/AgCl	CH_2Cl_2	$[NEt_4][ClO_4]$	50
$[Cr(CO)_4(PMe_3)_2]$	$0 \rightleftharpoons 1$	0.52	Ag/Ag⁺	MeCN	$[NBu_4][BF_4]$	41
$[Cr(CO)_4(dppe)]$	$0 \rightleftharpoons 1$	0.75	Ag/Ag⁺	CH_2Cl_2	$[NBu_4][BF_4]$	94
$[W(CO)_4(dppe)]$	$0 \rightleftharpoons 1$	0.95	Ag/Ag⁺	CH_2Cl_2	$[NBu_4][BF_4]$	94
$[Cr(CO)_4(dmpe)]$	$0 \rightleftharpoons 1$	0.46	Ag/Ag⁺	CH_2Cl_2	$[NEt_4][ClO_4]$	50
	$1 \rightleftharpoons 2$	1.08				
$[Cr(CO)_4\{CN(p\text{-}tolyl)\}]$	$0 \rightleftharpoons 1$	0.67	Ag/Ag⁺	CH_2Cl_2	$[NEt_4][ClO_4]$	50
$[Mo(CO)_4\{CN(t\text{-}Bu)\}]$	$0 \rightleftharpoons 1$	0.80	Ag/Ag⁺	CH_2Cl_2	$[NEt_4][ClO_4]$	50
$[Cr(CO)_4(NH_2CH_2CH_2NH_2)]$	$0 \rightleftharpoons 1$	0.14	Ag/Ag⁺	CH_2Cl_2	$[NEt_4][ClO_4]$	50
$[Cr(CO)_4(bipy)]$	$0 \rightleftharpoons 1$	0.61	Ag/Ag⁺	CH_2Cl_2	$[NEt_4][ClO_4]$	50
$[Mo(CO)_4(bipy)]$	$0 \rightleftharpoons -1$	-2.2	Ag/Ag⁺	Glyme	$[NBu_4][ClO_4]$	84
	$-1 \rightleftharpoons -2$	-2.8				
$[Cr(CO)_4\{PhN=C(Me)C(Me)=NPh\}]$	$0 \rightleftharpoons -1$	-1.92	Ag/Ag⁺	Glyme	$[NBu_4][ClO_4]$	´84
	$-1 \rightleftharpoons -2$	-2.28				
$[Cr(CO)_4\{(t\text{-}Bu)N=CHCH=N(t\text{-}Bu)\}]$	$0 \rightleftharpoons 1$	0.10	Ag/Ag⁺	MeCN	$[NEt_4][ClO_4]$	90
	$0 \rightleftharpoons -1$	-1.81				

Complex	Process	E	Reference electrode	Solvent	Electrolyte	Ref.
fac-[Cr(CO)$_3$(PMe$_3$)$_3$]	$0 \rightleftharpoons 1$	0.23	Ag/Ag$^+$	MeCN	[NBu$_4$][BF$_4$]	41
[Cr(CO)$_3${CN(p-ClC$_6$H$_4$)}$_3$]	$0 \rightleftharpoons 1$	0.51	Ag/Ag$^+$	CH$_2$Cl$_2$	[NEt$_4$][ClO$_4$]	58
fac-[Mo(CO)$_3$(PPh$_3$)(imid)$_2$][c]	$0 \rightleftharpoons 1$	0.01	Sce	CH$_2$Cl$_2$	[NBu$_4$][ClO$_4$]	52
mer,trans-[Mo(CO)$_3$(PPh$_3$)(imid)$_2$][c]	$0 \rightleftharpoons 1$	-0.24	Sce	CH$_2$Cl$_2$	[NBu$_4$][ClO$_4$]	52
mer,cis-[Mo(CO)$_3$(PPh$_3$)(imid)$_2$][c]	$0 \rightleftharpoons 1$	-0.25	Sce	CH$_2$Cl$_2$	[NBu$_4$][ClO$_4$]	52
[Cr(CO)$_3$L][d]	$0 \rightleftharpoons 1$	-0.38[e]	Sce	CH$_2$Cl$_2$	[NBu$_4$][ClO$_4$]	121
cis-[Mo(CO)$_2$(bipy)$_2$]	$0 \rightleftharpoons 1$	-0.54	Sce	MeCN		88
	$0 \rightleftharpoons -1$	1.64				
trans-[Cr(CO)$_2$(dppm)$_2$]	$0 \rightleftharpoons 1$	-0.61	Ag/AgCl	Acetone	[NEt$_4$][ClO$_4$]	101
	$1 \rightleftharpoons 2$	0.84				
cis-[Cr(CO)$_2$(dppm)$_2$]	$0 \rightleftharpoons 1$	0.01	Ag/AgCl	Acetone	[NEt$_4$][ClO$_4$]	101
trans-[Mo(CO)$_2$(dppm)$_2$]	$0 \rightleftharpoons 1$	-0.24	Ag/AgCl	Acetone	[NEt$_4$][ClO$_4$]	101
	$1 \rightleftharpoons 2$	0.95				
cis-[Mo(CO)$_2$(dppm)$_2$]	$0 \rightleftharpoons 1$	0.3	Ag/AgCl	Acetone	[NEt$_4$][ClO$_4$]	101
[Cr(CNPh)$_6$]	$0 \rightleftharpoons 1$	-0.32	Sce	CH$_2$Cl$_2$	[NBu$_4$][ClO$_4$]	61
	$1 \rightleftharpoons 2$	0.27				61
	$2 \rightleftharpoons 3$	0.98				61
[Mo(CNPh)$_6$]	$0 \rightleftharpoons 1$	-0.19	Sce	CH$_2$Cl$_2$	[NBu$_4$][PF$_6$]	66
	$1 \rightleftharpoons 2$[b]	0.48				
[Cr{CN(t-Bu)}$_6$]$^{2+}$	$2 \rightleftharpoons 3$	0.84	Sce	CH$_2$Cl$_2$	[NBu$_4$][PF$_6$]	65
	$2 \rightleftharpoons 1$	-0.28				
	$1 \rightleftharpoons 0$	-1.04				

[a] A calomel electrode, 1 M in LiCl, unless stated otherwise.
[b] Irreversible process at room temperature.
[c] imid = $\overline{\text{CN(Me)CH}}$=CHNMe.
[d] L = o-C$_6$H$_4${PhP(CH$_2$)$_3$}$_2$NMe   or   cis-2,10-diphenyl-6-methyl-6-aza-2,10-diphosphabicyclo-[9.4.0]pentadeca-1(11),12,14-triene.
[e] At -78°C.

7

hydroxycarbenes by such a mechanism is suggested as an alternative to hydridoacylmetal intermediates in Fischer–Tropsch catalysis (46).

The carbene complexes $[M\{C(OMe)R\}(CO)_5]$ (M = Cr, Mo, or W; R = aryl) are reduced to monoanions ($E_{1/2}$ = −1.25 to −1.60 V in MeCN) (47). At −50°C, chemical reduction gives the stable radical $[Cr\{C(OMe)Ph\}(CO)_5]^-$, the ESR spectrum of which shows small metal hyperfine splittings but relatively large couplings to the methyl group and to five inequivalent phenyl protons.

In the neutral complex the phenyl group lies perpendicular to the Cr–C(carbene)–O plane (48) but reduction causes a geometric change such that the p orbital on the carbene carbon atom conjugates with the phenyl ring. That is, the Ph–Cr–C(carbene)–O group tends toward coplanarity. Molecular orbital (MO) calculations on the neutral complex show that the highest occupied molecular orbitals (HOMOs) for the perpendicular and planar conformations are approximately equal in energy; the lowest unoccupied molecular orbital (LUMO) of the latter is stabilized by ∼2.5 eV (49).

Electrochemical studies on $[W\{C(OMe)aryl\}(CO)_5]$ also support the proposed structural changes. Ortho substitution of the aryl group, which causes crowding in the planar configuration, leads to more negative reduction potentials (e.g., aryl = mesityl, $E_{1/2}$ = −1.60 V). By contrast, complex **1**,

1

in which the aryl ring is constrained to lie in the W–C(carbene)–O plane, is reduced at a relatively positive potential ($E_{1/2}$ = −1.31 V) (48).

The one-electron oxidation potentials of the carbene complexes $[M(CXY)(CO)_5]$ (M = Cr or W; X = OR, SR, NRR′, etc.; Y = 2-furyl, Fc, alkyl, aryl, etc.) are highly dependent on X and Y (Table I). MO calculations suggest electron loss from an orbital almost entirely of metal and CO character (50).

The P-donor derivatives fac-$[Cr(CO)_3LL'_2]$ [L = C(OEt){CH(SiMe_3)_2} or $\overline{CNEtCH_2CH_2NEt}$ (L^{Et}), L′ = P(OPh)$_3$, L′$_2$ = dmpe or dppe] are oxidized by AgBF$_4$ to isolable radical cations such as $[Cr(CO)_3L(dmpe)][BF_4]$; photolytic substitution gives, for example, $[Cr(CO)_2L(PPh_3)(dmpe)]^+$ and $[Cr(CO)(L^{Et})(dmpe)_2]^+$ (51).

The oxidative electrochemistry of *mer,trans-*; *mer,cis-*; and *fac-*[Mo-(CO)₃L(carbene)₂] [L = CNPh, PEt₃, PPh₃, or pyr; carbene = $\overline{\text{CN(Me)CH=CHNMe}}$] involves a complex series of isomerizations *(52)*. The interconversion of *cis-* and *trans-*[Mo(CO)₄(carbene)₂] is more straightforward but differs from that of [Mo(CO)₂(dppe)₂] (Section II,C,6). The *trans*-tetracarbonyl is oxidized at 0.11 V to the *cis* cation which is reduced at this potential to *cis*-[Mo(CO)₄(carbene)₂] ($E^0 = 0.34$ V). Thus, redox-induced trans–cis isomerization can occur with no net current flow *(53)*.

The reversible oxidation of the carbyne complex [WBr(CO)₄(CFc)] is associated with the ferrocenyl group *(54)*. The cation [Cr(CO)₅(CNEt₂)]⁺ undergoes reductive ligand coupling with Li[acac] or LiAsMe₂ to give [(CO)₅Cr{μ-C(R)C(R)}Cr(CO)₅] (R = NEt₂) *(55)*.

3. Cyanides and Isocyanides

Reactions between [M(CN)(CO)₅]⁻ (M = Cr, Mo, or W) and arene-diazonium ions yield the functionalized isocyanide complexes [M(CO)₅-(CNR)] (R = α-tetrahydrofuranyl, CHCl₂, or CCl₃) via the mechanism shown in Scheme 1 *(56)*. The initial electron transfer also occurs electro-

$$[M(CN)(CO)_5]^- + [N_2C_6H_4X\text{-}\underline{p}]^+ \longrightarrow [M(CN)(CO)_5] + [N_2C_6H_4X\text{-}\underline{p}]$$

$$[N_2C_6H_4X\text{-}\underline{p}] \longrightarrow N_2 + C_6H_4X\text{-}\underline{p}$$

$$C_6H_4X\text{-}\underline{p} + RH \longrightarrow C_6H_5X\text{-}\underline{p} + R\cdot$$

$$R\cdot + [M(CN)(CO)_5] \longrightarrow [M(CO)_5(CNR)]$$

$$RH = \text{thf, } CH_2Cl_2, \text{ or } CHCl_3$$

SCHEME 1

chemically in CH₂Cl₂ (M = Cr) *(36, 50)*, and [Cr(CN)(CO)₅] can be independently synthesized *(57)*.

The voltametric half-wave potentials for the oxidation of [M(CO)₆₋ₙ(CNR)ₙ] [M = Cr, Mo, or W; n = 1–3; R = alkyl or aryl] (Table I) are dependent on M, n, and R *(50)*, and correlate with λ_{max} for the metal-to-isocyanide charge-transfer band in the electronic spectrum *(58)*.

Equation (3) quantifies the relationship between $E_{1/2}$ and the HOMO energies of $[Cr(CO)_{6-n}(CNR)_n]$ and $[Mn(CO)_{6-n}(CNR)_n]^+$ (Section II,D) in terms of the empirically determined parameters A, B, and C; n is the

$$E_{1/2} = A + Bn + Cx \tag{3}$$

number of ligands, L, and x is the number of those ligands which interact with the $d\pi$ orbital comprising the HOMO of the complex. There is not only good agreement between predicted and experimentally determined values of $E_{1/2}$, but also an explanation for the observed differences between the oxidation potentials of isomers ($n = 2$ or 3). For *cis*- and *trans*-$[M(CO)_4L_2]$ very similar $E_{1/2}$ values are predicted (59), as demonstrated for $[Mo(CO)_4\{P(n\text{-}Bu)_3\}_2]$ (Section II,C,6) but not for the analogous carbenes (Section II,C,2).

The complexes $[Cr(CNR)_6]$ are reversibly oxidized to mono-, di-, and trications (60–63), with the Hammett σ_p constants (R = aryl) correlating with E^0 for all three redox couples (64). In acetone, $AgPF_6$ and $[Cr(CNR)_6]$ (R = aryl) readily afford the mono- and dications (61), and alkyl analogs (R = t-Bu or Cy) can be prepared directly from CNR and chromium(II) salts in ethanol (65). Phosphine substitution of $[Cr(CNR)_6]^{2+}$ gives $[Cr(CNR)_5(dppe)]^{2+}$, *cis*-$[Cr(CNR)_4(dppe)]^{2+}$, or *trans*-$[Cr(CNR)_4(PR'_3)_2]^{2+}$ (R = R' = alkyl), all of which are reversibly oxidized to trications and reduced to monocationic and neutral species. The seven-coordinate dications $[Cr(CNR)_7]^{2+}$ dissociate to octahedral species on one-electron oxidation (65).

The redox chemistry of molybdenum and tungsten isocyanides is complicated by the increased stability of seven-coordinate species. CV shows that the reversible one-electron oxidation of $[M(CNPh)_6]$ (M = Mo or W) is followed by a second irreversible step, and $[M(CNPh)_7]^{2+}$ is formed either by bulk electrolysis or with $AgNO_3$ and excess CNR in acetone. Iodine and $[M(CNPh)_6]$ give $[MI(CNPh)_6]^+$ (66), and $[M(CNR)_7]^{2+}$, $[MX(CNR)_6]^+$ (M = Mo or W; X = Cl, I, or CN; R = alkyl), and $[MoI(CNR)_6]^+$ (R = aryl) are readily prepared by oxidizing $[M(CO)_3(CNR)_3]$ with I_2 or $[IClPh]Cl$ in the presence of CNR (67). The seven-coordinate complexes described above, $[W\{CN(t\text{-}Bu)\}_6(PR_3)]^{2+}$ (R = alkyl), and $[W\{CN(t\text{-}Bu)\}_5(dppe)]^{2+}$ (68), undergo quasi-reversible one-electron oxidation (66, 69), but on electrolysis $[Mo(CNR)_6L]^{2+}$ (L = CNR or PR'_3) gives (70) $[MoCl(CNR)_6]^{2+}$ rather than $[Mo(CNR)_6]^{3+}$ (68). CV of *trans*-$[M(CNR)_2(dppe)_2]$ (M = Mo or W, R = t-Bu or p-tolyl) shows the reversible formation of mono- and dications which may be prepared chemically with I_2 in benzene and Ag^+ in thf, respectively (71).

The nitrosyls $[M(CNR)_5(NO)]^+$ [M = Cr, R = Me, t-Bu, or p-ClC_6H_4 (60); M = Mo or W, R = Ph (66)] undergo one (M = Mo or W) or two (M = Cr) reversible one-electron oxidations, and $[Cr(CNR)_5(NO)]^{2+}$ may be prepared by $AgPF_6$ oxidation (R = p-ClC_6H_4) (60) or directly from $[Cr(CNR)_6]^{2+}$ and NO (R = alkyl) (72). Halide ions and $[Cr(CNR)_5$-$(NO)]^{2+}$ afford $[CrX(CNR)_4(NO)]^+$ (X = Cl or Br, R = alkyl) but amines give the reduced substitution products $[Cr(amine)(CNR)_4(NO)]^+$ (72).

4. Dinitrogen and Related Derivatives

Although dinitrogen complexes such as $[M(N_2)_2(dppe)_2]$ are not strictly organometallic, they closely resemble the carbonyls described in Section II,C,6. This analogy as well as the discovery of the syntheses of organo-nitrogen compounds via electron-transfer reactions have led us to include a brief description of the most recent developments.

Ferric chloride and $trans$-$[W(N_2)_2(PMePh_2)_4]$ in ethanol give $trans$-$[W(N_2)_2(PMePh_2)_4][FeCl_4]$ (73). The related complexes $[M(N_2)L(dppe)_2]^Z$ [M = Mo or W, Z = 0, L = NCR, R = alkyl or aryl (74); Z = −1, L = SCN, CN, or N_3 (75)] undergo two reversible one-electron oxidations, and $[Mo(SCN)(N_2)(dppe)_2]$ is isolable by reacting the corresponding anion with air (75). Equation (1) has also been used in the chemistry of molybdenum and tungsten dinitrogen complexes. For example, a solution of $[Mo(N_2)_2$-$(dppe)_2]$ and NH_3 in thf shows an oxidation wave at −0.71 V, as predicted for $[Mo(N_2)(NH_3)(dppe)_2]$ (75).

The diazoalkane complexes $[MF(NN=CH-p-XC_6H_4)(dppe)_2]^+$ are both oxidized (E^0 = 0.67–0.97 V) and reduced (E^0 = −1.02 to −1.82 V), and CPE in CH_2Cl_2 affords the dicationic and neutral, metal- and ligand-based radicals, respectively (76). On electrolysis at mercury, $[WF(NN=CH_2)(dppe)_2]^+$ undergoes reductive coupling of the diazomethane group, giving $[\{WF(dppe)_2\}_2(\mu-N_2CH_2CH_2N_2)]$ (77).

Under argon, $trans$-$[MoBr\{N\overline{N(CH_2)_4C}H_2\}(dppe)_2]^+$ **2** is irreversibly reduced in a two-electron step ($E_{1/2}$ = −1.62 V, in thf) and electrolysis yields $trans$-$[Mo\{N\overline{N(CH_2)_4C}H_2\}(dppe)_2]$ (**3**); subsequent reaction with HBr gives $[MoBr(NH)(dppe)_2]Br$ and piperidine (78). Under CO or N_2, the reduction of $[MoBr(NNR_2)(dppe)_2]^+$ requires four electrons and yields

$[Mo(CO)_2(dppe)_2]$ or *trans*-$[Mo(N_2)_2(dppe)_2]$ (**4**) and *N*-aminopiperidine. As **2** is readily made from **4**, the electrocatalytic synthesis of hydrazines from N_2 may be possible (*79*).

A range of interesting organonitrogen ligands is produced via the reaction between $[WBr(NNH_2)(dppe)_2]^+$ (**5**), $[IPh_2]^+$, and base. The initial steps of the reaction are shown in Scheme 2, but the final product depends on the

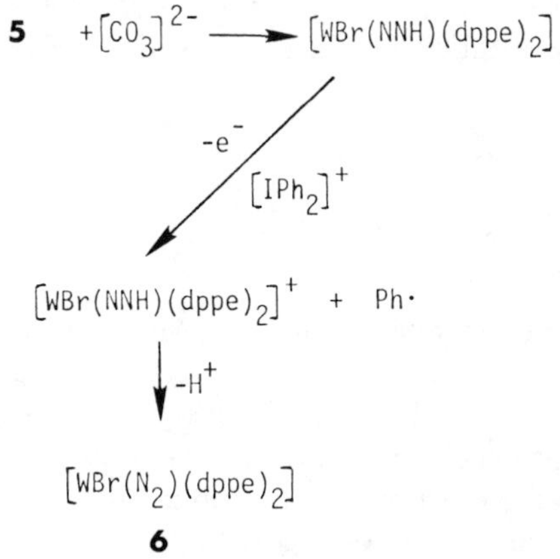

5 $+ [CO_3]^{2-}$ ⟶ $[WBr(NNH)(dppe)_2]$

$-e^-$

$[IPh_2]^+$

$[WBr(NNH)(dppe)_2]^+$ +   Ph·

$\downarrow -H^+$

$[WBr(N_2)(dppe)_2]$

6

SCHEME 2

solvent (*80–82*). For example, in CH_2Cl_2, chlorine atom abstraction by the phenyl radical is followed by $CHCl_2$ attack on the dinitrogen ligand of **6**; chloride loss and subsequent hydrolysis of $[WBr(NNCHCl)(dppe)_2]^+$ give $[WBr(NNCHO)(dppe)_2]$ (**7**) (*81*).

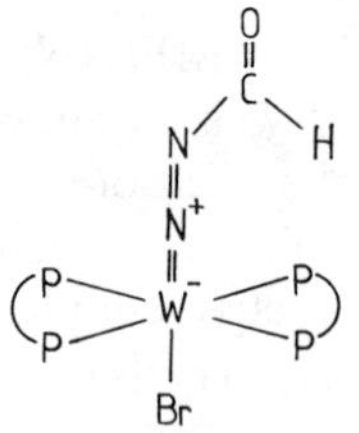

7; P–P = dppe

5. Amine, Pyridine, and Bipyridyl Complexes

The one-electron oxidation of $[M(CO)_5L]$ (M = Cr, L = NH_3, amine, or pyr; M = W, L = NEt_3) occurs at rather positive potentials (Table I)

(50), but trans-$[Mo(CO)_2(pyr)_4]^+$, prepared from $[Mo(pyr)_3(\eta^6\text{-toluene})]^+$ and CO, is reduced with difficulty ($E^0 = -1.03$ V in pyr) (83).

The bipy complexes $[M(CO)_4(bipy)]$ (M = Cr, Mo, or W) are reduced in two reversible one-electron steps (84, 85). The anion radicals $[M(CO_{4-n}(PPh_3)_n(bipy)]^-$ (n = 0, M = Cr, Mo, or W; n = 1 or 2, M = Mo) can be generated by CPE or alkali metal reduction (86, 87), and have ESR spectra very similar to that of $[bipy]^-$ but also showing metal and ^{31}P hyperfine coupling (85).

As expected, cis-$[Mo(CO)_2(bipy)_2]$ is reversibly reduced, and oxidized more readily than $[M(CO)_4(bipy)]$ (Table I); a second irreversible oxidation is also observed in MeCN ($E_{1/2} = 0.3$ V). The radical anion can be prepared by sodium amalgam or naphthalide reduction, and $AgBF_4$ or $AlCl_3$ gives $[\{Mo(CO)_2(bipy)_2\}_2][BF_4]_2$. In warm acetone the latter dissociates to trans-$[Mo(CO)_2(bipy)_2]^+$ which is further oxidized by Ag^+ in MeCN to yield $[Mo(CO)_2(NCMe)(bipy)_2]^{2+}$ (88).

The diazadiene (DAD) compounds $[M(CO)_4(DAD)]$ (M = Cr, Mo, or W; DAD = RN=CR'CR'=NR) also undergo two reversible reductions. Analysis of the IR carbonyl spectra of $[M(CO)_4\{PhN=C(Me)C(Me)=NPh\}]^Z$ (Z = 0, -1, and -2) suggests that charge transmission to and through the metal atom is via the σ framework of the molecule (89). The increased π-acceptor ability of DAD, relative to bipy, is demonstrated by the more positive reduction potentials of $[M(CO)_4(RN=CR'CR'=NR)]$ (R = alkyl or aryl, R' = H or Me) and by the increased ligand hyperfine couplings in the ESR spectra of $[M(CO)_4(DAD)]^-$ (85, 90). The complexes $[M(CO)_4(DAD)]$ are also oxidized, with the chromium cations being the most stable (90).

The binuclear complexes $[(CO)_5M(\mu\text{-L})M(CO)_5]$ (L = pyrazine) and $[(CO)_3M(\mu\text{-L})_3M(CO)_3]$ (L = pyridazine, M = Cr, Mo, or W) undergo one and three one-electron reductions, respectively, apparently associated with the bridging ligands L. Sodium reduction resulted in decomposition (91), but $[(CO)_5M(\mu\text{-L})M(CO)_5]^-$ (M = Mo or W, L = 4,4'-bipyridyl or pyrazine) can be prepared directly from $[M(CO)_6]$ and the anion radical L^-. The ESR spectra show the anionic complexes to be stabilized by hyperconjugation between the ligand LUMO (π^*) and the metal carbonyl σ^* orbitals (92).

6. P-, As-, and Sb-Donor Ligands

The $E_{1/2}$ values for the oxidation of $[M(CO)_{6-n}L_n]$ (L = P or As donor) are heavily dependent on n (Table I). In general, the cations $[M(CO)_{6-n}L_n]^+$

have been isolated only for $n = 4$, but $[Cr(CO)_4(dppe)][PF_6]$ (*93*) and $[Cr(CO)_4(dppm)]^+$ mixed with $[Cr(CO)_3(NO)(dppm)]^+$ (*94*) can be prepared using strong oxidants.

A study of the reactions between $[M(CO)_4(L-L)]$ (M = Cr, Mo, or W; L–L = dppm or dppe) and Ag^+ or $[NO]^+$ salts led (*94*) to a modification of the mechanism previously proposed (*95*) to account for the observation of both oxidation and substitution of metal carbonyls by nitrosonium salts (Section VIII). Scheme 3 shows that the intermediate $[ML_n(NO)]^+$ either

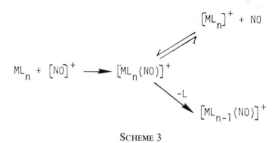

$$ML_n + [NO]^+ \longrightarrow [ML_n(NO)]^+$$

$$[ML_n]^+ + NO$$

$$\xrightarrow{-L} [ML_{n-1}(NO)]^+$$

SCHEME 3

reversibly dissociates to NO and the radical $[ML_n]^+$, or irreversibly loses L to yield the diamagnetic nitrosyl $[ML_{n-1}(NO)]^+$. The addition of NO gas to the reaction can cause preferential nitrosylation. Indeed, $[Cr(CO)_3(NO)(dppe)]^+$ can be prepared only by the isolation of $[Cr(CO)_4(dppe)]^+$ and subsequent reaction with excess nitric oxide.

Silver(I) ions and $[Cr(CO)_4(L-L)]$ yield $[Cr(CO)_4(L-L)]^+$ but, in spite of the favorable $E_{1/2}$ values, the molybdenum and tungsten analogs are not formed. Instead, $[Ag\{M(CO)_4(L-L)\}_2]^+$, which may be considered models for the intermediate $[ML_n(NO)]^+$ in Scheme 3, are isolated (*94*).

In CH_2Cl_2, $[M(CO)_5L]$ and $[M(CO)_4L_2]$ (M = Mo or W, L = $PPh_{3-n}Fc_n, n = 1–3$) are sequentially oxidized at the ferrocenyl centers. For example, CV of $[W(CO)_5(PFc_3)]$ shows three reversible waves at 0.64, 0.82, and 0.90 V; a fourth, irreversible wave ($E_{1/2} = 1.92$ V) is associated with the $W(CO)_5$ group. The similarities between the $E_{1/2}$ values for analogous penta- and tetracarbonyls lend support to ferrocenium ion formation (*96, 97*).

Complexes of the chelating ligand **8** behave rather differently. For $[Cr(CO)_4L]$ (L = **8**) the first of the two reversible oxidations ($E^0 = 0.54$

$$\begin{array}{c} \text{PPh}_2 \\ \text{Fe} \quad \text{CH}_2\text{NMe}_2 \end{array}$$

8

and 0.96 V in CH_2Cl_2) is thought to occur at chromium and the second at iron. Bulk electrolysis at 0.54 V leads to a considerable increase in $\tilde{\nu}(CO)$, but further oxidation at 0.96 V renders little or no further change in the IR carbonyl spectrum, and the blue-green color of the product is reminiscent of the ferrocenium ion. The molybdenum and tungsten analogs are irreversibly oxidized at the Group VI metal center prior to ligand oxidation at iron *(98)*.

The bis(phosphine) ligand $[Cr(\eta^6\text{-}C_6H_5PPh_2)_2]$ also chelates the $M(CO)_4$ group (M = Cr, Mo, or W), and oxidation of the product gives monocations whose ESR spectra show the single electron to be largely localized on the bis(arene)chromium moiety. The IR carbonyl bands are, however, shifted to higher energy by 10–30 cm^{-1} *(99)*.

Cis- and *trans-*$[Mo(CO)_4\{P(n\text{-}Bu)_3\}_2]$ are reversibly oxidized with potentials essentially independent of structure *(100)*, as predicted by MO theory *(59)*. CV showed rapid cis–trans isomerization but the close proximity of the two oxidation waves rendered difficult a detailed interpretation of the kinetics and thermodynamics of the reactions involved.

The isomerizations of *cis-* and *trans-*$[M(CO)_2(L-L)_2]^Z$ [M = Cr, Mo, or W; L–L = dppm *(101)* or dppe *(102, 103)*, Z = 0 or 1] are more fully defined. At $-75°C$, the one-electron oxidation of *cis-*$[Cr(CO)_2(L-L)_2]$ is reversible, but at room temperature it is less so and CV shows a new couple at a more negative potential. The molybdenum and tungsten analogs behave similarly but the initial electron-transfer process is incompletely reversible even at the lower temperature.

Chemical oxidation of *cis-*$[M(CO)_2(L-L)_2]$ by air in acidic thf *(104, 105)*, by I_2 *(106)*, or by mercury, Ag^+, or $[NO]^+$ salts *(105)* gives a high yield of *trans-*$[M(CO)_2(L-L)_2]^+$; CV clearly shows that the reduction of *trans-*$[Cr(CO)_2(L-L)_2]^+$ [L–L = dppm *(105)* or dppe *(102)*] gives rise to the product wave in the voltamogram of *cis-*$[Cr(CO)_2(L-L)_2]$.

The interconversion of *cis-* and *trans-*$[M(CO)_2(L-L)_2]^Z$ (Z = 0 and 1) is qualitatively described by Scheme 4. The isolation of *trans-*$[M(CO)_2(L-L)_2]$ *(107)* and double potential step chronoamperometry on *cis-*$[M(CO)_2(L-L)_2]$

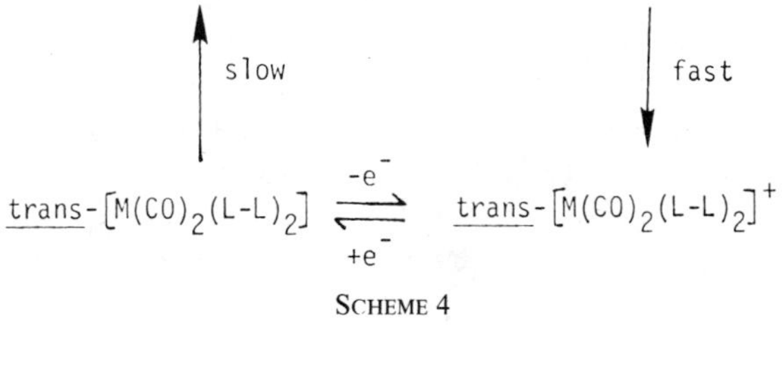

$$\text{cis-}[M(CO)_2(L-L)_2] \underset{+e^-}{\overset{-e^-}{\rightleftharpoons}} \text{cis-}[M(CO)_2(L-L)_2]^+$$

slow (↑)   fast (↓)

$$\text{trans-}[M(CO)_2(L-L)_2] \underset{+e^-}{\overset{-e^-}{\rightleftharpoons}} \text{trans-}[M(CO)_2(L-L)_2]^+$$

SCHEME 4

$[L-L = Ph_2P(CH_2)_nPPh_2, n = 1-3]$ (9) have led to the quantification of the Scheme. For example, cis-$[Cr(CO)_2(dppm)_2]^+$ isomerizes to the trans cation with a rate constant of 1.2 sec^{-1} (25°C, 0.1 M $[NEt_4][ClO_4]$ in acetone), and the conversion of trans-$[Cr(CO)_2(dppm)_2]$ to the cis isomer occurs with $k = 0.21$ sec^{-1}. Qualitatively, the cationic dppe complexes isomerize more rapidly than the dppm analogs whereas the order is reversed in the neutral state. The relative rate dependence on the metal (Cr > Mo > W) is the same for both $[M(CO)_2(L-L)_2]^Z$ ($Z = 0$ and 1). The negative entropies of activation and the independence of the rate on L–L and CO concentration and on the solvent suggest a nondissociative twist mechanism for the isomerization of both the neutral and cationic complexes (9). Less detailed studies had previously suggested otherwise (103).

The very different redox potentials for cis- and trans-$[M(CO)_2(L-L)_2]$ and the isomeric preference shown in each redox state have been rationalized by MO calculations on $[MoL_2(PH_3)_4]$. Strong π acceptors (e.g., L = CO) stabilize the neutral cis and the cationic trans complexes whereas the trans isomer is preferred in both cases when strong donors (e.g., L = O^{2-}) are present (108). Interestingly, cis-$[Mo(CO)(NO)(dppe)_2]^+$ is isomerized on oxidation (36) whereas $[ML(NO)(dppe)_2]^Z$ (M = Mo or W), with one strong π acceptor (NO) and one σ donor (L = halide, $Z = 0$; L = nitrile, $Z = 1$), retain the trans geometry on electron loss (36, 109).

Trans-$[Cr(CO)_2(dppm)_2]^+$ is also oxidized to trans-$[Cr(CO)_2(dppm)_2]^{2+}$ which isomerizes to the cis dication (101). The increased stability of cis-$[Cr(CO)_2(dppm)_2]^{2+}$ over the trans isomer is also revealed by the CV of cis-$[Cr(CO)_2(dppm)_2]$ at $-75°C$; two reversible waves show the formation of the dication directly from cis-$[Cr(CO)_2(dppm)_2]^+$.

The oxidation of $[M(CO)_2(L-L)_2]^+$ (M = Mo or W, L–L = dppm or dppe) is irreversible. Thus, attempts to prepare the dication, for example from $[NO][PF_6]$ and cis-$[Mo(CO)_2(dppe)_2]$ in CH_2Cl_2, gave the seven-coordinate Mo(II) complex $[MoF(CO)_2(dppe)_2][PF_6]$. In RCN (R = Me or Ph), $[NO][PF_6]$ affords $[Mo(CO)_2(NCR)(dppe)_2]^{2+}$ (110); cis-$[M(CO)_2(dppm)_2]$, $HClO_4$, and O_2 yield $[MH(CO)_2(dppm)_2]^+$ (M = Cr, Mo, or W) (105).

Although $AgBF_4$ and cis-$[M(CO)_2(dmpe)_2]$ (M = Cr or Mo) give trans-$[M(CO)_2(dmpe)_2][BF_4]$ as expected, silver(I) salts of coordinating anions react further to yield cis-$[MoX(CO)_2(dmpe)_2]X$ (X = NCS or NO_3) or zwitterionic $[MoX(CO)_2(dmpe)_2]$ (X = CO_3 or SO_4). Silver nitrite reacts similarly, but oxidation and anion coordination are followed by loss of CO_2 and formation of $[Mo(NO_2)(CO)(NO)(dmpe)_2]$ (111).

Reaction of tcne and cis-$[Cr(CO)_2(dmpe)_2]$ afford only trans-$[Cr(CO)_2(dmpe)_2][tcne]$, but the molybdenum analog reacts further to produce

cis-[Mo{$C_2(CN)_3$}$(CO)_2(dmpe)_2$][CN]. With 1,2,4,5-tetracyanobenzene (tcnb) and 1,3,5-trinitrobenzene (tnb), cis-[Mo$(CO)_2(dmpe)_2$] yields species formulated, respectively, as cis-[Mo{$C_6H_2(CN)_4$}$(CO)_2(dmpe)_2$][(tcnb)$_8$] and cis-[Mo{$C_6H_3(NO_2)_3$}$(CO)_2(dmpe)_2$][tnb] (112).

The most important reactions of cis-[M$(CO)_2(dmpe)_2$] are those with halocarbons (e.g., CCl_4, $PhCH_2Br$, CPh_3Cl, t-BuI, $CH_2{=}CHCH_2Br$), providing excellent evidence for the intermediacy of one-electron transfer in oxidative addition. The formation of cis-[MX$(CO)_2(dmpe)_2$]X (Scheme 5)

$$\underline{cis}-\left[Mo(CO)_2(dmpe)_2\right] + RX \longrightarrow \underline{trans}-\left[Mo(CO)_2(dmpe)_2\right]^+ + R\cdot + X^-$$

$$\underline{trans}-\left[Mo(CO)_2(dmpe)_2\right]X + RX \longrightarrow \underline{cis}-\left[MoX(CO)_2(dmpe)_2\right]X + R\cdot$$

SCHEME 5

depends on an outer-sphere electron transfer from cis-[M$(CO)_2(dmpe)_2$] to RX, and a second oxidative step involving inner-sphere atom transfer (M = Mo) (113).

7. O- and S-Donor Ligands

Photolysis of [M$(CO)_6$] with o-quinones (O–O) gives ligand-based anion radicals such as [M$(CO)_4$(O–O)]$^-$, [M = Mo, O–O = phenanthrene-quinone (pq); M = W, O–O = pq, 1,2-naphthoquinone, etc.] and [Cr$(CO)_2$-$(pq)_2$]$^-$, which are related to the bipy and DAD complexes in Section II,C,5. Carbonyl substitution affords, for example, [W$(CO)_3(AsPh_3(pq)$)]$^-$, [Mo$(CO)_2(PPh_3)_2(pq)$]$^-$, and [Cr$(CO)(PPh_3)(pq)_2$]$^-$, the ESR spectra of which reveal some delocalization over the metal (M = Cr) and the Group V donor ligands (114).

The first one-electron oxidation of [Mo$(CO)(S_2CNEt_2)_2L_2$] (L = PMe$_2$Ph or PMePh$_2$, L–L = dppe) is reversible (e.g., L–L = dppe, $E^0 = 0.24$ V in thf) but the second is only so for the dppe complex at fast cyclic voltametric rates ($E^0 = 1.03$ V). At $-10°C$, CPE gives [Mo$(CO)(S_2CNEt_2)_2(dppe)$]$^+$, characterized by the shift of $\tilde{v}(CO)$ on oxidation from 1780 to 1930 cm^{-1} (115).

8. *Pentacarbonyl Halide Derivatives*

The anions $[MX(CO)_5]^-$ [M = Cr (36, 50, 116), Mo, or W (116); X = Cl, Br, or I] are oxidized in two one-electron steps both of which are irreversible for molybdenum and tungsten. For chromium, the first electron transfer is reversible, as is the second at $-75°C$ (Table I); the radical $[CrI(CO)_5]$, generated using $KMnO_4$ in aqueous acetone at $-78°C$, shows the expected one-electron reduction wave in the cyclic voltamogram. The neutral chloride and bromide, prepared in solution by $[NO][PF_6]$ oxidation of $[CrX(CO)_5]^-$ (X = Cl or Br), are less stable (117) than the isolable iodide (118); all are thought to disproportionate to $[CrX(CO)_5]^-$ and $[CrX(CO)_5]^+$ which rapidly decompose (117).

Although electrochemistry provides no corroborative evidence, ^{60}Co γ irradiation of $[MI(CO)_5]^-$ (M = Cr, Mo, or W) in methyltetrahydrofuran gives radicals whose ESR spectra are consistent with the formulation $[MI(CO)_5]^{2-}$; the unpaired electron is assigned to a metal–halogen σ^* orbital (119).

Seven-coordinate $[MX_2(CO)_3L_2]$ (M = Mo or W, X = halide, L = PPh_3, $AsPh_3$, or $SbPh_3$) undergo irreversible two-electron reduction, and CPE under CO gives $[MX(CO)_5]^-$ (120).

D. *Manganese, Technetium, and Rhenium*

At 77 K, ^{60}Co γ irradiation of powdered $[Mn_2(CO)_{10}]$ yields a radical cation whose ESR spectrum shows hyperfine coupling to two magnetically inequivalent manganese atoms (122); fast scan CV shows $[Mn_2(CO)_{10}]^+$ to have some stability in CH_2Cl_2 (40). The binuclear cation is also implicated in the reaction between $[Mn_2(CO)_{10}]$ and tcne, which gives the radical $[Mn(CO)_5\{N{=}C{=}C(CN)C(CN)_2\}]$ (123), and in the formation of $[Mn_2(CO)_9\{N(O)R\}]^+$ by irradiating the decacarbonyl in the presence of nitrosodurene (NOR) as a spin trap (124). In MeCN (40), CF_3CO_2H (42), or benzene and molten $AlCl_3$-ethylpyridinium bromide (43), $[Mn_2(CO)_{10}]$ is irreversibly oxidized in a two-electron step giving, for example, $[Mn(CO)_5$-$(NCMe)]^+$.

Detailed studies (40, 84, 125–128) of the reduction of $[M_2(CO)_{10}]$ (M = Mn or Re) show an irreversible two-electron reduction at very negative potentials [e.g., M = Mn, $E_{1/2} = -1.65$ V in thf (40)]. However, ^{60}Co γ irradiation of $[M_2(CO)_8L_2]$ [M = Mn, L = CO (122) or PBu_3 (129); M = Re, L = CO (122)] generates $[M_2(CO)_8L_2]^-$ with a half-filled σ^* orbital of predominantly metal–metal d_{z^2} character.

Exhaustive electrolytic reduction of $[M_2(CO)_{10}]$ leads to the consumption of less than two electrons per molecule ($n \approx 1.8$) because of competing chemical reactions such as that in Eq. (4) (125). Reductive cleavage occurs at

$$[Mn_2(CO)_{10}] + [Mn(CO)_5]^- \longrightarrow [Mn_3(CO)_{14}]^- \qquad (4)$$

more negative potentials on carbonyl substitution, and the $E_{1/2}$ values of $[M_2(CO)_{10-n}(PPh_3)_n]$ (M = Mn or Re, M_2 = MnRe, n = 0 or 2) correlate with the energy of the electronic transition between the σ and σ^* metal–metal orbitals (126).

The anions $[M(CO)_5]^-$ undergo irreversible one-electron oxidation at relatively positive potentials (40) (e.g., M = Mn, $E_{1/2}$ = 0.04 V in MeCN), yielding $[M_2(CO)_{10}]$. The interconversion of the anion and the dimer is simply represented by Eq. (5) and yet the alternative shown in Scheme 6

$$[M_2(CO)_{10}] + 2e^- \rightleftharpoons 2[M(CO)_5]^- \qquad (5)$$

(M = Mn or Re, n = 5) is certainly possible for other binuclear carbonyls such as $[\{Cr(CO)_3Cp\}_2]$ (Section VII,C). It is noteworthy, then, that the radical $[Mn(CO)_5]$ (130) and its more stable P-donor derivatives are now well characterized.

$$[M_2(CO)_{2n}] \rightleftharpoons 2[M(CO)_n]$$

$$[M(CO)_n] \underset{-e^-}{\overset{+e^-}{\rightleftharpoons}} [M(CO)_n]^-$$

SCHEME 6

The anions $[M(CO)_{5-n}L_n]^-$ [M = Mn or Re, L = PPh_3, n = 1 or 2; L = P(OMe)_3, n = 3, etc.] are oxidized by $[C_7H_7]^+$ in methanol and thf to $[M(CO)_{5-n}L_n]$. The tetracarbonyl radicals dimerize but the more sterically hindered tricarbonyls yield mer,trans-$[MH(CO)_3L_2]$ by α-hydrogen abstraction from the solvent (131). The neutral compounds are also more labile to associative substitution than the anions, in the order $[Mn(CO)_5]$ > $[Mn(CO)_4L]$ > $[Mn(CO)_3L_2]$ (132). ESR spectroscopy defines the square-pyramidal geometry of species such as $[Mn(CO)_3L_2]$ (e.g., L = PBu_3); the basal ligands L are trans disposed and the half-filled orbital is parallel to the C_2-axis (133, 134).

MO calculations correctly predict that $[M(CO)_5]$ are oxidized (and reduced) more readily than $[M_2(CO)_{10}]$ (M = Mn or Re). Photolysis of the latter [or of $[\{W(CO)_3Cp\}_2]$, Section VII,C] in CH_2Cl_2 or MeCN with $[N(n\text{-}Bu)_4][ClO_4]$, a one-electron oxidant such as $[Fe(\eta\text{-}C_5Me_5)_2]^+$ or $[C_7H_7]^+$, and CCl_4 as a source of chlorine radicals gives both $[MCl(CO)_5]$ and $[M(CO)_5(OClO_3)]^+$ or $[M(CO)_5(NCMe)]^+$. The pentacarbonyl chloride results from a radical pathway but the cations are formed as in Scheme 7 *(14)*.

$$[M_2(CO)_{10}] \xrightarrow{\ h\nu \ } 2[M(CO)_5]$$

$$[M(CO)_5] \underset{+e^-}{\overset{-e^-}{\rightleftarrows}} [M(CO)_5]^+$$

$$[M(CO)_5]^+ + L \longrightarrow [M(CO)_5L]^+$$

$$L = [ClO_4]^- \text{ or NCMe}$$

SCHEME 7

The irreversible two-electron reduction of $[MX(CO)_5]$ [X = halide *(86, 127, 128)*, aryl *(127)*, SnR_3, or PbR_3, with R = Me, Et, or Ph *(135)*] yields $[M(CO)_5]^-$. However, ^{60}Co γ irradiation [M = Mn, X = Cl, Br, or I *(136)*; M = Re, X = Br or I *(119)*] gave the radical anions $[MX(CO)_5]^-$, whose ESR spectra revealed electron addition to a σ^* LUMO composed mainly of metal d_{z^2} and halide p_z orbitals. A second radical detected after ^{60}Co γ irradiation of $[MnX(CO)_5]$ may be high spin $[MnX(CO)_5]^+$ *(136)*. Interestingly, then, $[MnBr(CO)_5]$ is reversibly oxidized in CF_3CO_2H to an unstable monocation *(42)*.

The oxidative electrochemistry of the Group V-donor derivatives of $[MBr(CO)_5]$ is extensive, and *fac*-$[MX(CO)_3(L-L)]$ [M = Mn or Re, X = Cl or Br, L–L = dppm *(137)*, dppe, dpae, etc. *(138)*] undergo redox-induced isomerization according to Scheme 8. At $-35°C$, the oxidation of

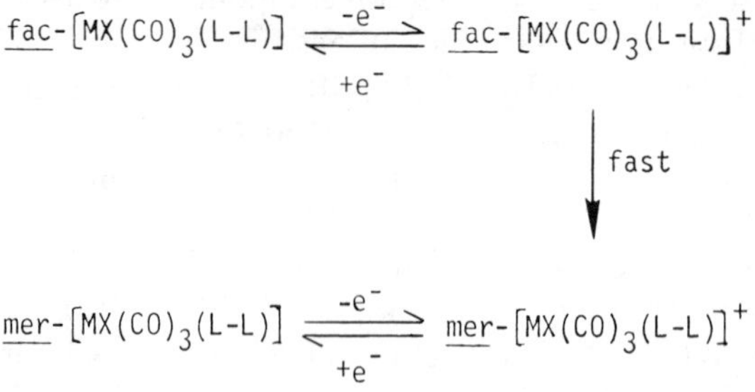

$$\underline{\text{fac}}\text{-}[MX(CO)_3(L-L)] \underset{+e^-}{\overset{-e^-}{\rightleftarrows}} \underline{\text{fac}}\text{-}[MX(CO)_3(L-L)]^+$$

$$\Big\downarrow \text{fast}$$

$$\underline{\text{mer}}\text{-}[MX(CO)_3(L-L)] \underset{+e^-}{\overset{-e^-}{\rightleftarrows}} \underline{\text{mer}}\text{-}[MX(CO)_3(L-L)]^+$$

SCHEME 8

the fac isomer is reversible (e.g., M = Mn, X = Cl, L–L = dppm, E^0 = 1.48 V vs Ag/AgCl) but in MeCN at room temperature the fac cation gives mer-$[MnX(CO)_3(L–L)]^+$, which is reduced by the solvent to otherwise inaccessible mer-$[MnX(CO)_3(L–L)]$ (e.g., X = Cl, L–L = dppm, E^0 = 0.95 V vs Ag/AgCl).

Complexes of monodentate donors, such as fac- and mer-$[MX(CO)_3L_2]$ [M = Mn or Re, X = Cl or Br, L = PPh_3, $P(OPh)_3$, $AsPh_3$, $SbPh_3$, etc.], behave similarly and the isomerizations can be effected chemically, for example by $[NO][PF_6]$ oxidation of fac-$[MnBr(CO)_3(SbPh_3)_2]$ (10). The products, which are particularly light sensitive for rhenium, have mer geometry rather than the fac structures originally assigned (139).

The thermodynamics and kinetics associated with Scheme 8 have been partially quantified. At 25°C, ΔH^* for the isomerization of fac-$[MnCl(CO)_3$-$(dppm)]^+$ to the mer cation is 56 (± 4) kJ mol^{-1}, and the rate of the reaction is 2.4 (± 0.2) sec^{-1} (138); the solvent independence of the rate implies a non-dissociative twist mechanism. For the neutral species, mer-$[MnX(CO)_3$-$(L–L)]$ does not convert to the fac isomer, and the complexes of mono-dentate ligands isomerize only over extended periods (10), probably by a dissociative process.

The mer,cis-dicarbonyls $[MnBr(CO)_2L_3]$ (L = phosphine or phosphite) are oxidized by $[NO][PF_6]$ to either mer,cis- (L = PMe_3) or mer,trans-[L = $P(OMe)_3$, PMe_2Ph, etc.] cations; the latter [L = $P(OR)_3$, R = Me or Et] are reduced at 0°C by hydrazine to mer,trans-$[MnBr(CO)_2L_3]$, which rapidly revert to the more thermodynamically stable mer,cis isomers (139). Similarly, NO_2 or $[NO][PF_6]$ oxidation of cis,cis-$[MnBr(CO)_2L(L–L)]$ (L = phosphine or phosphite, L–L = dppm or dppe) followed by hydrazine reduction of the unstable trans dications provides a route to trans-$[MnBr$-$(CO)_2L(L–L)]$ (140).

Although $[ReCl(CO)_3(PPh_3)_2[BF_4]$ has been isolated from $[ReCl(CO)_3$-$(PPh_3)_2]$ and $[OEt_3][BF_4]$ in CH_2Cl (141), fac- and mer-$[ReCl(CO)_3$-$(PMe_2Ph)_2]$ are oxidized in a two-electron step in MeCN (142). Similarly, $[TcCl(CO)_2(PMe_2Ph)_3]$ shows two ill-defined oxidation waves, and exhaustive electrolysis gave $[TcCl(CO)(NCMe)_2(PMe_2Ph)_3][ClO_4]_2$ (143).

The first one-electron reduction of trigonal-bipyramidal $[ReCl(CO)_2$-$(PPh_3)_2]$ is irreversible and thought to be associated with the formation of a square-pyramidal (sp) anion. A second, reversible step yields $[ReCl(CO)_2$-$(PPh_3)_2]^{2-}$, which retains the sp structure (144).

The dinitrogen complexes $[ReCl(N_2)(CO)_{4-n}L_n]$ (n = 2–4, L = P donor) and $[Re(N_2)(NCR)(dppe)_2]$ (R = Me or Ph) are reversibly oxidized in thf. The E^0 values range from 1.01 V (n = 2, L = PPh_3) to 0.28 V for $[ReCl(N_2)$-$(dppe)_2]$ and linearly correlate with $\tilde{v}(N_2)$; species with $E^0 > 0.8$ V are described as "electron poor" (*144*). Related isocyanide derivatives [ReCl-(CNR)(dppe)$_2$] (R = alkyl or aryl) are reversibly oxidized in two one-electron steps (e.g., R = Me, $E_1^0 = 0.08$ V, $E_2^0 = 0.99$ V). The linear correlation between E_1^0 and the Hammett σ_{p^+} parameter suggests the HOMO of the neutral complex to be conjugated to the aryl substituent. In addition, a consideration of the E^0 values in terms of the ligand parameter ρ_L (Section II,A) leads to the conclusion that the CNR ligands are bent (*145*).

The oxidation potential of [Mn(CO)$_6$]$^+$ is undoubtedly too positive to detect electrochemically, but carbonyl substitution can lead to the isolation of dications. The one-electron oxidation of *trans*-[Mn(CO)$_2$(dppe)$_2$]$^+$ is fully reversible ($E^0 = 1.08$ V vs Ag/AgCl in MeCN) (*102*) and the trans dication is readily isolable using oxidants such as bromine, KMnO$_4$, or SbCl$_5$ (*146*).

TABLE II

THE ONE-ELECTRON OXIDATION OF ISOCYANIDE DERIVATIVES OF
[M(CO)$_6$]$^+$ (M = Mn OR Re)

Complex	E^0 (V)^a	Reference
[Mn(CO)$_5$(CNMe)]$^+$	2.65^b	*151*
cis-[Mn(CO)$_4$(CNMe)$_2$]$^+$	2.14	*151*
cis-[Mn(CO)$_4$(CNPh)$_2$]$^+$	2.28	*63*
[Mn(CO)$_3$(CNMe)$_3$]$^+$	1.65^b	*151*
fac-[Mn(CO)$_3$(CNPh)$_3$]$^+$	2.12	*151*
mer-[Mn(CO)$_3$(CNPh)$_3$]$^+$	1.98	*151*
fac-[Re(CO)$_3$(CNMe)$_3$]$^+$	2.21^c	*149*
cis-[Mn(CO)$_2$(CNPh)$_4$]$^+$	1.70	*63*
trans-[Mn(CO)$_2$(CNPh)$_4$]$^+$	1.54	*63*
cis-[Re(CO)$_2$(CNMe)$_4$]$^+$	1.70 (2.6)	*149*
[Mn(CO)(CNMe)$_5$]$^+$	0.79^b (1.99)	*151*
[Mn(CO)(CNPh)$_5$]$^+$	1.28 (2.21)	*63*
[Re(CO)(*p*-CNtolyl)$_5$]$^+$	1.41 (2.26)	*149*
fac-[Mn(CO)(CNMe)$_3$(dppe)]$^+$	0.83	*148*
mer-[Mn(CO)(CNMe)$_3$(dppe)]$^+$	0.91	*148*
[Mn(CNMe)$_6$]$^+$	0.38^b	*151*
[Mn(*p*-CNC$_6$H$_4$NO$_2$)$_6$]$^+$	1.31	*147*
[Mn(pyr)(*p*-CNtolyl)$_5$]$^+$	0.62 (1.6)	*150*
[Mn(CNMe)$_4$(dppe)]$^+$	0.46 (1.43)	*148*

^{a} Versus sce in CH$_2$Cl$_2$ with [NBu$_4$][ClO$_4$] as base electrolyte unless indicated otherwise. Where appropriate the potential for the second wave is given in parentheses.
^{b} In MeCN, with [NEt$_4$][ClO$_4$].
^{c} Irreversible.

Isomerization also occurs within this series of complexes; cis-[Mn(CO)$_2$-{PPh(OMe)$_2$}$_4$]$^+$ is converted to the trans form via [NO][PF$_6$] oxidation followed by hydrazine reduction (139).

The electron-transfer properties of [M(CO)$_{6-n}$(CNR)$_n$]$^+$ (M = Mn or Re, R = alkyl or aryl, n = 1–6) (63, 147–151) shown in Table II are very similar to those of [Cr(CO)$_{6-n}$(CNR)$_n$] (Section II,C,3). The increment in positive charge, however, renders the cations much less readily oxidized. Studies of the two series have tended to be made in parallel, and MO calculations referred to in Section II,C,3 also apply here. The calculated HOMO energies of [Mn(CO)$_{6-n}$(CNMe)$_n$]$^+$ (n = 0–6) correlate with E^0 for the one-electron oxidation (152), and the calculated dependence of E^0 on isomer geometry (59) is displayed by fac- and mer-[Mn(CO)$_3$(CNPh)$_3$]$^+$ (139).

Surprisingly few of the dicationic derivatives have been synthesized, but nitric acid oxidation gives [Mn(CO)$_{6-n}$(CNR)$_n$]$^{2+}$ (n = 5 or 6, R = Me) (151) from the appropriate monocations. The latter are also irreversibly oxidized to unstable trications (147, 151).

One derivative of [Re(CO)$_6$]$^+$ is worthy of particular note, namely [Re(CO)$_3$(NCMe)(o-phen)]$^+$. Irreversible reduction occurs at −1.2 V, but on UV photolysis the excited state generated is reduced at ca. 1.5 V. The enormous change in $E_{1/2}$ leads to a photosubstitution reaction with PPh$_3$ which depends on the generation of a labile 19-electron intermediate (Scheme 9), and for which the quantum yield is far greater than one. The

[Re(CO)$_3$(NCMe)(o-phen)]$^+$ $\xrightarrow{h\nu}$ *[Re(CO)$_3$(NCMe)(o-phen)]$^+$

*[Re(CO)$_3$(NCMe)(o-phen)]$^+$ + PPh$_3$ → [Re(CO)$_3$(NCMe)(o-phen)] + [PPh$_3$]$^+$

[Re(CO)$_3$(NCMe)(o-phen)] + PPh$_3$ → [Re(CO)$_3$(PPh$_3$)(o-phen)] + MeCN

*[Re(CO)$_3$(NCMe)(o-phen)]$^+$ + [Re(CO)$_3$(PPh$_3$)(o-phen)]

↓

[Re(CO)$_3$(NCMe)(o-phen)] + [Re(CO)$_3$(PPh$_3$)(o-phen)]$^+$

SCHEME 9

final electron-transfer step is facilitated by the ready oxidation of $[Re(CO)_3$-$(PPh_3)(o\text{-phen})]^0$ ($E_{1/2} \approx -1.2$ V) (15).

Sodium reduction of $[Mn(CO)_3(NO)]$ in thf at 200 K gives $[Mn(CO)_3$-$(NO)]^-$ identified by ESR spectroscopy (153).

E. *Iron, Ruthenium, and Osmium*

The electrochemistry of the iron carbonyls is complex and in some respects ill defined. The pentacarbonyl $[Fe(CO)_5]$ is oxidized in most common, non-aqueous electrochemical solvents in an irreversible but diffusion-controlled process (4, 154, 155). The monocation $[Fe(CO)_5]^+$ shows some stability in a mixture of benzene and molten $AlCl_3$-ethylpyridinium bromide (43), and the oxidation is reversible in CF_3CO_2H at fast cyclic voltametric scan rates (42).

The final product of the electrolytic reduction of $[Fe(CO)_5]$ in aprotic solvents is $[Fe_2(CO)_8]^{2-}$ (40, 154, 155), but the mechanism of its formation is not agreed upon. Polarography (155, 156) suggests an initial two-electron transfer, but other authors claim a diffusion-controlled one-electron reduction at platinum, mercury, or glassy carbon (40, 154). Interestingly, equimolar quantities of $[Fe(CO)_5]$ and sodium naphthalide yield $[Fe_2(CO)_8]^{2-}$ whereas a 1:2 ratio of the reagents gives $[Fe(CO)_4]^{2-}$ (154). It is also notable that electrolytic reduction of $[Fe(CO)_5]$ at mercury requires one electron while the presence of water leads to $[FeH(CO)_4]^-$ and the uptake of two electrons. The hydride may form via the protonation of $[Fe(CO)_4]^-$ followed by one-electron reduction of $[FeH(CO)_4]$ (154), although the oxidation of $[FeH(CO)_4]^-$ is irreversible at mercury (155).

The chemical reduction of $[Fe(CO)_5]$ with alkali metals or alloys in very dry thf is also complex. The product contains up to four paramagnetic species, namely $[Fe_3(CO)_{12}]^-$, which is also formed by the reversible one-electron reduction of $[Fe_3(CO)_{12}]$, and what were thought to be $[Fe_2(CO)_8]^-$, $[Fe_3(CO)_{11}]^-$, and $[Fe_4(CO)_{13}]^-$. The last three species, identified by ESR spectroscopy of the ^{57}Fe- and ^{13}CO-labeled derivatives, can also be generated by Ag^+ or $[FeCp_2]^+$ oxidation of the corresponding dianions in thf at $-80°C$ (157). However, CV shows that $[Fe_3(CO)_{11}]^{2-}$ is neither oxidized nor reduced, casting some doubt on the existence of $[Fe_3(CO)_{11}]^-$ (158). Further discussion of the electrochemistry of $[M_3(CO)_{12}]$ (M = Fe, Ru, or Os), and of closely related hydrido clusters will be presented elsewhere (31).

The Group V-donor derivatives $[Fe(CO)_{5-n}L_n]$ ($n = 1$ or 2, L = P-, As-, or Sb-donor) are oxidized in a one-electron step at platinum or mercury (*159, 160*) (e.g., L = PPh_3, $n = 1$, $E_{1/2} \approx 1.2$ V; $n = 2$, $E^0 = 0.63$ V vs Ag/AgCl in acetone at Pt). The reversibility of the process depends on L, n, the solvent, and, significantly, the electrode material. In general, the oxidation of the tetracarbonyls is irreversible, and MeCN is reactive toward $[Fe(CO)_{5-n}L_n]^+$ ($n = 1$ or 2). At a mercury electrode, the electron transfer is markedly more reversible, and the formation of "mercury-stabilized" radical cations, related to the silver adducts described below, cannot be discounted (*160*).

The reversible, one-electron oxidation of *trans*-$[Fe(CO)_3(PPh_3)_2]$ in CH_2Cl_2 is clearly defined (*159, 160*), and the air-sensitive salt *trans*-$[Fe(CO)_3(PPh_3)_2][PF_6]$ (9) is isolable using $AgPF_6$ or $[N(p-C_6H_4Br)_3]$-$[PF_6]$. By contrast, the reaction between $AgPF_6$ and $[Fe(CO)_3(PPh_3)_2]$ in toluene yields the 2:1 adduct $[Ag\{Fe(CO)_3(PPh_3)_2\}_2]^+$, which dissolves in CH_2Cl_2 to give silver metal and 9. The formation of an analogous adduct may well explain the observation of the ESR spectrum of $[Fe(CO)_3(AsPh_3)_2]^+$ (10) when Ag^+ salts and $[Fe(CO)_3(AsPh_3)_2]$ react in CH_2Cl_2. Although the neutral arsine complex undergoes a completely irreversible one-electron oxidation at platinum, the slow decomposition of $[Ag\{Fe(CO)_3(AsPh_3)_2\}_2]^+$ (Scheme 10) would provide a low but steady-state concentration of 10 (*159*).

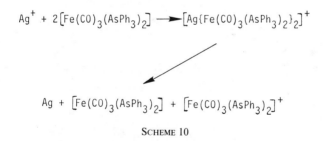

$$Ag^+ + 2[Fe(CO)_3(AsPh_3)_2] \longrightarrow [Ag\{Fe(CO)_3(AsPh_3)_2\}_2]^+$$

$$Ag + [Fe(CO)_3(AsPh_3)_2] + [Fe(CO)_3(AsPh_3)_2]^+$$

SCHEME 10

The reaction between $[Fe(CO)_3(AsPh_3)_2]$ and CCl_4 also involves the intermediacy of 10, and $[Fe(CO)_3(SbPh_3)_2]^+$ has been detected in the reaction between $[Fe(CO)_3(SbPh_3)_2]$ and $CHBr_3$ (*161*).

The stoichiometric reactions of 9 with halogens and halide ions and infrared stopped flow kinetic studies clearly show the importance of one-electron transfer in the oxidative elimination reactions between $[Fe(CO)_3L_2]$ and X_2 (X = Cl, Br, or I). The major steps in the mechanism are shown in

Scheme 11 (*159*), but the intermediate formed in the reaction between **9** and I^- is probably incorrectly formulated as the six-coordinate, 19-electron complex $[FeI(CO)_3(PPh_3)_2]$ (*159*). Its ESR spectrum shows g and A_P values similar to those of **9**, and it is more likely a five-coordinate, 17-electron radical, perhaps the iodoacyl $[Fe(COI)(CO)_2(PPh_3)_2]$ (*162*).

$$[Fe(CO)_3L_2] + I_2 \longrightarrow [Fe(CO)_3L_2]^+ + I^\cdot + I^-$$

$$[Fe(CO)_3L_2]^+ + I^- \rightleftharpoons [FeI(CO)_3L_2]$$

$$[Fe(CO)_3L_2]^+ + [FeI(CO)_3L_2] \longrightarrow [Fe(CO)_3L_2] + [FeI(CO)_3L_2]^+$$

$$[FeI(CO)_3L_2]^+ + I^- \longrightarrow [FeI_2(CO)_3L] + L$$

$$L = PPh_3$$

SCHEME 11

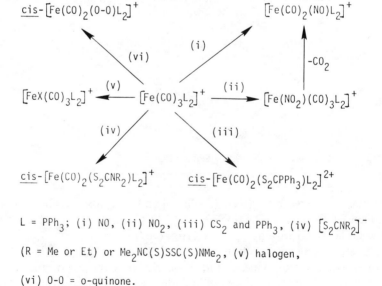

$L = PPh_3$; (i) NO, (ii) NO_2, (iii) CS_2 and PPh_3, (iv) $[S_2CNR_2]^-$

(R = Me or Et) or $Me_2NC(S)SSC(S)NMe_2$, (v) halogen,

(vi) O–O = $\underline{o}$-quinone.

SCHEME 12

The reactions of **9** also provide routes to otherwise inaccessible Fe(II) carbonyls; radical coupling occurs with NO, NO_2, $Me_2NC(S)SSC(S)NMe_2$ (*162*), and halogens (*160*), and 1,2-diketones or o-quinones (O–O) give paramagnetic $[Fe(CO)_2(O–O)(PPh_3)_2]^+$ (Scheme 12). The latter and the ruthenium analogs $[Ru(CO)_2L(PPh_3)(o\text{-chloranil})][PF_6]$ [L = PPh$_3$, P(OPh)$_3$, or AsPh$_3$] (*163, 164*) are best formulated as metal(II) complexes of semiquinone radical anions. Their ESR spectra are similar to those of $[O–O]^-$, although small hyperfine splittings to the metal (M = Ru) and Group V-donor atoms are observed. The spin-trapping of **9** by t-BuNO has also been accomplished (*162*).

Although P donors do not substitute the CO ligands of **9**, the paramagnetic product formed by oxidizing $[Fe(CO)_3(dppe)]$ in the presence of PPh_2Me is formulated as $[Fe(CO)_2(PPh_2Me)(dppe)]^+$. The monocarbonyl $[Fe(CO)(dppe)_2][BPh_4]$ has also been prepared via the electrolytic oxidation of $[Fe(CO)(dppe)_2]$ (*165*).

Reaction of the carbene complexes $[Fe(CO)_3LL']$ [L = $\overline{CNMe(CH_2)_2N}$-Me, L' = PR$_3$, AsPh$_3$, or P(OPh)$_3$] with AgBF$_4$ readily affords the isolable monocations, and the related binuclear dication $[\{Fe(CO)_3L\}_2(\mu\text{-dppe})]^{2+}$ contains two noninteracting iron(I) centers. The IR carbonyl spectra of $[Fe(CO)_3LL']^+$ are somewhat different from that of **9**, leading to the assignment of a distorted trigonal-bipyramidal structure. The carbene cations also differ from **9** in undergoing carbonyl substitution with P donors. ESR spectroscopy suggests nearly square-pyramidal geometry for $[Fe(CO)\text{-}L_2L'_2]^+$ and $[Fe(CO)L\{P(OPh)_3\}_3]^+$, and probably for the dicarbonyls $[Fe(CO)_2LL'_2]^+$ (*166*).

A few derivatives of the unknown dications $[M(CO)_6]^{2+}$ (M = Fe, Ru, or Os) are involved in electron-transfer reactions. *Trans*-$[Fe(CO)_2(dppe)_2]^{2+}$ is reduced in an irreversible two-electron step to five-coordinate $[Fe(CO)_2\text{-}(dppe)_2]$ (*165*), and Ce(IV) oxidation of $[Os(CO)(NH_3)_5]^{2+}$ under N_2 gives $[\{Os(CO)(NH_3)_4\}_2(\mu\text{-}N_2)]^{4+}$ via $[Os(CO)(NH_3)_5]^{3+}$. The oxidation of the dicationic amine complex by $[IrCl_6]^{2-}$ may provide a salt of the trication, namely $[Os(CO)(NH_3)_5][IrCl_6]$ (*167*). Each of the complexes $[Fe(CNCy)_3\text{-}\{PPh(OEt)_2\}_3]^{2+}$ and $[FeX(CNR)_2\{PPh(OEt)_2\}_3]^+$ (X = halide, R = aryl) is oxidized and reduced in reversible one-electron and irreversible two-electron processes, respectively (*168*).

The mononitrosyl $[Fe(CO)_3(NO)]^-$ is oxidized at mercury in two one-electron steps. The first involves the formation of $[Fe(CO)_3(NO)]$ and the second the oxidation of $[Hg\{Fe(CO)_3(NO)\}_2]$ (*169*). The two one-electron reductions of $[Fe(CO)_{2-n}(PPh_3)_n(NO)_2]$ (n = 0–2) give mono- and dianions which decompose at room temperature (*170*). However, the ESR spectrum of $[Fe(CO)_2(NO)_2]^-$ can be detected on ^{60}Co γ irradiation of $[Fe(CO)_2$-$(NO)_2]$ at $-196°C$ and shows the unpaired electron to be largely confined to the nitrosyl ligand (171). The reduction of $[Fe(PPh_3)_2(NO)_2]$ was previously assumed to involve electron addition to the NO groups (170) despite the absence of ^{14}N hyperfine splitting.

The complexes $[Fe(L–L)(NO)_2]^Z$ (L–L = bipy, o-phen, or di-2-pyridyl ketone) form an extensive redox series for which $Z = +1$ to -2. All of the electron-transfer steps are fully reversible in 1,2-dimethoxyethane (e.g., L–L = bipy, $E^0 = -0.56$, -2.14, and -2.78 V vs Ag/AgCl), and all of the ions may be generated by exhaustive electrolysis of $[Fe(L–L)(NO)_2]$ at the appropriate potentials. The cations and the monoanions have ESR spectra typical of metal- and chelate–ligand-based radicals, respectively; the latter are similar to those of uncomplexed $[L–L]^-$ (172). It is noteworthy that $[Fe(NO)_2(\overline{CNMeCH_2CH_2NMe})_2][BF_4]$ can be isolated via $AgBF_4$ oxidation of the neutral complex. Again, no ^{14}N coupling was observed in the ESR spectrum (166).

F. Cobalt, Rhodium, and Iridium

In CH_2Cl_2 (135) or EtOH (173), $[Co_2(CO)_8]$ undergoes irreversible two-electron reduction to $[Co(CO)_4]^-$, which, in turn, is irreversibly oxidized to $[Co_2(CO)_8]$ at platinum or to $[Hg\{Co(CO)_4\}_2]$ at mercury (173). In common with $[Mn_2(CO)_{10}]$ and $[Mn(CO)_5]^-$ (Section II,D), $E_{1/2}$ for the reduction of the dimer $[-0.9$ V vs $Ag/AgClO_4$ in CH_2Cl_2 (135)] is much more negative than $E_{1/2}$ for the oxidation of the anion $[-0.2$ V vs $Ag/AgClO_4$ in glyme (174)]. The mercurial $[Hg\{Co(CO)_4\}_2]$ is also reduced in a two-electron step to $[Co(CO)_4]^-$ (135, 173), and $E_{1/2}$ for both $[Co_2(CO)_6L_2]$ and $[Hg\{Co(CO)_3L\}_2]$ (L = phosphine or phosphite) correlates linearly with the half neutralization potential of L (34).

The irreversible one-electron reduction of $[Co(CO)_4(SnR_3)]$ (R = Me or Ph) gives $[Co(CO)_4]^-$ and SnR_3 (135), but ESR spectroscopy provides evidence for the formation of $[Co(CO)_4(PbPh_3)]^-$ when the neutral complex is ^{60}Co γ irradiated at 77 K (175). One-electron reduction is also implicated in the formation of $[Co(CO)_3L_2][SnCl_3]$ (L = PBu_3, PPh_3, or $AsPh_3$) from $[Co(CO)_4(SnCl_3)]$ and L (Scheme 13) (176).

Although $[Co(CO)_4]$ is undetectable during the oxidative dimerization of $[Co(CO)_4]^-$, there is some evidence for its formation when the anion

$$[Co(CO)_4(SnCl_3)] + L \rightleftharpoons [Co(CO)_4(SnCl_3L)]$$

$$[Co(CO)_4(SnCl_3L)] \longrightarrow [Co(CO)_4] + SnCl_3L$$

$$[Co(CO)_4] + L \longrightarrow [Co(CO)_3L] + CO$$

$$[Co(CO)_3L] + [Co(CO)_4(SnCl_3L)] \longrightarrow [Co(CO)_3L]^+ + [Co(CO)_4(SnCl_3L)]^-$$

$$[Co(CO)_3L]^+ + L \longrightarrow [Co(CO)_3L_2]^+$$

$$[Co(CO)_4(SnCl_3L)]^- \longrightarrow [Co(CO)_4] + SnCl_3^- + L$$

SCHEME 13

reacts with alkyl iodides (177), and phosphine or phosphite substitution leads to isolable radicals.

The anions $[Co\{P(OR)_3\}_4]^-$ (R = Me or Et), prepared via the irreversible two-electron reduction of $[Co\{P(OR)_3\}_5]^+$, are reversibly oxidized to $[Co\{P(OR)_3\}_4]$. The last crystallize from MeCN on exhaustive oxidation of the anions, or can be isolated by mixing equivalent quantities of $[Co\{P(OR)_3\}_4]^-$ and $[Co\{P(OR)_3\}_5]^+$ (178); electrolytic reduction of cobalt(II) perchlorate in the presence of the appropriate ligand provides a route to $P(OPh)_3$ (179) and $P(O\text{-}i\text{-}Pr)_3$ (180) analogs. The phosphine complex $[Co(PMe_3)_4]$ reacts with alkali metals to give air-sensitive $M[Co(PMe_3)_4]$ (M = Li, Na, or K) which reduce naphthalene or benzophenone to their radical anions (181). Carbon monoxide and $[CoL_4]^-$ [L = PMe_3 (181) or $P(OPh)_3$ (179)] afford $[Co(CO)_{4-n}L_n]^-$ (n = 2–4).

The magnetic moments of $[Co\{P(OR)_3\}_4]$ are frequently lower than expected for one unpaired electron, and partial dimerization to $[Co_2\{P(OR)_3\}_8]$ has been invoked (178). However, it has been shown recently that $[Co\{P(OMe)_3\}_4]$ and the independently synthesized compound $[Co_2\{P(OMe)_3\}_8]$ do not interconvert, and the dimer is thought to form only when $[Co\{P(OMe)_3\}_4]^-$ is added to $[Co\{P(OMe)_3\}_4]^+$ (182).

The cations $[Co\{P(OR)_3\}_4]^+$ (R = Me or Et) are isolable at $-10°C$ and are paramagnetic with two unpaired electrons; the $P(OMe)_3$ derivative is an olefin-isomerization catalyst (*183*). The more stable, chelated analog $[Co(dppe_2)]^+$ is reversibly reduced in two one-electron steps (*184, 185*), but the adduct $[Co(CO)(dppe)_2]^+$ shows a reversible two-electron wave ($E^0 = -1.3$ V) and may be electrolytically converted to $[Co(CO)(dppe)_2]^-$ at $-30°C$ under argon (*185*).

The bulk electrolytic reduction of $[Rh(dppe)_2]^+$ affords $[RhH(dppe)_2]$ (*185, 186*) but the mechanism of the reaction is disputed. The two separate pathways shown in Scheme 14 have been proposed, based on different

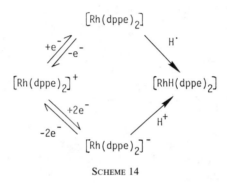

SCHEME 14

interpretations of apparently similar results. At a hanging mercury drop, CV in MeCN (*186*) or PhCN (*187*) shows a reversible wave ($E^0 = -2.1$ V vs $Ag/AgNO_3$) thought to correspond to the formation of $[Rh(dppe)_2]$; subsequent hydrogen atom transfer from the solvent or supporting electrolyte would give $[RhH(dppe)_2]$. However, polarography in MeCN is said to involve a reversible *two*-electron process, with the neutral hydride formed via protonation of $[Rh(dppe)_2]^-$ (*188*).

Very recent studies seem to support the second proposal, with CV at a mercury on gold electrode showing a peak separation of 30 mV, as expected for a reversible two-electron reduction. Exhaustive electrolysis in anhydrous MeCN at $-35°C$ gave a solution of the anion which showed a two-electron oxidation wave at -2.1 V, as might be expected. At $-75°C$, $[Rh(dppe)_2]^-$ decomposes with a half-life of about 10^3 seconds, and $[RhH(dppe)_2]$ precipitates from solution (*184*).

Interestingly, the reduction of $[Rh(dppe)_2]^+$ in a mixture of PhCN and cyclohexane gave low yields of cyclohexene and dicyclohexyl, implying C—H bond activation by the radical [Rh(dppe)$_2$] (*187*). However, the more recent electrochemical results (*184*) outlined above suggest that this, and a related reaction involving the isomerization of Ph-*t*-Bu to Ph-*i*-Bu (*187*), may need to be reinterpreted in terms of the intermediacy of [Rh(dppe)$_2$]$^-$.

The five-coordinate hydrides [MH(dppe)$_2$] (M = Co, Rh, or Ir) are also electroactive (*189*). At platinum in a mixture of MeCN and toluene, [CoH(dppe)$_2$] shows two one-electron waves, the first of which is reversible ($E^0 = -1.08$ V vs Ag/Ag$^+$). Electrolysis at -0.5 V leads to the characterization of [CoH(dppe)$_2$][BPh$_4$], and oxidation at the second irreversible wave ($E_{1/2} = 0.2$ V) gives a stable solution of [CoH(dppe)$_2$(NCMe)]$^{2+}$, which is reduced back to [CoH(dppe)$_2$] in a two-electron process. The phosphite analogs [CoHL$_4$][L = P(OMe)$_3$ or P(OEt)$_3$] also show reversible one-electron waves [e.g., L = P(OMe)$_3$, $E_0 = -0.58$ V vs Ag/Ag$^+$] and the monocations may be generated electrolytically. The second irreversible oxidation leads to the formation of [CoL$_4$]$^+$.

At $-35°$C, [MH(dppe)$_2$] (M = Rh or Ir) behave similarly to the cobalt complexes, but electrolysis at the second wave gives [M(dppe)$_2$]$^+$ and [MH$_2$(dppe)$_2$]$^+$ via the reaction between [MH(dppe)$_2$]$^{2+}$ and [MH-(dppe)$_2$]. The first oxidation of [MH(CO)(PPh$_3$)$_3$] is also reversible, and the cation [RhH(CO)(PPh$_3$)$_3$]$^+$ can be generated in ClCH$_2$CH$_2$Cl at $-35°$C. Under these conditions, the unstable dication can also be formed (*189*).

The hydrides [IrHL$_4$] (L = PPh$_3$ or AsPh$_3$) precipitate from MeCN–toluene mixtures when [IrHCl$_2$L$_3$] is electrochemically reduced in the presence of the ligand L (*190*); a similar reduction of [RhClL$_3$] (L = PPh$_3$ or PMePh$_2$) was thought (*191*) to give [RhL$_4$] but the products are probably [RhHL$_4$] (see above).

The adducts [Ir(X$_2$)(dppe)$_2$]$^+$ (X = O, S, or Se) show two reduction waves in MeCN (e.g., X = S, $E_{1/2} = -1.75$ and -2.05 V vs Ag/AgNO$_3$). The first involves dissociation of [X$_2$]$^-$ from the unstable radical [Ir(X$_2$)-(dppe)$_2$], and the second the reversible reduction of [Ir(dppe)$_2$]$^+$ [cf. [Rh(dppe)$_2$]$^+$ above]. MO calculations support the addition of the first electron to an orbital antibonding with respect to the X$_2$ ligand (*192*).

The potentials for the irreversible two-electron reductions of [MX(CO)L$_2$] [M = Rh or Ir, X = inorganic monoanion (*193–195*), alkyl, or aryl (*196*), L = Group V donor, $E_{1/2} = -1.8$ to -2.7 V] correlate with the electron-donor ability of X (*195*), and electrolytic reduction in the presence of L affords [M(CO)L$_3$]$^-$. The one-electron oxidation of [IrX(CO)L$_2$] (X = halide, L = phosphine) is irreversible at platinum, possibly giving metal–metal bonded products (*197*). By contrast, [IrH(CO)(PPh$_3$)$_2$] is reversibly oxidized ($E^0 = 0.07$ V, in MeCN) to the mononuclear monocation (*39*), and [IrCl(CO)(PMe$_2$Ph)$_2$]$^+$ is a proposed intermediate in the oxidative addition of MeCHClNO$_2$ to [IrCl(CO)(PMe$_2$Ph)$_2$] (198). At mercury, the oxidation of [IrCl(CO)(PPh$_3$)$_2$] gives the polynuclear species [Hg{IrCl(CO)-(PPh$_3$)$_2$}$_n$]$^{2+}$ (n = 3 or 4) and [(NO$_3$)$_2$Hg{IrCl(CO)(PPh$_3$)$_2$}$_2$] (199).

The oxidative addition of RX to [MY(CO)L$_2$], followed by the electro-chemical reduction of [MXYR(CO)L$_2$], provides a novel route to d^8 alkyl and aryl complexes. For example, [RhCl(CO)(PPh$_3$)$_2$] and CPh$_3$Br afford [Rh(CH$_3$)(CO)(PPh$_3$)$_2$] via reductive elimination of ClBr from [RhClBr-(CPh$_3$)(CO)(PPh$_3$)$_2$] (200). The d^6, six-coordinate adducts are reduced in an irreversible two-electron process, and for [IrH$_n$Cl$_{3-n}$(CO)(PPh$_3$)$_2$] (n = 0–2), there is a linear correlation between $E_{1/2}$ and $\tilde{v}$(CO) (201).

The five-coordinate cation [Co(CO)$_2$(PMe$_3$)$_3$]$^+$ is polarographically reduced in an irreversible two-electron process ($E_{1/2}$ = −1.68 V, vs Ag/Ag$^+$), and exhaustive electrolysis gives [{Co(CO)$_2$(PMe$_3$)$_2$}$_2$] (194). As mentioned above, [Co{P(OR)$_3$}$_5$]$^+$ undergoes a similar electron-transfer reaction, but affords [Co{P(OR)$_3$}$_4$]$^-$ (178).

Tcne (E^0 = 0.22 V) and [Co(CNMe)$_5$]$^+$ (E^0 = 0.10 V) give high yields of [Co(CNMe)$_4$(η^2-tcne)]$^+$ by an oxidative substitution reaction involving initial one-electron transfer; the intermediate, [tcne]$^-$, was detected by flow ESR spectroscopy (202). The dication [Co(CNPh)$_5$]$^{2+}$ is isolable as a per-chlorate salt (203), but solutions of the blue MeNC analog give red crystals of dimeric [Co$_2$(CNMe)$_{10}$][ClO$_4$]$_4$ (204). No doubt the monomer–dimer equilibrium interfered with the electrochemical study of [Co(CNMe)$_5$]$^{2+}$ in water (205).

The mixed-ligand complexes trans-[CoL$_2$(CNR)$_3$][PF$_6$] (L = phosphine, R = alkyl or aryl) also undergo electrochemical oxidation, complicated by the reactivity of the dications. Nevertheless, reaction with AgPF$_6$ allowed the isolation of [Co(PPh$_3$)$_2$(CN-t-Bu)$_3$][PF$_6$]$_2$. The significantly different E^0 values of [Co(dppe)(CN-p-tolyl)$_3$]$^+$ (E^0 = 0.24 V) and [Co(PMePh$_2$)$_2$-(CN-p-tolyl)$_3$]$^+$ (E^0 = 0.39 V) probably reflect the dependence of oxidation potential on geometry (206).

The polarographic reduction of [Co(CO)$_3$(NO)] involves the reversible formation of unstable [Co(CO)$_3$(NO)]$^-$ (E^0 = −1.28 V in thf). The derivatives [Co(CO)$_{3-n}$L$_n$(NO)] (n = 1 or 2, L = PPh$_3$, AsPh$_3$, dppe, etc.) behave similarly although an increase in n serves to lower the reduction potential markedly (e.g., L = AsPh$_3$, n = 1, E^0 = −1.47 V; n = 2, E^0 = −2.07 V) (207). The E^0 values correlate linearly with $\tilde{v}$(NO) for the neutral compounds, although the relatively small changes in the stretching frequency suggest electron addition to a delocalized LUMO rather than one predominantly NO ligand based (208). Complexes of bipy-like ligands, [Co(CO)(NO)(L–L)] (L–L = bipy, o-phen, or di-2-pyridyl ketone), are reduced in two fully reversible stages to mono- and dianions (e.g., L–L =

bipy, $E^0 = -2.26$ and -2.81 V vs Ag/Ag$^+$). At $-50°$C in 1,2-dimethoxy-ethane, the ESR spectrum of [Co(CO)(NO)(L–L)]$^-$ shows hyperfine coupling to cobalt and to the nitrogen atoms and two protons of L–L, but not to the nitrosyl group (172).

Surprisingly, the electrochemical oxidation of [Co(CO)$_{3-n}$L$_n$(NO)] (L = P donor) has not been studied despite the irreversible formation of [Co(bipy)(CO)(NO)]$^+$ ($E_{1/2} = -0.50$ V vs Ag/Ag$^+$) and the isolation of, for example, [Co{N(C$_2$H$_4$PPh$_2$)$_3$}(NO)]^Z ($Z = 0$ and 1) (209).

The dinitrosyl [Co{P(OEt)$_3$}$_2$(NO)$_2$]$^+$ undergoes irreversible one-electron oxidation and two one-electron reductions in MeCN. The formation of the neutral radical is accompanied by a decrease in $\tilde{v}$(NO) of ca. 200 cm^{-1} (210). The unsubstituted analog [Co(CO)$_2$(NO)$_2$] may be generated by photolyzing [Co(CO)$_3$(NO)] with NO at 195 K. Its ESR spectrum is consistent with the single electron essentially confined to the $2p_z$ orbitals of the two nitrogen atoms (153).

G. Nickel, Palladium, and Platinum

The redox properties of [Ni(CO)$_4$] are ill defined (84) and no products were characterized following the irreversible one-electron reduction of [Ni(CO)$_{4-n}${P(OMe)$_3$}$_n$] ($n = 0$–3) (34). However, representatives of each member of the series [NiL$_4$]^Z ($Z = 0$, 1, or 2; L = P donor) are now well documented.

In MeCN, [M(dppe)$_2$]$^{2+}$ (M = Pd or Pt) are reduced in a single two-electron step to [M(dppe)$_2$] (e.g., M = Pd, $E^0 = -0.59$ V) (211). By contrast, [Ni(dppe)$_2$]^Z ($Z = 0$ and 1) are formed from the dication in sequential one-electron processes ($E_1^0 = -0.23$ V, $E_2^0 = -0.52$ V vs Ag/AgCl) (211, 212). Bulk reduction of [Ni(dppe)$_2$]$^{2+}$ at -0.25 V ($-40°$C) gave the monocation whose ESR spectrum suggests a half-filled $d_{x^2-y^2}$ orbital and a square planar structure (212). Interestingly, both [Ni(CO){N(C$_2$H$_4$AsPh$_2$)$_3$}]-[BPh$_4$] (213) and [Ni(PMe$_3$)$_4$][BPh$_4$] (214) have been fully characterized and the X-ray structure of the latter revealed a slightly distorted tetrahedral array.

The preparation of [Ni(PMe$_3$)$_4$]$^+$ by the addition of PMe$_3$ to [NiXMe-(PMe$_3$)$_2$] (X = halide) (215) is particularly noteworthy in suggesting that the reductive elimination of alkyl or aryl halides from Ni(II) phosphine complexes via 17-electron intermediates may be important in catalytic cycles

(214). One-electron transfer is certainly involved in the reverse process and Scheme 15 shows the proposed mechanism for the oxidative elimination

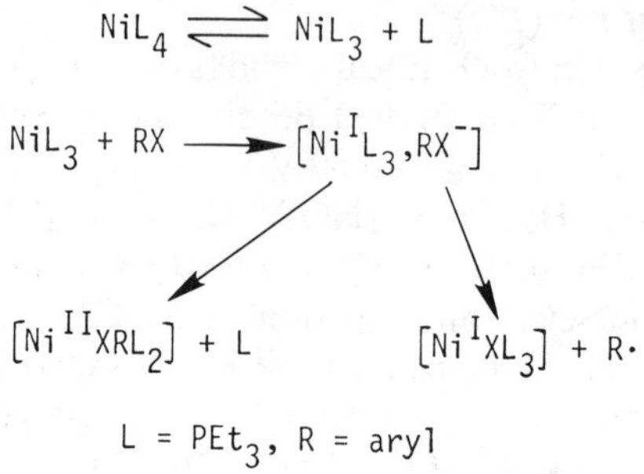

$$NiL_4 \rightleftharpoons NiL_3 + L$$

$$NiL_3 + RX \longrightarrow [Ni^IL_3, RX^-]$$

$$[Ni^{II}XRL_2] + L \qquad\qquad [Ni^IXL_3] + R\cdot$$

$$L = PEt_3, \quad R = aryl$$

Scheme 15

reaction between aryl halides, RX, and NiL_4. The ion pair $[Ni^IL_3, RX^-]$ can collapse to the Ni(II) complex $[NiXRL_2]$, or fragmentation of $[RX]^-$ and aryl diffusion from the solvent cage can give the Ni(I) species $[NiXL_3]$ *(216)*. The detailed studies of the possible electrocatalyzed reductive coupling of alkyl or aryl halides by $[Ni(PPh_3)_4]$ will not be discussed further *(21, 217–219)*.

III

ALKYL AND ARYL COMPLEXES

Very few σ alkyl or aryl complexes, apart from those also containing carbonyl (Section II) or cyclopentadienyl (Section VII) ligands, have been studied by electrochemical methods. Such methods should prove invaluable in the characterization of early transition metal aryls for which extensive redox series apparently exist.

The anions $[V(Mes)_3R]^-$ [R = Mes or NPh_2] are oxidized in air to $[V(Mes)_3R]$ *(220–222)*, and both $[V(dmop)_4]^-$ and the dianion may be prepared from Li[dmop] and $[VBr_3(thf)_3]$ and $[VCp_2]$, respectively *(223)*. Similarly, $[Cr(Mes)_4]^-$ *(222)* and oxygen give $[Cr(Mes)_4]$ *(224)*, and $Li_2[Cr(Mes)_4]$ can be made from $[Cr(Mes)_2(thf)_3]$ and Li[Mes] *(225)*. For molybdenum, an additional oxidation state is stabilized. Air oxidation of $[Mo(Mes)_4]^-$ affords $[Mo(Mes)_4]$ *(226)* which reacts with halogens to give $[Mo(Mes)_4][X_3]$ (X = Br or I) *(227)*.

Sodium naphthalide and $[W\{(CH_2)_2\text{-}o\text{-}C_6H_4\}_3]$ $(E^0 = -1.68 \text{ V})$ in thf yield the corresponding monoanion *(228)*, and 1,2-diketonate complexes such as $[CrPh(acac)_2]$ are also reversibly reduced, with E^0 correlating with the Hammett parameters of the acetylacetonate substituents (229).

The complexes cis-$[FeR_2(bipy)_2]^Z$ (R = alkyl, Z = 0, 1, or 2) provide an interesting insight into the effect of redox state on the mechanism of metal–alkyl bond cleavage. The neutral species, which are reversibly oxidized at -1.03 V with a second irreversible one-electron step at about 0.2 V, decompose via β elimination. By constrast, the dications, when formed by electrolytic oxidation at 0.8 V, provide good yields of R–R coupled compounds. The monocation (R = Et), which can be isolated via exhaustive electrolysis at -0.4 V in thf, undergoes homolytic fragmentation; the products are typical of those formed via cage reactions of alkyl radicals (230).

Sodium or magnesium amalgam reduction of $[RuCl_2(PPh_3)_4]$ in MeCN affords the ortho-metallated complex $[RuH(C_6H_4PPh_2)(NCMe)(PPh_3)_2]$ (231), previously incorrectly formulated as $[Ru(\eta^2\text{-}NCMe)(NCMe)(PPh_3)_4]$ (232).

In MeCN, in the presence of PPh_3, trans-$[NiBrPh(PPh_3)_2]$ (11) is reduced in two one-electron steps. The first involves the formation of $[NiPh(PPh_3)_3]$ which decomposes to biphenyl by homolytic coupling; the second gives $[NiPh(PPh_3)_3]^-$ which also yields biphenyl but via heterolytic coupling with 11. Both reductive processes also afford $[Ni(PPh_3)_4]$ which reacts with PhBr to regenerate 11. In principle, electrocatalytic C—C bond formation is possible (233).

The one-electron oxidation of $[PtR_2(PMe_2Ph)_2]$ is irreversible in MeCN (R = Me, $E_{1/2}$ = 0.72 V; R = Et, $E_{1/2}$ = 0.40 V vs $Ag/AgNO_3$). Chemical oxidation requires two equivalents of $[IrCl_6]^{2-}$, with the radical cation $[PtR_2(PMe_2Ph)_2]^+$ giving $[PtCl_2Me_2(PMe_2Ph)_2]$ and $[PtEt(NCMe)(PMe_2Ph)_2]^+$ by competing pathways (Scheme 16) (234). The related

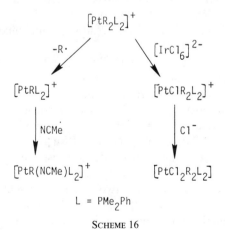

$$L = PMe_2Ph$$

SCHEME 16

metallocyclobutane complex $[Pt(CH_2CH_2CH_2)(bipy)]$ (**12**) is synthesized by the electrolytic reductive elimination of halogen from $[PtCl_2$-$(CH_2CH_2CH_2)(bipy)]$ ($E_{1/2} = -1.35$ V), a route which may have general applicability. Compound **12** is oxidized irreversibly, but the radical anion ($E^0 = -1.68$ V) may be generated electrolytically or by sodium metal reduction. Carbonylation of **12** gives **13** and **14**, both of which are also reversibly reduced to monoanions (*235*).

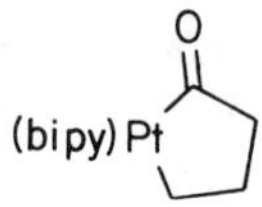

(bipy)Pt

13

(bipy)Pt

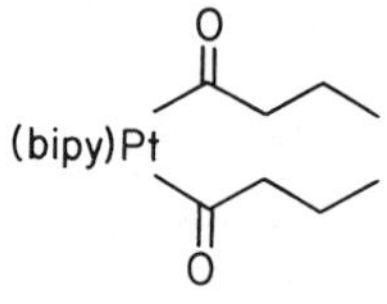

14

IV

η²-ALKENE COMPLEXES

The one-electron reduction of $[Fe(CO)_4(\eta^2\text{-alkene})]$ (alkene = dimethyl-maleate, cinnamaldehyde, etc.) was initially thought (*89*) to afford the radical anions $[Fe(CO)_4(\eta^2\text{-alkene})]^-$, but it is now clear (*236*) that the paramagnetic products are $[Fe(CO)_3(\eta^2\text{-alkene})]^-$; the ESR spectra are consistent with a single electron localized largely on the metal atom but also weakly interacting with the ligand protons. The radical anions are substitutionally labile, with $P(OMe)_3$ readily giving fluxional $[Fe(CO)_2$-$\{P(OMe)_3\}(\eta^2\text{-alkene})]^-$ (*236*).

The chemical reduction of $[Fe(CO)_3(\eta^4\text{-diene})]$ (diene = bd, chd, etc.) also yields 17-electron anions, $[Fe(CO)_3(\eta^2\text{-diene})]^-$, in which electron attachment leads to decomplexation of one of the alkene bonds (*236*) (cf. Section VI; diene = cot). Negative-ion mass spectrometry also shows that the bd and chd derivatives readily form monoanions (*237, 238*), and at $-196°C$ ^{60}Co γ irradiation of $[Fe(CO)_2(PPh_3(\eta^4\text{-chd})]$ gives (*239*) a species with an ESR spectrum very similar to those mentioned above (*236*).

The heterodiene complex $[Fe(CO)_3(\eta^4\text{-benzylideneacetone})]$ is somewhat different in that polarography reveals two reversible, one-electron reductions ($E^0 = -0.88$ and -1.61 V in thf). The stable monoanion, whose ESR spectrum shows detachment of the ketonic carbonyl group, and the unstable dianion can be generated by CPE at -1.0 and -1.8 V, respectively. In the presence of crotyl bromide the first reduction process is irreversible due to the formation of $[Fe(CO)_2(\eta^3\text{-}1\text{-}Me\text{-}C_3H_4)_2]$. In dmf the monoanion gives paramagnetic $[Fe(CO)_3(dmf)_n]^-$, which reacts with $[C_7H_7]^+$ to yield $[Fe(CO)_3\text{-}(\eta^4\text{-}C_{14}H_{14})]$ (240).

V

η^3-ALLYL COMPLEXES

The polarographic reduction of $[FeX(CO)_3(\eta^3\text{-}C_3H_4R)]$ (X = halide or NO_3, R = H or Ph) occurs in two one-electron steps resulting in the formation of $[Fe(CO)_3(\eta^3\text{-}C_3H_4R)]$ and then $[Fe(CO)_3(\eta^3\text{-}C_3H_4R)]^-$ (241). Chemical reduction of $[FeBr(CO)_2(PPh_3)(\eta^3\text{-}allyl)]$ yields similar products, namely $[Fe(CO)_2(PPh_3)(\eta^3\text{-}allyl)]$ with zinc dust, and the anion with sodium amalgam (242).

The allyltricarbonyl radicals equilibrate with dimeric $[\{Fe(CO)_3(\eta^3\text{-}allyl)\}_2]$, in contrast to fluxional $[FeL_3(\eta^3\text{-}C_8H_{13})]$ [L = P(OMe)$_3$, C_8H_{13} = cyclooctenyl] which is formed (243) by sodium amalgam reduction of $[FeL_3(\eta^3\text{-}C_8H_{13})]^+$ (244). X-Ray studies (245) show the last, formally a 16-electron cation, to be stabilized by a short Fe$\cdots$H interaction (1.874 Å) which is lengthened on reduction (2.77 Å) (246). The monoanion $[FeL_3(\eta^3\text{-}C_8H_{13})]^-$ has not been detected but isoelectronic $[CoL_3(\eta^3\text{-}C_8H_{13})]$ shows no such stabilizing feature; MO calculations have allowed the structural changes to be rationalized (246).

The electrochemistry of $[RuBr(CO)_3(\eta^3\text{-}allyl)]$ differs from that of iron in showing only one irreversible one-electron reduction. The product, thought to be $[\{Ru(CO)_3(\eta^3\text{-}allyl)\}_2]$, shows no tendency to dissociate or to reduce to $[Ru(CO)_3(\eta^3\text{-}allyl)]^-$ (247).

The two-electron reductions of $[Fe(CO)_2(NO)(\eta^3\text{-}C_3H_4R)]$ (R = Cl, Br, or Me) (248) and $[Co(CO)_2L(\eta^3\text{-}allyl)]$ (L = phosphine or phosphite) (249) result in metal–allyl bond cleavage. For the cobalt complexes there is a linear correlation between $E_{1/2}$ and the half neutralization potential of L. The rhodium complex $[Rh\{P(OMe)_3\}_2(\eta^3\text{-}allyl)]$ reacts with $AgPF_6$ to give $[Rh\{P(OMe)_3\}_2(\eta^4\text{-}C_6H_{10})]$ in which the two C_3 fragments have been oxidatively coupled (250).

p-Benzoquinones and $[\{NiBr(\eta^3\text{-}2\text{-}Me\text{-}C_3H_4)\}_2]$ give allyl-substituted dihydroquinones in a reaction in which one-electron transfer precedes C—C bond formation (251). Analogous palladium complexes $[\{PdX(\eta^3\text{-}allyl)\}_n]$ (X = halide, Ph, $n = 2$; X = acac, $n = 1$) are polarographically reduced in two one-electron processes. The first involves Pd—X cleavage and the second yields palladium metal and propane (252).

VI

η^4-BONDED COMPLEXES

The monoanions $[MnL(\eta^4\text{-bd})_2]^-$ [L = CO or P(OMe)$_3$] may be prepared by the sodium amalgam reduction of the neutral radicals. The latter are structurally similar to $[Mn(CO)_{5-n}L_n]$ (Section II,D) but electronically very different; ESR spectroscopy, supported by MO calculations, shows $[MnL(\eta^4\text{-bd})_2]$ to be square pyramidal with L in the apical position and the half-occupied orbital perpendicular to the C_2 axis (134).

In most cases, the one-electron reduction of $[Fe(CO)_3(\eta^4\text{-diene})]$ gives the η^2-bonded alkene anions $[Fe(CO)_3(\eta^2\text{-diene})]^-$ (Section IV). However, $[Fe(CO)_3(\eta^4\text{-cot})]$ is reversibly reduced in two one-electron steps (84) [e.g., $E^0 = -1.24$ and -1.71 V in dmf (253)] with little structural change on anion formation (254). Electrolytic reduction (255) or potassium in thf (253) gives the monoanions $[Fe(CO)_2L(\eta^4\text{-}C_8H_8)]^-$ (L = CO or PPh$_3$), whose ESR spectra are consistent with the unpaired electron based largely on the cot ligand (255). In the presence of protons the dianion yields $[Fe(CO)_3(\eta^4\text{-cyclooctatriene})]$ (253).

The asymmetric benzoferrole complex **15** undergoes two one-electron reductions in thf ($E^0 = -0.89$ and -1.30 V vs Ag/AgCl); the first is reversible but the second is only so at fast cyclic voltametric scan rates. Electrolytic reduction of **15** at -1.1 V gives the monoanion whose ESR spectrum shows coupling to four near-equivalent protons; reduction at -1.5 V yields the unstable dianion.

By contrast, the symmetric isomer **16** is reversibly reduced in one two-electron step ($E^0 = -1.14$ V) and the dianion, when generated at -1.3 V, is stable. MO calculations suggest that the anions of **15** and **16** are stabilized

15   16   17

TABLE III

THE ONE-ELECTRON OXIDATION OF $[M(CO)_{3-n}L_n(\eta^4\text{-DIENE})]$

M	n	L	E^0 (V)^a	Diene	Reference
Fe	1	PPh$_3$	0.60 (R)	C$_4$Ph$_4$^{b}	260
Fe	1	P(OMe)$_3$	0.76 (R)	C$_4$Ph$_4$	260
Fe	2	P(OMe)$_3$	0.22 (R)	C$_4$Ph$_4$	260
Ru	2	P(OMe)$_3$	0.36 (R)	C$_4$Ph$_4$	260
Ru	3	P(OMe)$_3$	0.02 (R)	C$_4$Ph$_4$	260
Fe	1	PPh$_3$	0.22 (R)	Nbd	257
Ru	1	P(OMe)$_3$	0.65 (I)	Chd	259
Ru	1	P(OMe)$_3$	0.72 (I)	Dmbd^c	259
Fe	1	P(OCH$_2$)$_3$CMe	0.50 (I)	Chpt	257
Fe	0	—	0.88 (I)	Cot	258
Fe	1	P(OMe)$_3$	0.48 (I)	Cot	258
Fe	2	P(OMe)$_3$	0.12 (I)	Cot	258
Fe	3	P(OMe)$_3$	−0.22 (I)	Cot	258

^{a} Versus a calomel electrode, 1 M in LiCl, in CH$_2$Cl$_2$ with [NEt$_4$][ClO$_4$] as base electrolyte. R, Reversible; I irreversible.

^{b} C$_4$Ph$_4$, Tetraphenylcyclobutadiene.

^{c} 2,3-Dimethylbutadiene.

by extensive delocalization over the organic ligand. The simple ferrole complex **17** is reduced in one irreversible, one-electron step (256).

The complexes $[M(CO)_{3-n}L_n(\eta^4\text{-diene})]$ (M = Fe or Ru) (Table III) usually give short-lived radical cations (257–259), but $[M(CO)_{3-n}(\eta^4\text{-}C_4Ph_4)]$ (M = Fe or Ru, L = P donor, n = 1–3, C$_4$Ph$_4$ = tetraphenyl-cyclobutadiene) and $[Fe(CO)_2(PPh_3)(\eta^4\text{-nbd})]$ are reversibly oxidized, and silver(I) salts afford $[Fe(CO)_2(PPh_3)(\eta^4\text{-}C_4Ph_4)][PF_6]$ and $[Fe(CO)\{P(OMe)_3\}_2(\eta^4\text{-}C_4Ph_4)][BF_4]$ (**18**). Complex **18** is also formed in the reactions of $[Fe(CO)\{P(OMe)_3\}_2(\eta^4\text{-}C_4Ph_4)]$ with tcne and AgNO$_3$, which give $[Fe(CO)(\eta^2\text{-tcne})\{P(OMe)_3\}(\eta^4\text{-}C_4Ph_4)]$ and $[Fe(CO)\{P(OMe)_3\}\text{-}(NO)(\eta^4\text{-}C_4Ph_4)]^+$, respectively (260). A comparison of the X-ray structures of $[Fe(CO)\{P(OMe)_3\}_2(\eta\text{-}C_4Ph_4)]^Z$ (Z = 0 and 1) has revealed little change in the iron–ring bonding (261), implying electron loss from a predominantly metal-based orbital.

The cyclobutadieneruthenium analogs undergo second irreversible oxidations at potentials 0.4–0.5 V positive of the first. The radical cations have not been detected but are implicated in the formation of $[RuI(CO)_{3-n}\text{-}\{P(OMe)_3\}_n(\eta^4\text{-}C_4Ph_4)]^+$ (n = 2 or 3) from the neutral complexes and iodine (260).

Although the oxidation of the cyclobutadieneiron complexes gives metal-based radicals, the chpt and cot analogs behave rather differently. Silver(I) salts and $[Fe(CO)_3(\eta^4\text{-chpt})]$ give $[Fe(CO)_3(\eta^5\text{-}C_7H_9)]^+$, probably via hydrogen atom addition to $[Fe(CO)_3(C_7H_8)]^+$ (257), but the oxidation of $[Fe(CO)_3(\eta^4\text{-cot})]$ is far more interesting. In CH_2Cl_2, Ag^+ or $[N(p\text{-Br-}C_6H_4)_3][PF_6]$ gives high yield of $[Fe_2(CO)_6(\eta^5,\eta'^5\text{-}C_{16}H_{16})]^{2+}$ (19) via the mechanism shown in Scheme 17 (258).

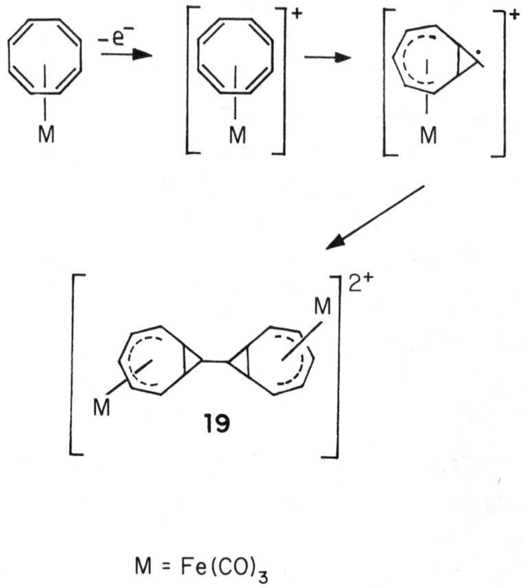

M = Fe(CO)₃

SCHEME 17

Complex 19 is the precursor to novel polycyclic hydrocarbon complexes, reacting with hydride or halide ions (258) and phosphines (262), as shown in Scheme 18. Complexes 20 and 21 also undergo oxidation, with $[FeCp_2]^+$, giving high yields of 22 and 23 via stereospecific cyclopropane ring opening. Further C—C bond formation occurs when 23 is reduced to 24 by sodium amalgam (Scheme 18) (263).

Very few other η^4-diene complexes have been studied, apart from those also containing the cyclopentadienyl ligand (Section VII). In MeCN at platinum (264) or mercury (265), $[M(L-L)(\eta^4\text{-diene})]^+$ (M = Rh or Ir, L–L = bipy, o-phen, 2,2'-bipyrazine, etc.; diene = 1,5-cod or nbd) are reduced in two reversible, one-electron stages [e.g., $[Rh(bipy)(1,5\text{-cod})]^+$, $E^0 = -1.26$ and -1.73 V]; the chelate, L–L, is lost at more negative potentials and mercury-containing products are formed via irreversible

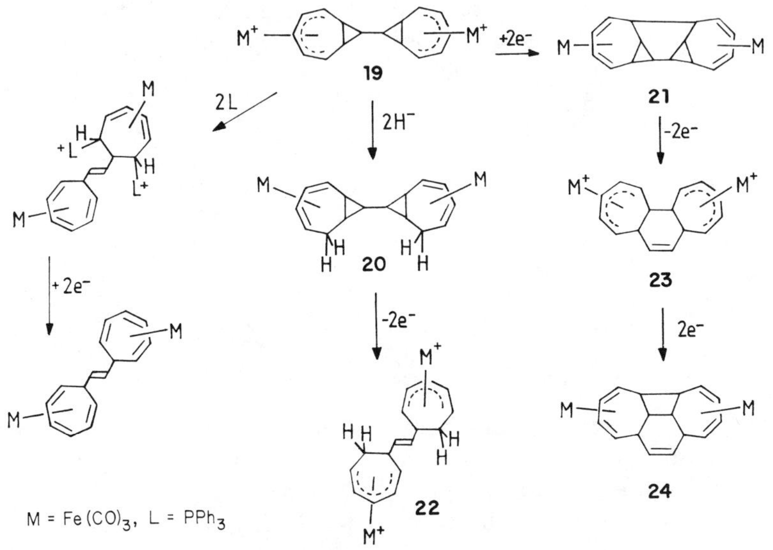

SCHEME 18

oxidation (265). The neutral radicals, generated by CPE, have ESR spectra consistent with the unpaired electron in a π^* orbital of L–L (264) rather than on the metal (265); the spectrum of [Rh(bipy)(η^4-nbd)] shows hyperfine coupling to two nitrogen and four hydrogen atoms.

Sodium naphthalide or amalgam is said to give the anion radicals of [Ni(dq)(η^4-1,5-cod)] (266) and [Ni(dq)$_2$] (267).

VII

η^5-CYCLOPENTADIENYL COMPLEXES

A. Titanium, Zirconium, and Hafnium

The reduction of [MX$_2$Cp$_2$] (M = Ti, Zr, or Hf; X = halide, alkoxide, alkyl, etc.) has been studied extensively but the reactions involved are not fully understood. A wealth of electrochemical data exists, particularly for the dihalides, but their interpretation is controversial and it is quite clear that the isolation and full characterization of some of the postulated inter-mediates are necessary.

The zirconium complexes [ZrCl$_2$(η-C$_5$H$_4$R)$_2$] [R = H (268), Me, Et, or SiMe$_3$ (269)] are thought to be reversibly reduced to the corresponding monoanions (e.g., R = H, $E^0 = -1.7$ V in thf), but [TiX$_2$Cp$_2$] (X = Cl or

Br) show three polarographic waves (270) [e.g., $X = Cl$, $E_{1/2} = -0.80$, -2.10, and -2.38 V in thf (271)]. Two sequences have been proposed to account for the reduction reactions of the titanium compounds (Schemes 19 and 20). The first (Scheme 19), in which halide ion is lost after the addition

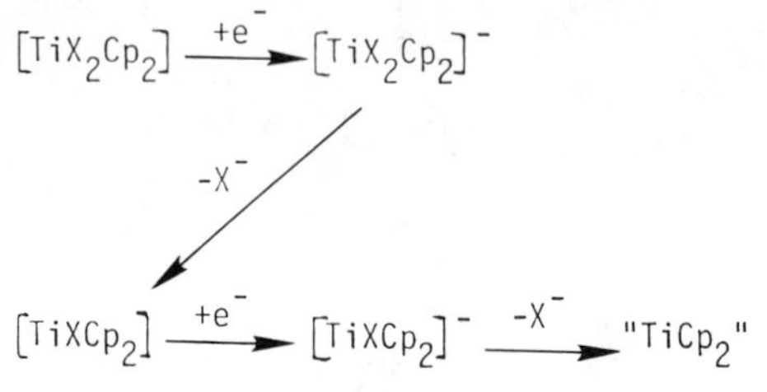

$$[TiX_2Cp_2] \xrightarrow{+e^-} [TiX_2Cp_2]^-$$

$$-X^-$$

$$[TiXCp_2] \xrightarrow{+e^-} [TiXCp_2]^- \xrightarrow{-X^-} \text{"}TiCp_2\text{"}$$

SCHEME 19

of each of two electrons, was generally accepted (86, 272–276) until recently and seems to be supported by the electrochemistry of $[TiClCp_2]$. The monochloride shows two reduction waves (-2.12 and -2.42 V) which are very similar to the second and third waves of $[TiCl_2Cp_2]$, and in the presence of Cl^- the oxidation of $[TiClCp_2]$ leads to the formation of the dichloride (277).

Recent studies, however, have striven to show that the mechanism shown in Scheme 20 is correct. The first reduction wave of $[TiX_2Cp_2]$ ($X = Cl$ or

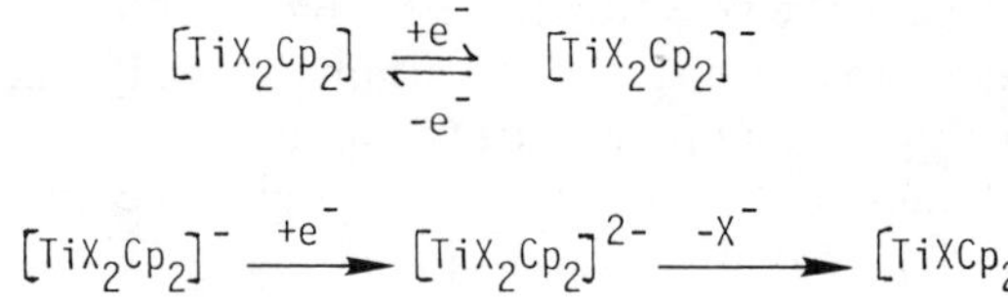

$$[TiX_2Cp_2] \underset{-e^-}{\overset{+e^-}{\rightleftharpoons}} [TiX_2Cp_2]^-$$

$$[TiX_2Cp_2]^- \xrightarrow{+e^-} [TiX_2Cp_2]^{2-} \xrightarrow{-X^-} [TiXCp_2]^-$$

SCHEME 20

Br) is said to be fully reversible (267, 278, 279), even in dmf (278) where the results of earlier work (271) are claimed to be erroneous. After exhaustive one-electron reduction of $[TiX_2Cp_2]$ ($X = Cl$ or Br), no free chloride ion was detectable in the solution (278), and a one-electron oxidation was observed and attributed to the reformation of the neutral compound from $[TiX_2Cp_2]^-$. The ESR spectra of the reduced species were also assigned to the monoanion (268), but others believe they are identical to those of $[TiXCp_2]$ (271). In this respect it is noteworthy that [60]Co γ irradiation of $[TiCl_2Cp_2]$ in methyltetrahydrofuran at 77 K gives what is thought to be $[TiCl_2Cp_2]^-$; the free electron is assigned to the d_{z^2} orbital. Warming the product to 120 K gives $[TiClCp_2]$, also identified by ESR spectroscopy (280).

Further studies designed to distinguish between Schemes 19 and 20 are also inconclusive. Addition of PMe_2Ph to the reduced solution of $[TiX_2Cp_2]$ led to a ^{31}P doublet splitting in the ESR spectrum, and to the conclusion that the product was $[TiX_2(PMe_2Ph)Cp_2]^-$ (279). However, it is clear that the radical present is not the anion but rather $[TiX(PMe_2Ph)Cp_2]$ (*271*). Be that as it may, this study provides no evidence against the formation of $[TiX_2Cp_2]^-$; halide replacement by PMe_2Ph is, of course, possible.

The reduction of $[TiCl_2Cp_2]$ under CO is also cited as evidence for $[TiCl_2Cp_2]^-$ in that the product, $[Ti(CO)_2Cp_2]$, cannot be prepared from $[TiClCp_2]$ (*279*). However, the intermediacy of $[TiCl_2(CO)Cp_2]^-$ based on analogy with $[TiCl_2(PMe_2Ph)Cp_2]^-$ (see above) is unlikely.

The electrochemistry of other cyclopentadienyltitanium compounds is also complicated, not only by the loss of halide or a similar ligand but also by ring–metal cleavage. The anion $[TiCl_3Cp]^-$, with a half-occupied $d_{x^2-y^2}$ orbital, has been detected on ^{60}Co γ irradiation of $[TiCl_3Cp]$ at 77 K (*280*), and the alkoxides $[Ti(OR)_2Cp_2]$ and $[Ti(OR)_3Cp]$ are reduced to unstable monoanions (*270, 281*). The latter usually lose Cp^- or OR^- to give paramagnetic $[Ti(OR)_2Cp]$, but complex **25** is sufficiently stable to be detected directly by ESR spectroscopy.

25

Related complexes of S donors, namely $[Ti(SPh)_2Cp_2]$ (*86*) and $[Ti(L-L)Cp_2]$ [L–L = tdt (*86*) or mnt (*282*)], are also reduced, the first irreversibly and the second to a more stable monoanion. The monocyclopentadienyl complex $[Ti(mnt)_2Cp]^-$ undergoes reversible one-electron oxidation and reduction (*282*).

The reaction of $[TiCl_2Cp_2]$ with $NaMR_2$ gives paramagnetic species, formulated as $[Ti(MR_2)_2Cp_2]^-$ (MR_2 = NHPh, PPh_2, $AsPh_2$, etc.) on the basis of ESR spectroscopy (*283*). Presumably, ring–metal cleavage to give $[Ti(MR_2)_2Cp]$ is not ruled out. There is, however, good evidence for $[TiH_2Cp_2]^-$ which, when prepared from $[TiClCp_2]$ and RMgCl (R = Et or i-Pr) (*284*) or from $[TiCl_2Cp_2]$ and sodium or lithium naphthalide under argon (*285*), shows hyperfine coupling to the two hydrides and the 10 ring protons.

The one-electron reduction of $[TiR_2Cp_2]$ (R = Me or Bz) in thf is thought to result in ring cleavage (*281*) rather than carbanion loss (*286*); CV shows that these species and $[ZrR_2Cp_2]$ (R = Bz or $CHPh_2$) (*269*) react with sodium naphthalide to give $[MR(\eta\text{-}C_5H_4R')_2]$ (R = alkyl, M = Ti or Zr,

$R' = H$; $M = Hf$, $R' = i\text{-}Pr$ *(269)*. A second report of the reduction of $[TiR_2Cp_2]$ ($R = n\text{-}Bu$ or CH_2SiMe_3, $R_2 = CH_2SiMe_2XSiMe_2CH_2$, $X = CH_2$ or O), which ascribed *(287)* the observed ESR spectra to $[TiR_2Cp_2]^-$, may be in error. However, complex **26** is reversibly reduced ($M = Ti$,

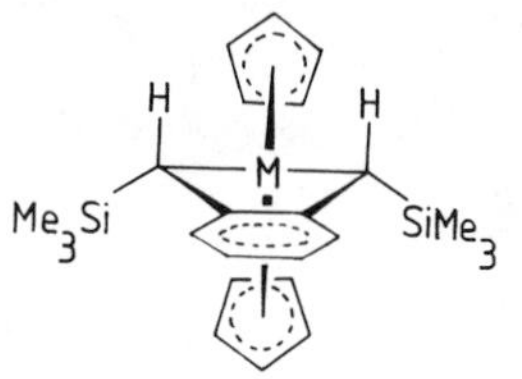

26

$E^0 = -1.46$; $M = Zr$, $E^0 = -2.02$; $M = Hf$, $E^0 = -2.26$ V in thf) and the radical anion is stable for seconds at room temperature *(288)*.

The electrochemical one-electron reduction of $[MClR(\eta\text{-}C_5H_4R')_2]$ [$M = Ti$, $R' = H$, $R = Me$, Ph, or C_6F_5 *(275)*; $M = Zr$, $R' = Me$, $t\text{-}Bu$, or $SiMe_3$, $R = CH(SiMe_3)_2$ *(269)*] is thought to result in chloride loss from the titanium compounds, but there is good evidence for ring cleavage in the zirconium complexes. However, chemical reduction of $[ZrClRCp_2]$ by sodium amalgam in thf has yielded a variety of products depending on R, such as the relatively stable chiral monoanion $[ZrCl\{CH(SiMe_3)(o\text{-}tolyl)\}\text{-}Cp_2]^-$ *(289)*, the side-on bound dinitrogen complex $[Zr\{CH(SiMe_3)_2\}\text{-}(N_2)Cp_2]$ *(290)*, and compound **27** *(291)*.

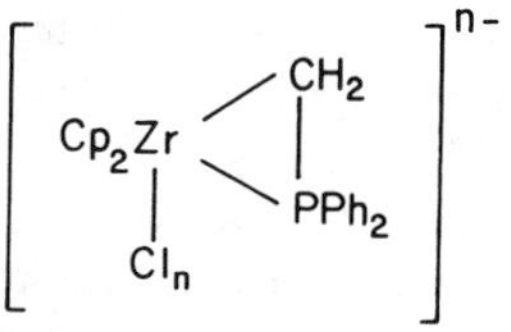

27

Both $[Ti(bipy)Cp_2]$ *(292)* and its monocation *(293)* are isolable. The former is diamagnetic but has a low-lying triplet excited state in which one electron is located in a bipy π^* orbital and the second on the $TiCp_2$ group *(292)*. Related *o*-phen complexes undergo extremely interesting reactions in which ligand C—H bond activation occurs due to intramolecular electron transfer from the metal. Thus, $[Ti(CO)_2Cp_2]$ and 3,4,7,8-tetramethyl-1,10-

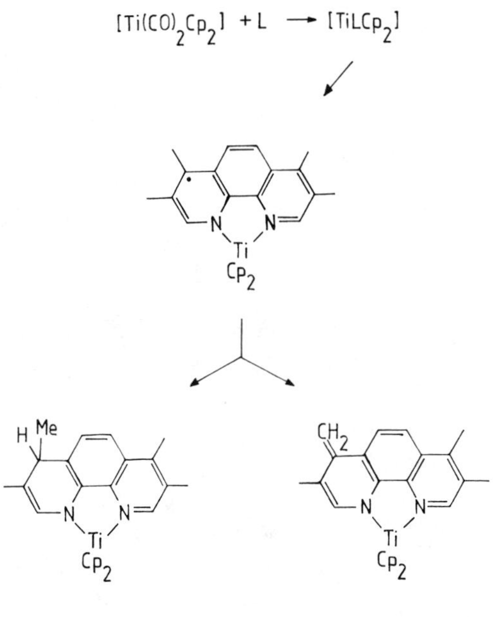

$$[Ti(CO)_2Cp_2] + L \longrightarrow [TiLCp_2]$$

28   29

L = 3, 4, 7, 8 - tetramethyl - 1, 10 - phenanthroline

SCHEME 21

phenanthroline give paramagnetic **28** and **29**, as shown in Scheme 21. MO calculations predict the initial electron transfer, showing a large contribution to the HOMO by the p_x orbital of the carbon atom at the 4-position (*294*).

B. *Vanadium, Niobium, and Tantalum*

In thf, $[V(CO)_4Cp]$ undergoes reversible one-electron reduction ($E^0 = -1.97$ V) but electrolysis in the presence of water gives $[VH(CO)_3Cp]^-$ with the consumption of two electrons (*295*). The hydride is probably formed via carbonyl loss from $[V(CO)_4Cp]^-$, reduction of $[V(CO)_3Cp]^-$, and protonation of the known dianion $[V(CO)_3Cp]^{2-}$. CPE of $[V(CO)Cp_2]$ under CO also yields $[V(CO)_3Cp]^{2-}$, requiring three electrons per molecule.

One-electron reduction affords $[V(CO)Cp_2]^-$, which is carbonylated to $[V(CO)_4Cp]$; the reaction then follows the pathway described above (295).

The formation of monocations from $[VLCp_2]$ [L = CO, $E^0 = -0.3$ V (295); L = CNCy, $E^0 = -0.57$ V (296)] is reversible at fast cyclic voltametric scan rates. Similarly, $[V(CO)_{4-n}L_nCp]$ ($n = 1$, L = PR_3; $n = 2$, L_2 = dppe) undergo one-electron oxidation, reversibly for the phosphine derivatives (e.g., R = Ph, $E^0 = 0.25$ V) but not for those of phosphites (e.g., R = OPh, $E_{1/2} = 0.42$ V). The radical cations $[V(CO)_{4-n}L_nCp]^+$ ($n = 1$, L = PPh_3 or PPh_2Me; $n = 2$, L_2 = dppe) can be generated in CH_2Cl_2, using $[N_2\text{-}p\text{-}F\text{-}C_6H_4]^+$ as the oxidant, and characterized by ESR spectroscopy; the air-sensitive salts $[V(CO)_3(PPh_3)Cp][PF_6]$ and $[V(CO)_2\text{-}(dppe)Cp][PF_6]$ are isolable via $[NO][PF_6]$ oxidation in methanol–toluene (297).

The electrochemistry of $[VCl_2Cp_2]$, like that of $[TiCl_2Cp_2]$ (Section VII,A), is the subject of some controversy. In 1,2-dimethoxyethane (86) or acetone (298), two polarographic one-electron reduction waves are observed and a mechanism similar to that in Scheme 19 was proposed. However, more recent studies in thf suggest the first process ($E^0 = -0.29$ V) to be reversible with chloride loss only after formation of $[VCl_2Cp_2]^{2-}$ ($E_{1/2} = -1.54$ V). A third one-electron wave corresponds to the reduction of vanadocene (Section X,B) (299).

In thf, equimolar quantities of AgCl and $[VCp_2]$ rapidly and quantitatively yield $[VClCp_2]$, presumably via one-electron oxidation of vanadocene (Section X,B). A second equivalent of AgCl gives $[VCl_2Cp_2]$, implying, in turn, the initial formation of $[VClCp_2]^+$. The second oxidation is also effected by organic electron acceptors so that, for example, $[VBrCp_2]$ and CPh_3Cl give $[VClBrCp_2]$ (300). Similarly, $[VMeCp_2]$ and CuCl or AgCl give $[VClMeCp_2]$ (301). The paramagnetic dialkyls $[M(CH_2SiMe_3)_2Cp_2]$ (M = Nb or Ta) are oxidized by Ag(I) salts to monocations which are deprotonated by $Li[N(SiMe_3)_2]$ to the carbenes $[M(CH_2SiMe_3)(CHSiMe_3)\text{-}Cp_2]$ (302).

The chelated derivatives $[V(L-L)Cp_2]^Z$ (L–L = monoanion, $Z = 1$; L–L = dianion, $Z = 0$), containing 1,1-, 1,2-, or 1,3-dithio, dioxo, or oxothio ligands are reversibly reduced in acetone (298, 303, 304). The reduction potential is not only dependent on Z but also on the donor atom and substituents [e.g., L–L = mnt, $Z = 0$, $E^0 = -0.56$ V; L–L = S_2CNEt_2, $Z = 1$, $E^0 = -0.41$ V; L–L = $S_2P(O\text{-}i\text{-}Pr)_2$, $Z = 1$, $E^0 = -0.12$ V vs Ag/AgCl]. Most of the reduction products rapidly decompose via L–L loss, but sodium amalgam and $[V(S_2CX)Cp_2]^+$ give $[V(S_2CX)Cp_2]$ (X = NBu_2 or OBu) (298). The paramagnetic dication $[Ta(dmpe)(\eta\text{-}C_5H_4Me)_2]^{2+}$ is readily prepared from the monocation and $[CPh_3][BF_4]$ (38).

C. *Chromium, Molybdenum, and Tungsten*

In common with other binuclear carbonyls (Sections II,D, II,F, and VII,E), $[\{M(CO)_3Cp\}_2]$ are reduced in a two-electron step to $[M(CO)_3Cp]^-$ [M = Cr, $E_{1/2}$ = -0.18 V (*84, 135, 305, 306*); M = Mo, $E_{1/2}$ = -0.92 V; M = W, $E_{1/2}$ = -1.08 V vs Ag/AgCl in PhCN (*305*)]; the reduction potential correlates linearly with the energy of the $\sigma-\sigma^*$ transition in the electronic spectrum (M = Cr, Mo, or W; M_2 = CrMo, CrW, or MoW) (*305*). The monoanions are also formed by the irreversible two-electron reduction of $[Hg\{M(CO)_3Cp\}_2]$ (M = Cr, Mo, or W) (*84, 86, 135, 306*), or from $[MX(CO)_2LCp]$ (M = Mo or W, X = halide, alkyl, aryl, $SnMe_3$, $PbPh_3$; L = CO or PPh_3) (*135, 306, 307*). Although the monoanions are eventually reoxidized to the dimers, the initial one-electron transfer is fully reversible for $[Cr(CO)_3Cp]^-$ (E^0 = -0.17 V vs Ag/AgCl in PhCN); the radical $[Cr(CO)_3Cp]$ is also present in low concentration when $[\{Cr(CO)_3-Cp\}_2]$ is dissolved in PhCN (*308*).

CV also shows a second, irreversible oxidation wave for $[M(CO)_3Cp]^-$ (e.g., M = Cr, $E_{1/2}$ = 0.87 V vs Ag/AgCl) (*305*), and the cations $[M(CO)_3-Cp]^+$ (M = Cr or Mo) are thought to be involved in the formation of $[MMe(CO)_2Cp]^-$ when $[MMe(CO)_3Cp]$ is photolyzed (Scheme 22) (*309*).

$$[MMe(CO)_3Cp] \rightleftharpoons [MMe(CO)_2Cp] + CO$$

$$[MMe(CO)_2Cp] \rightleftharpoons [M(CO)_2Cp] + Me\cdot$$

$$[M(CO)_2Cp] + CO \rightleftharpoons [M(CO)_3Cp]$$

$$[M(CO)_3Cp] + [MMe(CO)_2Cp] \longrightarrow [M(CO)_3Cp]^+ + [MMe(CO)_2Cp]^-$$

SCHEME 22

Similarly, UV irradiation of $[\{W(CO)_3Cp\}_2]$ in the presence of a one-electron oxidant gives products derived from both $[W(CO)_3Cp]$ radicals and $[W(CO)_3Cp]^+$ (see also Section II,D) (*14*).

The alkyne complexes $[Mo(MeC_2Me)(CO)LCp]^+$ (e.g., L = PPh_3, E^0 = -0.89 V) and $[Mo(MeC_2Me)\{P(OMe)_3\}_2Cp]^+$ (E^0 = -1.27 V), formally substitution products of $[Mo(CO)_3Cp]^+$, are reversibly reduced to neutral radicals; the latter are implicated in the formation of allyl derivatives from the cations and $NaBH_4$. The bis(alkyne) compound $[Mo(CO)(MeC_2Me)_2Cp]^+$ is irreversibly reduced, but $[Mo(SMe)(MeC_2Me)\{P(OMe)_3\}Cp]$ is oxidized in a reversible one-electron reaction at 0.02 V and a second irreversible step at about 0.6 V (310).

The hydride $[MoH(CO)_3Cp]$ forms a monocation which rapidly yields $[\{Mo(CO)_3Cp\}_2]$ by proton dissociation (39). Analogous alkyl cations are intermediates in the oxidatively induced carbonyl insertion reactions of $[MR(CO)_3Cp]$ (311). Thus, $[M(CH_2\text{-}p\text{-}F\text{-}C_6H_4)(CO)_3Cp]$ (M = Mo or W), LiCl, and Ce(IV) in methanol give methyl-p-fluorophenyl acetate (Scheme 23); the mechanism is discussed more fully in Section VII,E.

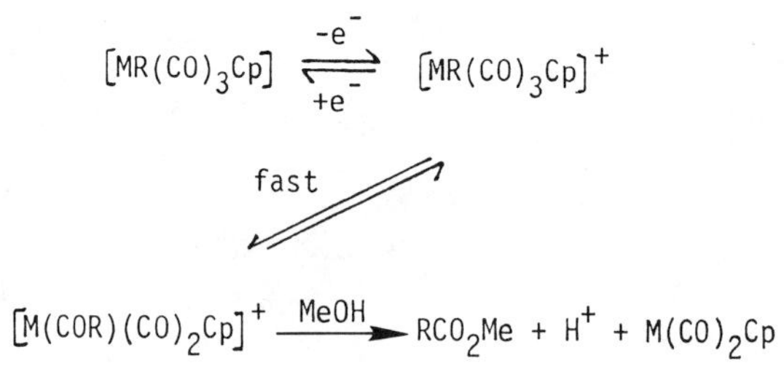

$$[MR(CO)_3Cp] \underset{+e^-}{\overset{-e^-}{\rightleftharpoons}} [MR(CO)_3Cp]^+$$

fast

$$[M(COR)(CO)_2Cp]^+ \xrightarrow{MeOH} RCO_2Me + H^+ + M(CO)_2Cp$$

SCHEME 23

Iodine and $[Mo(dppe)_2Cp]^+$ give the isolable paramagnetic dication (312), and related dithiolene complexes $[M(S\!-\!S)_2Cp]^Z$ (M = Mo or W) form a three-membered redox series (Z = 0, −1, and −2). For example, the monoanionic mnt derivative is reversibly oxidized and reduced (M = Mo, E^0 = 0.78 and −1.42 V) (282, 313). The electrochemistry of $[Mo(CO)_2(XYCNMe_2)Cp]$ (XY = S_2, Se_2, SeO, etc.) and the tricarbonyls $[Mo(CO)_3\{X(Y)CNMe_2\}Cp]$ (XY = SO and SeO) was no doubt complicated by the thermal decarbonylation of the latter to the former (314).

Both $[WH_2Cp_2]$ and $[WHPhCp_2]$ give monocations ($E_{1/2}$ = −0.35 and −0.09 V, respectively), but $[WH_2Cp_2]^+$ has a lifetime of less than 1 msec and rapidly yields $[W_2(\mu\text{-}H)H_2Cp_2]^+$ (39); the molybdenum analog $[MoH_2Cp_2]^+$ is proposed as an intermediate in the reactions of $[MoH_2Cp_2]$ with α-dicarbonyls such as camphorquinone (315). The phenyl complex $[WHPhCp_2]^+$ is more persistent ($t_{1/2} \approx$ 20 seconds) and its ESR spectrum can be detected at −45°C via flow electrolysis in MeCN (39).

The oxidation of $[WCl_2Cp_2]$ [E^0 = 0.36 V] and $[WMe_2Cp_2]$ [E^0 = −0.42 V] yields radical cations which are much more stable than the hydrides described above (39). At −78°C, $[CPh_3]^+$ and the dimethyl complex afford $[WMe_2Cp_2]^+$, but at higher temperatures hydrogen abstraction by the CPh_3 radical gives $[WH(C_2H_4)Cp_2]^+$. Loss of the α-hydrogen atom from a methyl group was confirmed by the formation of $[WH(C_3H_6)Cp_2]^+$ from $[WMeEtCp_2]$ and $[CPh_3]^+$, and the intermediate carbene was trapped by PMe_2Ph as $[W(CH_2PMe_2Ph)EtCp_2]^+$ (316).

Although the appropriate potentials have not been reported, $[ML_2Cp_2]$ (M = Mo or W, L = anionic N, O, or S donor, or halide) are chemically oxidized by HNO_3 or H_2O_2; in some cases the monocations have been isolated as $[PF_6]^-$ salts. Electrolytic oxidation also generated the radicals which were characterized by ESR spectroscopy (317). However, the four-line spectrum assigned to an intermediate was shown (318) to be due to ClO_2, formed by decomposition of the base electrolyte salt $[N-n-Bu_4][ClO_4]$. The importance of controlled-potential electrolysis for the generation of redox products is clear.

The cationic nitrosyls $[W(NO)_2LCp]^+$ [L = PPh_3 or $P(OR)_3$, R = Me or Et] are reduced in two one-electron steps (e.g., L = PPh_3, $E^0 = -0.18$ V, $E_{1/2} = -1.68$ V), the second of which is irreversible. Hydrazine, zinc, or alkoxide ion gives the radicals $[W(NO)_2LCp]$ which show $\tilde{v}(NO)$ to be lowered by 160–175 cm^{-1} on reduction. MO calculations support the implication that the unpaired electron is localized largely on the NO ligands. The singly occupied orbital is significantly anti-bonding between the two nitrosyl groups with the N—W—N angle in $[W(NO)_2\{P(OPh)_3\}Cp]$ (102.7°) greater than that in $[WCl(NO)_2Cp]$ (92°C) (319).

In CH_2Cl_2, the one-electron oxidations of $[MoI(NO)(L–L)Cp]^-$ (L–L = mnt, pfdt, or tcbdt; $E^0 = 0.4–0.6$ V), $[Mo(NO)\{P(OPh)_3\}(mnt)Cp]$ ($E^0 = 1.09$ V), and $[MoI(NO)(S_2CNR_2)Cp]$ (R = Me or Et, $E^0 \approx 1.3$ V) are reversible, but that of $[MoI(NO)\{S_2CC(CN)_2\}Cp]^-$ is not (320).

D. Manganese and Rhenium

The electrochemical reduction of $[M(CO)_3Cp]$ [M = Mn (84, 321) or Re (86)] has only recently been further quantified. At platinum in thf, the manganese complex exhibits a reversible wave at -2.34 V followed by a second irreversible process at -2.82 V. Although exhaustive electrolysis at the potential of the first wave resulted in the passage of 2.4 electrons per molecule, the air-sensitive product is formulated as $[NBu_4][Mn(CO)_3Cp]$ (322).

The ketones $[Mn(CO)_3(\eta^5-C_5H_4COR)]$ (R = Me, aryl, furyl, pyrrolyl, etc.) show two cathodic polarographic waves, and reduction in the presence of protons yields the corresponding alcohol (323). At $-80°C$, however, the radical anions $[Mn(CO)_3\{\eta\text{-}C_5H_3R(COPh)\}]^-$ (R = H, 2-Me, or 3-Me) can be generated using potassium. The room temperature ESR spectra show small couplings to the metal atom ($A_{55_{Mn}}$ = 7.68 G), splitting due to the ortho and para hydrogen atoms of the phenyl group, and g values close to 2.00. The anions are therefore similar to those of arenechromium derivatives (Section VIII) in that the single electron is confined largely to the organic ligand (324).

The one-electron oxidation of $[Mn(CO)_3Cp]$ in CF_3CO_2H is reversible (E^0 = 0.79 V), with a second irreversible step at a much more positive potential ($E_{1/2}$ = 1.98 V); the metal-based radical cation ($A_{55_{Mn}}$ = 92.5 G, g = 2.015) can be generated by CPE (42). The oxidation potentials of representative examples of $[Mn(CO)_{3-n}L_n(\eta\text{-}C_5R_4R')]$ are given in Table IV; while there is the usual dependence of E^0 on L, R, or R', the effect of a change in n is most marked (12, 325–329). The disubstituted complexes $[Mn(CO)L_2(\eta\text{-}C_5R_5)]$ (L = phosphine) can be oxidized chemically by $AgPF_6$ (325), arenediazonium ions (326), the organic π acids tcne and chloranil, and acceptors such as $[MCl_2Cp_2]$ (M = Ti, Zr, or Hf) and $[TiCl_3Cp]$ (330) to give isolable salts, e.g., $[Mn(CO)(dppe)Cp][PF_6]$

TABLE IV

The One-Electron Oxidation of $[Mn(CO)_{3-n}L_n(\eta\text{-}C_5R_4R')]$

n	L	R	R'	E^0(V)	Reference[a] electrode	Solvent	Base electrolyte	Reference
0	—	H	H	0.79	Sce	CF_3CO_2H	$[NBu_4][BF_4]$	42
1	$P(OPh)_3$	H	H	0.92		CH_2Cl_2	$[NEt_4][ClO_4]$	326
1	PPh_3	H	H	0.59		CH_2Cl_2	$[NEt_4][ClO_4]$	326
1	PPh_3	Me	Me	0.37		CH_2Cl_2	$[NEt_4][ClO_4]$	326
2	PPh_3	H	H	−0.09	Sce	CH_2Cl_2	$[NBu_4][ClO_4]$	325
2	1/2 dppe	H	H	-0.50^b	$Ag/AgClO_4$	MeCN	$[NEt_4][BF_4]$	328
1	CN-t-Bu	H	Me	0.54	Sce	MeCN	$[NEt_4][ClO_4]$	12
3	CNMe	H	H	-0.30^c	Sce	CH_2Cl_2	$[NBu_4][ClO_4]$	327
1	C(OMe)Ph	H	Me	0.45	Sce	CH_2Cl_2	$[NEt_4][ClO_4]$	329
1	NH_3	H	H	0.29	Ag/AgCl	thf	$[NBu_4][ClO_4]$	322
1	N_2H_4	H	H	0.31	Ag/AgCl	thf	$[NBu_4][ClO_4]$	322
1	$N_2H_2{}^d$	H	H	0.35	Ag/AgCl	thf	$[NBu_4][ClO_4]$	322
1	NCMe	H	Me	0.19	Sce	MeCN	$[NEt_4][ClO_4]$	12

[a] A calomel electrode 1 M in LiCl, unless stated otherwise.
[b] Second oxidation wave at 0.79 V.
[c] Irreversible waves at 0.72 and 1.80 V.
[d] $[\{Mn(CO)_2(\eta\text{-}C_5H_5)\}_2(\mu\text{-}N_2H_2)]$.

(325, 326). Silver(I) salts and the dicarbonyls (L = P donor) yield red or purple solutions of the radical cations $[Mn(CO)_2L(\eta\text{-}C_5R_5)]^+$, characterized by the large shifts of $\tilde{v}(CO)$ to higher wave number (ca. 100–150 cm^{-1}). Unlike the monocarbonyls, the dicarbonyl cations were too easily reduced for their ESR spectra to be detected. Oxidation by $[N(p\text{-Br-}C_6H_4)_3][SbCl_6]$ also gave $[Mn(CO)_2L(\eta\text{-}C_5R_5)]^+$, but further reaction occurred to produce $[MnX(CO)_2L(\eta\text{-}C_5R_5)]^+$ (X = Cl); halogens and $[Mn(CO)_2L(\eta\text{-}C_5R_5)]$ afford analogous cations (X = Cl, Br, or I) directly (326).

The one-electron oxidation of $[Mn(CO)_2LCp]$ [L = PhC$_2$Ph (328), isocyanide (12), or carbene (329)] has been briefly noted, and I$_2$ and $[Mn(CO)_2\{C(OMe)Fc\}(\eta\text{-}C_5H_4Me)]$ yield a manganese- rather than an iron-based radical cation (329). Derivatives of simple nitrogen donors, such as $[Mn(CO)_2LCp]$ (L = N$_2$, N$_2$H$_4$, or NH$_3$) and $[\{Mn(CO)_2Cp\}_2$-$(\mu\text{-}N_2H_2)]$, not only undergo reversible oxidation but also irreversible reduction at very negative potentials (e.g., L = N$_2$, $E_{1/2}$ = -2.01 V in thf) (322). In the presence of oxygen, $[Mn(CO)_2(NH_2\text{-}m\text{-}Me\text{-}C_6H_4)Cp]$ and H$_2$O$_2$ give $[Mn(CO)_2(NH\text{-}m\text{-}Me\text{-}C_6H_4)Cp]$. The complex is described as a trapped aminyl radical but MO calculations suggest the half-occupied orbital to be 70–80% Mn(CO)$_2$Cp group in character; the ESR spectrum shows coupling to the metal ($A_{55_{Mn}}$ = 50 G) but not to the NH group (331).

Perhaps the most interesting aspect of the electrochemistry of $[Mn(CO)_{3-n}$-$L_n(\eta\text{-}C_5R_5)]$ is the occurrence of catalytic, oxidative substitution reactions. Partial oxidation of $[Mn(CO)_2(NCMe)(\eta\text{-}C_5H_4Me)]$ (E^0 = 0.19 V) in the presence of PPh$_3$ rapidly gives quantitative yields of $[Mn(CO)_2(PPh_3)$-$(\eta\text{-}C_5H_4Me)]$ via the mechanism shown in Scheme 24. The reaction, which occurs with very high current efficiency and long kinetic chain length, is

$$[M(NCMe)] \underset{+e^-}{\overset{-e^-}{\rightleftharpoons}} [M(NCMe)]^+$$

$$[M(NCMe)]^+ + PPh_3 \longrightarrow [M(PPh_3)]^+ + MeCN$$

$$[M(PPh_3)]^+ + [M(NCMe)] \longrightarrow [M(PPh_3)] + [M(NCMe)]^+$$

$$M = Mn(CO)_2(\eta\text{-}C_5H_4Me)$$

Scheme 24

TABLE V

THE ONE-ELECTRON TRANSFER REACTIONS OF $[Mn(NO)LL'(\eta-C_5H_4R)]^Z$

L	L'	R	Z	Process	E^0 (V)[a]	Reference
CO	PPh$_3$	H	1	$1 \rightleftharpoons 2$	1.62	332
(PMe$_2$Ph)$_2$		H	1	$1 \rightleftharpoons 2$	0.90	332
o-phen		Me	1	$1 \rightleftharpoons 2$	0.89	332
S$_2$CNMe$_2$		Me	0	$0 \rightleftharpoons 1$	0.38	282
S$_2$CSC(t-Bu)		H	0	$0 \rightleftharpoons 1$	0.53	332
S$_2$C=C(CN)$_2$		H	−1	$-1 \rightleftharpoons 0$	0.20	333
Mnt		Me	−1	$-1 \rightleftharpoons 0$	0.09	282
Tcbdt		Me	0	$0 \rightleftharpoons 1$	0.91	282
Tcbdt		Me	0	$0 \rightleftharpoons -1$	−0.32	282
Tdt		Me	0	$0 \rightleftharpoons 1$	0.54	282
Tdt		Me	0	$0 \rightleftharpoons -1$	−0.52	282

[a] Versus a calomel electrode $1\,M$ in LiCl, in CH_2Cl_2 with $[NEt_4][ClO_4]$ as base electrolyte.

facilitated by the ability of $[Mn(CO)_2(PPh_3)(\eta-C_5H_4Me)]^+$ ($E^0 = 0.52$ V) to oxidize the neutral nitrile complex. The mechanism in Scheme 24 also applies to other monometallic systems in which a simple donor such as MeCN, acetone, or thf is substituted by a π-acceptor ligand (12). Poly-nuclear carbonyls, by contrast, undergo catalytic *reductive* substitution (13, 31).

Representative examples of the redox-active nitrosyl complexes $[Mn(NO)$-$LL'(\eta-C_5H_4R)]^Z$ are given in Table V. The E^0 values depend on L, L', R, and, most markedly, Z, and correlate linearly with $\tilde{v}(NO)$ (332). The anions $[Mn(NO)(S_2CX)Cp]^-$ [X = N(CN), C(CN)CO$_2$Et, or C(CN)CONH$_2$] and the neutral compounds $[Mn(NO)(S_2CNR_2)Cp]$ (R = Me or Et) are chemically oxidized to radicals by iodine (333). The complexes $[Mn(NO)$-$(L–L)(\eta-C_5H_4R)]$ (L–L = mnt, tdt, or tcbdt; R = H or Me) are part of a three-membered redox series (Table V) for which the E^0 values are very dependent on the sulfur-ligand substituent; both neutral and anionic species are isolable (282).

E. *Iron and Ruthenium*

The irreversible two-electron reduction of $[\{Fe(CO)_2Cp\}_2]$ (84, 174, 334) (e.g., $E_{1/2} = -1.48$ V in thf) gives $[Fe(CO)_2Cp]^-$, which, in turn, is reoxidized in a one-electron step [e.g., $E_{1/2} = -1.18$ V in thf (334)] to the dimer. Electrolytic oxidation of the anion at mercury gives $[Hg\{Fe(CO)_2-Cp\}_2]$ (84) via $[Hg\{Fe(CO)_2Cp\}_3]^-$ (334).

ESR spectroscopy provides indirect evidence for the intermediacy of $[Fe(CO)_2Cp]$ in the reactions of $[Fe(CO)_2Cp]^-$ with alkyl iodides, allyl or benzyl bromides, or $[C_7H_7]^+$; electron transfer yields detectable organic radicals (177). Interestingly, $[Fe(cod)Cp]$, isoelectronic with $[Fe(CO)_2Cp]$, can be isolated from the reaction between CPh_3Cl and $[LiFe(cod)Cp]$ (335).

The oxidation of $[\{Fe(CO)_2Cp\}_2]$ is irreversible and the potential is very electrode dependent (e.g., $E_{1/2} = 0.19$ V at carbon, 1.0 V at platinum, in MeCN). CPE (336) or reaction with iron(III) salts (337) in the presence of a ligand L provides a good route to $[Fe(CO)_2LCp]^+$ (L = acetone, NCMe, P donor, etc.). Coulometry reveals that the oxidative cleavage reaction requires the loss of two electrons per dimer (336). However, kinetic studies of the reaction with $[RuCl_2(bipy)_2]^+$ or $[\{Fe(CO)Cp\}_4]^+$ show the rate-determining step to involve one-electron transfer (338).

Although $[\{Fe(CO)_2Cp\}_2]^+$ remains undetectable by CV, the derivative $[Fe_2(\mu\text{-}CO)_2(\mu\text{-}dppe)Cp_2]$ is reversibly oxidized in a one-electron step ($E^0 = 0.10$ V in CH_2Cl_2) which is followed by a second irreversible process at 0.95 V (339). The monocations $[Fe_2(\mu\text{-}CO)_2(\mu\text{-}L)Cp_2]^+$ (L = Ph_2PRPh_2, R = CH_2, C_2H_2, C_2H_4. or NEt) are readily isolable using I_2 or silver(I) salts; further oxidation by Ag^+ gives $[Fe(CO)_2LCp]^+$, probably via the unstable binuclear dication (340).

The complexes $[Fe(CO)_{3-n}L_nCp]^+$ ($n = 0\text{-}3$, L = PPh_3, NCMe, or CNMe) ($84, 86, 341$) undergo irreversible one-electron reduction. For $n = 0$ or 1, a second polarographic wave is observed, corresponding to the reductive cleavage of $[\{Fe(CO)_2Cp\}_2]$; dimer formation is not detected for the more highly substituted cations (341).

Many of the neutral complexes $[FeX(CO)_{2-n}L_nCp]$ are reversibly oxidized at potentials sufficiently negative to allow the chemical isolation of radical cations ($325, 342\text{-}346$). Representative examples are given in Table VI. The E^0 values for $[FeBr(CNR)_2Cp]$ (R = aryl), which correlate linearly with the Hammett σ parameters for R, are relatively low. However, chemical or electrolytic oxidation yields only $[Fe(CNR)_3Cp]^+$, which are, in turn, oxidized at a potential ~ 0.9 V more positive than the neutral bromides (342). The analog $[Fe(PMe_3)_3Cp]^+$ ($E^0 = 0.71$ V in CH_2Cl_2) and $[NO][PF_6]$ afford the isolable dication (344).

Although $[Fe(SPh)L_2Cp]^+$ [L = PMe_3 (344) or $P(OPh)_3$ (346), L_2 = dppe (343)] are readily prepared by Ag^+ oxidation, $[Fe(SPh)(CO)_2Cp]$ and $[NO][PF_6]$ in acetone yield diamagnetic $[Fe_2(CO)_4(\mu\text{-}PhSSPh)Cp_2]^{2+}$

TABLE VI

THE ONE-ELECTRON OXIDATION OF $[FeX(CO)_{2-n}L_nCp]$

X	n	L	E^0 (V)[a]	Chemical oxidant[b]	Reference
I	1	CNPh	0.95		325
Br	2	CNPh	0.51[c]		342
I	1	P(OPh)$_3$	0.96		342
I	2	P(OPh)$_3$	ca. 0.6	[NO][PF$_6$] in C$_6$H$_6$	342
SnCl$_3$	2	Dppe	0.90		343
CN	2	Dppe	0.54	AgPF$_6$ in acetone	343
NCS	2	Dppe	0.32	AgPF$_6$ in acetone	343
Cl	2	Dppe	0.08	AgPF$_6$ in acetone	343
SnMe$_3$	2	Dppe	0.07	AgPF$_6$ in acetone	343
SnPh$_3$	2	PMe$_3$	-0.02	AgBF$_4$ in acetone	344
H	2	Dppe	-0.08		343
SPh	2	Dppe	-0.25	AgPF$_6$ in acetone, H$^+$	343
Me	2	Dppe	-0.26	AgPF$_6$ in acetone	343
SPh	2	PMe$_3$	-0.37^d	AgBF$_4$ in acetone	344
S$_2$O$_3$^{e}	2	Dppe	-0.37^f		343

[a] Versus sce, in CH$_2$Cl$_2$ with [NBu$_4$][ClO$_4$] as base electrolyte, unless stated otherwise.
[b] Specified if radical cation isolated.
[c] In MeCN.
[d] Second irreversible oxidation at 1.06 V.
[e] Isolated as the neutral radical.
[f] Reduction. An irreversible oxidation occurs at 0.77 V.

(30). The complexes [Fe(SPh)(CO)LCp] react similarly with [NO][PF$_6$] (L = phosphite) or AgPF$_6$ (L = phosphine), but in solution the products exist as an equilibrium mixture of blue, 17-electron, monomeric cations and red, diamagnetic dimers. Normally the electron would be lost from a metal-

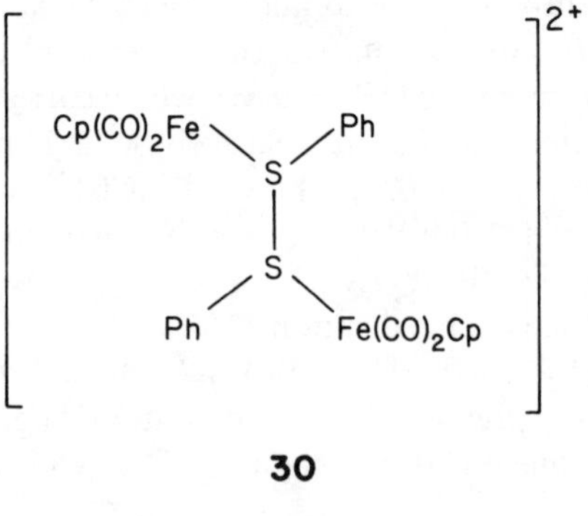

30

based orbital of $[FeX(CO)_{2-n}L_nCp]$, but the oxidation of the thiolato complexes probably involves a lone pair on sulfur (346).

The electrochemistry of $[Fe(CO)_2\{X(Y)CNMe_2\}Cp]$ and $[Fe(CO)(XYCNMe_2)Cp]$ (X, Y = O, S, or Se) (347) is similar to that of cyclopentadienylmolybdenum analogs (Section VII,C). The oxidation of $[FeL(RNNNR)Cp]$ (R = aryl, L = CO or phosphite) is fully reversible ($E^0 \approx$ 0.3–0.6 V vs Ag/AgI in CH_2Cl_2) but $[NO]^+$ and the carbonyl produce $[Fe(CO)(NO)(RNNNR)Cp]^+$ (348).

The reduction of $[FeR(CO)LCp]$ [R = alkyl, Ph, allyl, C_5H_5, or Fc; L = CO, PPh_3, or $P(OPh)_3$] was described as an irreversible two-electron process giving $[Fe(CO)LCp]^-$ and the carbanion; there are linear correlations between the reduction potentials measured ($E_{1/2} \approx -2.0$ V in MeCN) and the pK_a values and Taft σ^* inductive constants of RH (307, 349, 350). More recently, however, $[FeX(CO)_2Cp]$ (X = halide, $SnCl_3$, or $GeCl_3$) have been shown to exhibit two polarographic reduction waves, the first giving X^- and $[Fe(CO)_2Cp]$, and the second due to $[\{Fe(CO)_2Cp\}_2]$ (351). In the presence of PPh_3, the reduction of $[FeMe(CO)_2Cp]$ gives $[Fe(COMe)(CO)(PPh_3)Cp]$. Electron addition to a σ^* orbital weakens the iron–methyl bond and facilitates acyl formation; the phosphine stabilizes the coordinatively unsaturated anion $[Fe(COMe)(CO)Cp]^-$. If a small electrolytic pulse is passed through a solution of $[FeMe(CO)_2Cp]$ and PPh_3 at -1.8 V, electrocatalytic formation of $[Fe(COMe)(CO)(PPh_3)Cp]$ is observed. Neither precursor ($E_{1/2} = -1.90$ V) nor product ($E_{1/2} = -1.95$ V) is reduced at the applied potential (352).

The oxidative cleavage of $[FeR(CO)_2Cp]$ and related species (Section VII,C) by methanolic Ce(IV) in the presence of LiCl gives the acetate RCO_2Me. The proposed mechanism (311), involving initial one-electron oxidation followed by rapid carbonyl insertion and subsequent nucleophilic cleavage (Scheme 23), has recently been largely substantiated. The alkyls $[MR(CO)_{2-n}L_nCp]$ (M = Fe or Ru, n = 0 or 1, L = phosphine or phosphite) (353–356) are irreversibly oxidized [e.g., n = 0, $E_{1/2} \approx 1.3$ V; n = 1, $E_{1/2} \approx 0.8$ V vs Ag/AgCl in CH_2Cl_2 (353)], and the product wave in the cyclic voltammogram may be assigned to the quasi-reversible oxidation of $[Fe(COMe)(CO)LCp]$ (355). At $-78°C$, Ce(IV) or copper(II) triflate and $[FeMe(CO)_2Cp]$ in CH_2Cl_2–MeCN afford a deep green solution of the *acyl* radical cation $[Fe(COMe)(CO)(NCMe)Cp]^+$, and related species such as $[Fe(COMe)(CO)LCp]^+$ (L = CO or PPh_3) may be generated directly from the appropriate neutral complex; the phosphine cation is relatively stable ($t_{1/2} = 177$ seconds at 20°C) (355). The anions $[Fe(CN)(COR)(CO)Cp]^-$ (R = Me or Ph) are reversibly oxidized in MeCN at a relatively negative potential ($E^0 = 0.10$ V), and the neutral radical [Fe(CN)(COMe)(CO)Cp], generated by CPE, undergoes thermal decomposition to give acetone (*357*).

The oxidative cleavage of [FeR(CO)$_2$Cp] (R = alkyl) by CuX$_2$ (X = Cl or Br) to give RX and [FeX(CO)$_2$Cp] was thought to proceed via the mechanism shown in Scheme 25 (*354*). However, the rapid formation of acyl

$$[FeR(CO)_2Cp] + CuX_2 \longrightarrow [FeR(CO)_2Cp]^+ + [CuX_2]^-$$

$$[FeR(CO)_2Cp]^+ + [CuX_2]^- \longrightarrow [Fe(CO)_2Cp] + RX + CuX$$

$$[Fe(CO)_2Cp] + CuX_2 \longrightarrow [FeX(CO)_2Cp] + CuX$$

SCHEME 25

cations on oxidation suggests that RX may result from nucleophilic displacement at the α carbon of the acyl group (*355*).

In contrast to the alkyl complexes described above, the Ag$^+$ oxidation of [FeR(CO)$_2$Cp] (R = allyl or CH$_2$C≡CPh) leads to regiospecific dimerization; C—C bond formation involves the most highly substituted carbon atom of R (Scheme 26). An important extension of this reaction

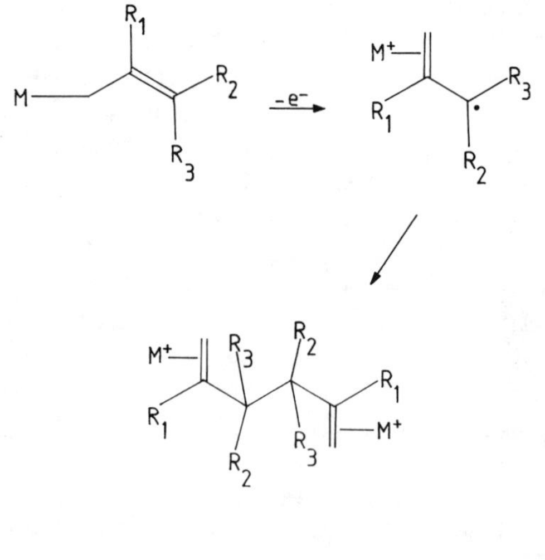

M = Fe(CO)$_2$Cp

SCHEME 26

involves the oxidative transfer of allyl ligands. Thus [Fe{σ-CH$_2$C(Me)= CH$_2$}(CO)$_2$Cp] and 1,3-diphenylisobenzofuran gave, on Ag(I) oxidation and subsequent reaction with NaI, compound **31** (*358*).

31

F. *Cobalt, Rhodium, and Iridium*

Sodium amalgam and [Co(CO)$_2$Cp] in 1,2-dimethoxyethane give paramagnetic [{Co(CO)Cp}$_2$]$^-$ (*31*) but CV in MeCN shows the first step in the reaction to be the formation of [Co(CO)$_2$Cp]$^-$ ($E^0 = -2.14$ V vs Ag/AgClO$_4$); decarbonylation and nucleophilic attack of [Co(CO)Cp]$^-$ on [Co(CO)$_2$Cp] yield the binuclear product. The brief electrochemical study of [Co(CO)$_2$Cp] in MeCN also showed a reversible oxidation at -0.06 V, but neither the radical cation nor [Co(CO)$_2$Cp]$^-$ were further characterized (*359*). However, the two ESR spectra observed on ^{60}Co γ irradiation of [Co(CO)$_2$Cp] in acetone at 77 K are assigned to the radical anion with a half-occupied d_{yz} σ^* orbital, and to the radical cation with the free electron in a d_{xz} orbital (*360*).

The monocarbonyl [Co(CO)(PPh$_3$)Cp] is thought to be reduced to a polymeric cyclopentadienylcobalt complex (*359*) but its oxidation, and that of related species, is better defined. In CH$_2$Cl$_2$ or thf, [Co(CO)LCp] (L = PPh$_3$ or PCy$_3$) is reversibly oxidized, and [FeCp$_2$]$^+$ gives [Co(CO)(PCy$_3$)-Cp]$^+$ or [Co(PPh$_3$)$_2$Cp]$^+$ (*361*); the latter is also isolable directly from [CoL$_2$Cp] (L = phosphine or phosphite) and AgBF$_4$ in thf (*362*). Carbonyl substitution of [Co(CO)LCp] by L (L = P donor) does not occur thermally, but one-electron oxidation weakens the Co—CO bond and renders the metal more susceptible to nucleophilic attack. Further oxidation of [Co(CO)$_{2-n}$L$_n$Cp]$^+$ is not observed in CH$_2$Cl$_2$ (*361*), and an air-stable, yellow complex isolated by addition of AgBF$_4$ to [Co(PPh$_3$)$_2$Cp]$^+$ in thf (*362*) cannot be the dication [Co(PPh$_3$)$_2$Cp]$^{2+}$.

On the basis of the potentials for the oxidation of [Co(CO)(PPh$_3$)Cp] ($E^0 = 0.13$ V) and the reduction of [Co(PPh$_3$)$_2$Cp]$^+$ ($E^0 = -0.71$ V), E^0

for the dicarbonyl should be ~ 1.0 V in CH_2Cl_2 (cf. in MeCN, above). Thus, as expected, $[Co(CO)_2Cp]$ and $[FeCp_2]^+$ do not react. In the presence of PPh_3, however, high yields of $[Co(PPh_3)_2Cp]^+$ rapidly form via the mechanism shown in Scheme 27. The cation $[Co(CO)(PPh_3)(\eta\text{-}C_5Me_5)]^+$

$$[Co(CO)_2Cp] + PPh_3 \; \underset{\longleftarrow}{\overset{\longrightarrow}{\;\;\;}} \; [Co(CO)(PPh_3)Cp] + CO$$

$$[Co(CO)(PPh_3)Cp] \; \underset{+e^-}{\overset{-e^-}{\rightleftarrows}} \; [Co(CO)(PPh_3)Cp]^+$$

$$[Co(CO)(PPh_3)Cp]^+ + PPh_3 \longrightarrow [Co(PPh_3)_2Cp]^+ + CO$$

SCHEME 27

results similarly from $[Co(CO)_2(\eta\text{-}C_5Me_5)]$, but in this case $[Co(CO)(PPh_3)(\eta\text{-}C_5Me_5)]$ cannot be made directly from the dicarbonyl and PPh_3 (361).

The complexes $[Co(CO)_{2-n}L_nCp]^+$ undergo radical–radical coupling and substitution reactions rather similar to those of $[Fe(CO)_3(PPh_3)_2]^+$ (Section II,E). The o-quinone derivatives $[Co(O\text{-}O)LCp]^+$ (e.g., O–O = o-chloranil, L = PCy_3) (361) and the rhodium analogs (363) show ESR spectra consistent with the single electron confined mainly to the semiquinone ligand.

The electrochemistry of $[Rh(CO)_{2-n}L_nCp]$ is very different from that of the cobalt analogs; CV in CH_2Cl_2 or thf shows the one-electron oxidation of $[Rh(CO)(PPh_3)Cp]$ to be irreversible ($E_{1/2} = 0.43$ V) with the lifetime of the radical cation no greater than 0.5 seconds at room temperature (364). The addition of Ag^+ to $[Rh(PPh_3)_2Cp]$ in thf (250), or of $[FeCp_2]^+$ or $[N_2\text{-}p\text{-}FC_6H_4]^+$ to $[Rh(CO)(PPh_3)Cp]$ in CH_2Cl_2 (364), gives the fulvalene complexes $[Rh_2(CO)_{4-n}L_n(\eta^5,\eta'^5\text{-}C_{10}H_8)]^{2+}$ (32, n = 2 or 4). The dications are reduced by sodium amalgam to $[Rh_2(CO)_{4-n}(\eta^5,\eta'^5\text{-}C_{10}H_8)]$ (33),

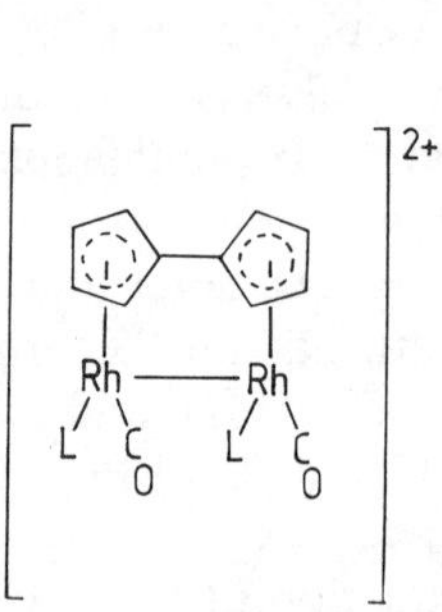

32 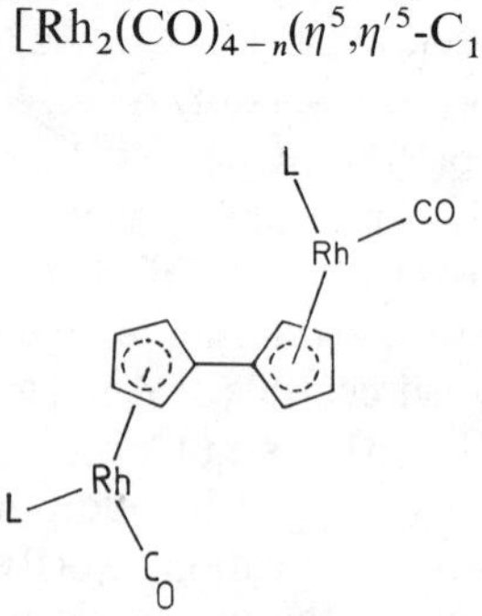   33

which are reoxidized by $[N_2\text{-}p\text{-}FC_6H_4]^+$, $[FeCp_2]^+$, or H^+. Complex **32** is also generated by CPE of $[Rh(CO)(PPh_3)Cp]$, and CV shows that the interconversion of **32** and **33** involves a diffusion-controlled, reversible, two-electron transfer ($E^0 = 0.01$ V) (*364*).

The mechanism of the oxidative dimerization of $[Rh(CO)_{2-n}L_nCp]$ is not simple. MO calculations (*250*) show the HOMO of the neutral complex to be predominantly metal based with, however, significant C_5H_5-ligand character. Thus, the dimerization of the radical cation may involve C—C bond formation as the first step. Complex **32** then results from elimination of H_2, or via the loss of two protons and subsequent oxidation of **33** (Scheme 28). The alternative, in which metal–metal bond formation precedes ring coupling, seems less likely (*250, 364*).

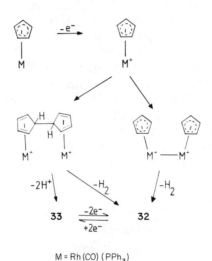

$$M = Rh(CO)(PPh_3)$$

SCHEME 28

The metal–metal bond of **32** was invoked to account for the observed diamagnetism, and such a bond, requiring the two Rh(CO)L groups to be located on the same (cis) side of the $C_{10}H_8$ ligand, has been found crystallographically. Surprisingly, in view of the electrochemical reversibility of the interconversion of **32** and **33**, X-ray studies show $[Rh_2(CO)_2(PPh_3)_2\text{-}(\eta^5,\eta'^5\text{-}C_{10}H_8)]$ to have the trans structure (*365*).

The reaction between $[Rh(CO)(PPh_3)Cp]$ and $AgPF_6$ in toluene does not give **32**, but rather the air-stable, 2:1 adduct $[Ag\{Rh(CO)(PPh_3)Cp\}_2][PF_6]$ (*364, 366*), which has a nearly linear Rh–Ag–Rh backbone. The complex acts as a stable source of the highly reactive radical cation $[Rh(CO)(PPh_3)Cp]^+$

(Scheme 29). Thus, NO or NO_2 gives $[Rh(NO)(PPh_3)Cp]^+$ via radical–radical coupling; nitric oxide and $[Rh(CO)(PPh_3)Cp]$ do not react (366).

$$[Ag\{Rh(CO)(PPh_3)Cp\}_2]^+ \equiv Ag + [Rh(CO)(PPh_3)Cp] + [Rh(CO)(PPh_3)Cp]^+$$

$$\downarrow NO/CH_2Cl_2$$

$$[Rh(NO)(PPh_3)Cp]^+ + [Rh(CO)(PPh_3)Cp] + Ag$$

SCHEME 29

The cobalt(III) complexes $[Co(NCMe)L_2Cp]^{2+}$ (L = NCMe, L_2 = bipy) are reduced at mercury; the bipy complex shows two waves ($E^0 = -0.13$ and 0.62 V in MeCN) corresponding to the formation of $[Co(bipy)Cp]^Z$ (Z = 0 and 1) (367). The triazenido compounds $[CoL(RNNNR)Cp]^+$ (R = aryl, L = phosphine or phosphite) are reversibly reduced ($E^0 \approx 0.0$ V vs Ag/AgCl in acetone) (348), and the neutral radical (L = PPh_3) may be prepared from $[CoCl(PPh_3)Cp]$ and $[RNNNR]^-$ (368); the dithiolenes $[M(L-L)Cp]$ [M = Co, L–L = pfdt (84) or mnt (282); M = Rh, L–L = pfdt (84)] give monoanions characterized by ESR spectroscopy.

The one-electron oxidation of 34 (M = Fe, $E^0 = -0.46$ V; M = Co, $E^0 = 1.02$ V) is associated more with the octahedral metal M than the

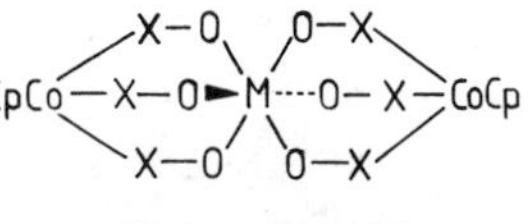

34; X = P(OEt₂)

cyclopentadienylcobalt fragment (369). The iron(III) monocation is simply prepared by oxidation in air (370), and the cobalt(III) analog is stable to reduction despite the very positive value of E^0 (369).

The bis(ethylene) complex $[Rh(\eta-C_2H_4)_2Cp]$ has been reported briefly to yield a radical cation (84) but the electron-transfer reactions of η^4-diene cobalt analogs are better defined. In thf or MeCN, the one-electron oxidation of $[Co(\eta^4$-diene)$(\eta-C_5R_5)]$ [R = H, diene = cyclopentadiene, cyclobutadiene, benzocycloheptatriene (371), cod (371, 372), or cot (372); R = Me, diene = cyclopentadiene (371) or cod (372)] is usually irreversible. In CH_2Cl_2, however, the process is more nearly reversible and the C_5Me_5 complexes show second oxidation waves due to the formation of unstable dications. Bulk electrolysis generally yields $[CoCp_2]^+$ and the free diene, but at $-15°C$ $[Co(cod)(\eta\text{-}C_5Me_5)]$ ($E^0 = 0.07$ V) gives a radical cation whose ESR spectrum is consistent with a d^7 metal configuration (372).

The polarographic reduction of η^4-diene complexes was first noted for $[M(\eta^4\text{-}1,4\text{-quinone})Cp]$ (M = Rh or Ir, quinone = duroquinone or 2,6-di(t-butyl)quinone), which show one two-electron wave (M = Ir), or two one-electron waves (M = Rh) at potentials more negative than those of the free quinone (373). The cyclopentadienone compounds $[M(\eta^4\text{-}C_5R_4O)Cp]$ (M = Co or Rh, R = Ph or C_6F_5) undergo reversible one-electron reduction to radical anions (e.g., M = Co, R = Ph, $E^0 = -1.46$ V in thf). The cobalt complexes, more stable than those of rhodium, are generated by alkali metal reduction at 100 K and show ESR spectra characteristic of metal-based d^9 species (374).

The reductive electrochemistry of $[Co(\eta^4\text{-diene})(\eta\text{-}C_5R_5)]$ (diene = cot, R = H or Me; diene = cod, R = H) is more complex yet provides an excellent example of redox-induced isomerization (Scheme 30). The cot

$$[Co(\eta^4\text{-}1,5\text{-diene})(\eta\text{-}C_5R_5)] \rightleftharpoons [Co(\eta^4\text{-}1,3\text{-diene})(\eta\text{-}C_5R_5)]$$

$$-e^- \updownarrow +e^- \qquad\qquad -e^- \updownarrow +e^-$$

$$[Co(\eta^4\text{-}1,5\text{-diene})(\eta\text{-}C_5R_5)]^{\cdot-} \rightleftharpoons [Co(\eta^4\text{-}1,3\text{-diene})(\eta\text{-}C_5R_5)]^-$$

SCHEME 30

complexes exist in solution as equilibrium mixtures of the 1,5- and 1,3-bonded isomers, with the former predominant. On one-electron reduction, both give $[Co(1,3\text{-cot})(\eta\text{-}C_5R_5)]^-$, which is reversibly oxidized to $[Co(1,3\text{-cot})(\eta\text{-}C_5R_5)]$ (e.g., R = H, $E^0 = -1,82$ V in thf). The anion is further reduced to a dianion ($E_{1/2} = -2.50$ V) and both are protonated to $[Co(\eta^4\text{-cyclooctatriene})(\eta\text{-}C_5R_5)]$.

The neutral 1,5-cod complex is reduced at a very negative potential, but in dmf the process is reversible ($E^0 = -2.35$ V). On electrolytic reduction in thf, however, $[Co(\eta^4\text{-}1,3\text{-cod})Cp]^-$ ($E^0 = -1.58$ V) is formed. The neutral 1,3-cod complex isomerizes only slowly to $[Co(\eta^4\text{-}1,5\text{-cod})Cp]$ so that reduction of the latter at -2.6 V, followed by oxidation at -1.4 V, provides a synthetic route to the otherwise inaccessible compound $[Co(\eta^4\text{-}1,3\text{-cod})Cp]$.

The rate constants for the various transformations have been estimated by electrochemical and NMR spectral methods (Table VII). Qualitatively, the isomerization of the cod complex necessarily requires hydrogen migration and is therefore slower than that of the cot derivative. For both, the anions isomerize more rapidly than the neutral species (11).

TABLE VII

RATE CONSTANTS FOR THE ISOMERIZATIONS OF
$[Co(\eta^4\text{-DIENE})Cp]^Z$

Diene	Z	Diene isomerization	Rate constant $(\sec^{-1})^a$
Cot	0	$1,3 \to 1,5$	30^b
Cot	-1	$1,5 \to 1,3$	$\geqslant 10^2$
Cod	0	$1,3 \to 1,5$	3×10^{-5}
Cod	-1	$1,5 \to 1,3$	0.1

[a] At room temperature, unless stated otherwise. Data from ref. *11*.

[b] At 375 K.

The ESR spectra of the anions are particularly interesting. Both isomers of $[Co(\eta^4\text{-cod})Cp]^-$ are metal-based radicals but $[Co(\eta^4\text{-1,3-cot})Cp]^-$ in frozen thf shows a more nearly isotropic spectrum with the singly occupied orbital possessing little cobalt d character. MO calculations are also consistent with localization of the free electron on the unbonded diene unit of $[Co(\eta^4\text{-1,3-cot})Cp]^-$ [cf. $[Fe(CO)_3(\eta^4\text{-cot})]^-$, Section VI], but on the metal of $[Co(\eta^4\text{-cod})Cp]^-$ (*255*).

G. *Nickel and Palladium*

The complexes $[NiL_2Cp]^+$ (L = phosphine, phosphite, dppe, das, cod, nbd) (*375–378*) are in one-electron steps, and the bis(ylide) $[Ni(CH_2PPh_3)_2\text{-}Cp]^+$ is also oxidized ($E^0 = -0.21$ V vs Ag/AgCl) (*378*). In the presence of a 10-fold excess of PPh_3, the reduction of $[Ni(PPh_3)_2Cp]^+$ is fully reversible, but in the absence of free ligand the equilibrium shown in Eq. (6) is reached.

$$2[NiL_2Cp] \rightleftharpoons [NiCp_2] + [NiL_4] \qquad (6)$$

Indeed, the neutral radicals can be prepared directly from nickelocene and $[NiL_4]$ (*378*). The ESR spectrum of $[Ni(\eta^4\text{-cod})Cp]$, generated by electrolytic reduction of the cation ($E^0 = -0.46$ V in CH_2Cl_2) at $-10°C$, is consistent with a metal-based radical (*375*), and the 19-electron configuration was verified by the X-ray structure of the analog $[Ni(bipy)Cp]$ (*378*).

The palladium complexes $[PdL_2(\eta\text{-}C_5Ph_5)]^+$ (L_2 = dppe, bipy, or cod) are not only reduced to neutral radicals (e.g., L_2 = cod, $E^0 = -0.47$ V) but are also reversibly oxidized to dications (e.g., L_2 = bipy, $E^0 = 1.18$ V). The paramagnetic C_5Ph_5 compounds are considerably more robust than their Cp analogs; $[Pd(\eta^4\text{-cod})(\eta\text{-}C_5Ph_5)]$ is stable at $-10°C$ in CH_2Cl_2 but the reduction of $[Pd(\eta^4\text{-cod})Cp]^+$ $(E_{1/2} = -0.74$ V) is completely irreversible (379).

The mononuclear dithiolene complex $[Ni(pfdt)Cp]$ undergoes one-electron reduction (84) but the related bimetallic complex $[Ni_2(\mu\text{-}C_2S_4)\text{-}(\eta\text{-}C_5Me_5)_2]$ (35) is oxidized to a monocation $(E^0 = 0.09$ V) and reduced to a monoanion $(E^0 = -0.92$ V) and a dianion $(E^0 = -1.43$ V) (380). The nitrosyl $[Ni(NO)Cp]$ is reduced $(E_{1/2} = -1.45$ V vs Ag/AgCl in dmf) to an unstable monoanion (381).

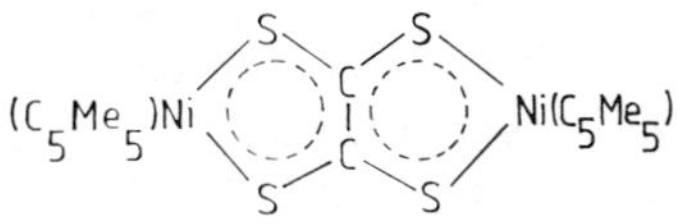

$$(C_5Me_5)Ni \left\langle \begin{array}{c} S \\ S \end{array} \begin{array}{c} C \\ \| \\ C \end{array} \begin{array}{c} S \\ S \end{array} \right\rangle Ni(C_5Me_5)$$

35

H. *Lanthanides and Actinides*

The oxidation of $[Yb(dme)(\eta\text{-}C_5Me_5)_2]$ by $[FeCp_2][PF_6]$ gives $[Yb(dme)(\eta\text{-}C_5Me_5)_2][PF_6]$ (382). CV shows the $[UClCp_3]$ (383), $[UCl_2\text{-}(\eta_2\text{-}C_5Me_5)_2]$ (384), and $[UCl_2(CNCy)(\eta\text{-}C_5Me_5)_2]$ (385) are reversibly reduced to monoanions. The salt $Na[UCl_2(thf)(\eta\text{-}C_5Me_5)_2]$ is isolable via sodium amalgam reduction in thf and undergoes reversible reoxidation to the neutral precursor; $[UCl(thf)(\eta\text{-}C_5Me_5)_2]$ is irreversibly reduced at about -0.7 V (384).

VIII

η^6-ARENE COMPLEXES

Apart from the many derivatives of $[Cr(CO)_3(\eta\text{-}C_6H_6)]$ (see below) and the sandwich complexes described in Section X, very few arene compounds have been studied electrochemically. The synthesis of $[V(CO)_3(\eta^6\text{-}C_6Ph_6)]$ (386) is somewhat surprising in view of the two-electron oxidation of $[V(CO)_3\text{-}(\eta^6\text{-mesitylene})]^-$ $(E_{1/2} = -0.92$ V in MeCN) (387).

The tricarbonyls $[Cr(CO)_3(\eta\text{-}C_6H_{6-n}Me_n)]$ $(n = 0-6)$, $[Cr(CO)_3(\eta\text{-}C_6H_5R)]$ (R = i-Pr, NH_2, OMe, etc.), and the isomers of $[Cr(CO)_3\{\eta\text{-}C_6H_4(NH_2)Me\}]$ are oxidized in two one-electron steps. The first is nearly reversible with $E_{1/2}$, ranging from 0.36 to 0.80 V, correlating with $E_{1/2}$ for the free arene and with the ionization potentials of the complex and the ligand (388). The ferrocenyl derivatives $[Cr(CO)_3(\eta\text{-}C_6H_5R)]$ (R = Fc or CH_3Fc) also show two oxidation waves but the first is associated with the iron center (389).

Carbonyl substitution leads to large shifts in $E_{1/2}$ to more negative potentials, facilitating the isolation of paramagnetic monocations. The alkyne complexes $[Cr(CO)_2(\eta^2\text{-}PhC_2Ph)(\eta^6\text{-}C_6H_{6-n}Me_n)][PF_6]$ ($n = 5$ or 6) are readily prepared from the neutral species ($E^0 = -0.2$ V in CH_2Cl_2) and $[NO][PF_6]$ or $AgPF_6$ in methanol–toluene mixtures, and show ESR spectra at room temperature ($g = 1.99$, $A_{53_{Cr}} = 15$ G) (390). The phosphine and phosphite derivatives $[Cr(CO)_2L(\eta\text{-}C_6Me_6)]$ are oxidized at more positive potentials ($E^0 = 0.4-0.7$ V) but the radical cations (L = $PPh_{3-n}Me_n$, $n = 0-2$) may be generated in CH_2Cl_2 using $[N_2\text{-}p\text{-}F\text{-}C_6H_4]^+$; $[Cr(CO)_2\text{-}(PPh_3)(\eta\text{-}C_6Me_6)][PF_6]$ is isolable. The ESR spectra are very different from those of the alkyne analogs, and are observed only at low temperature (e.g., L = PPh_3, $g = 2.041$, $A_{31_P} = 31.3$ G) (95). Tcne and $[Cr(CO)_2L(\eta\text{-}C_6Me_6)]$ afford $[Cr(CO)_2(\eta^2\text{-}tcne)(\eta\text{-}C_6Me_6)]$ via one-electron transfer followed by displacement of L from the cation by $[tcne]^-$ (391).

A mechanism, recently modified and discussed further in Section II,C,6, has been proposed to account for the observation of both CO substitution and one-electron oxidation in the reactions between $[Cr(CO)_2L(\eta\text{-}C_6Me_6)]$ and $[NO]^+$. Rather than depending simply on the relative E^0 values, the products arise by the decomposition of the intermediate $[Cr(NO)(CO)_2\text{-}L(\eta\text{-}C_6Me_6)]^+$, in which the NO group functions as a one-electron donor; carbonyl displacement gives the cationic nitrosyl but loss of NO yields the radical cation (95).

At $-43°C$, $[Cr(CO)_2(dam)]$ (dam = $Ph_2AsCH_2AsPh_2$), in which dam chelates via one arsenic atom and the η^6-bound aryl group, is reversibly oxidized at platinum ($E^0 = 0.60$ V vs Ag/AgCl); a second irreversible process occurs at ~ 1.27 V. The reversible reactions at mercury (Scheme 31) are rather different (392).

Arenechromiumtricarbonyl radical cations are also stabilized by Group IVB ring substituents. In propylene carbonate, $[Me_2SnPh_2\{Cr(CO)_3\}_n]$ ($n = 1$ or 2) and $[MeSnPh_3\{Cr(CO)_3\}_3]$ are reversibly oxidized ($E^0 \approx 0.8$ V vs Ag/AgCl), and coulometry reveals one-electron transfer for each $Cr(CO)_3$ group; the resulting cationic sites are essentially noninteracting (393).

$$2[M(CO)_2(dam)] + 2Hg \rightleftharpoons 2[HgM(CO)_2(dam)]^+ + 2e^-$$

$$2[HgM(CO)_2(dam)]^+ \rightleftharpoons [Hg\{M(CO)_2(dam)\}_2]^{2+} + Hg$$

SCHEME 31

The electrochemistry of $[Mo(CO)_3(\eta\text{-arene})]$ remains undefined but $[MoL_3(\eta\text{-}C_6H_5R)]^+$ (R = H or Me, L = pyr or 1-methylimidazole) are reduced to unstable $[MoL_3(\eta\text{-}C_6H_5R)]$ (e.g., L = pyr, $E^0 = -0.96$ V) (83).

The two-electron reduction of $[Cr(CO)_3(\eta\text{-arene})]$ occurs at very negative potentials (e.g., arene = C_6H_5Me, $E_{1/2} = -2.25$ V in MeCN) (84, 394, 395). The process is irreversible for substituted benzene derivatives but stable dianions, in which the arene is bonded as an η^4-diene, can be generated by electrolytic reduction of the naphthalene analogs (395).

When electron delocalization can occur onto suitable arene substituents, ligand-based radical anions can be identified. For example, $[Cr(CO)_3(\eta^6\text{-L})]$ (L = PhCO-t-Bu or Ph_2CO) are reversibly reduced in two steps to mono- and dianions (e.g., L = Ph_2CO, $E^0 = -1.45$ and -1.72 V in dmf) (396). The monoanions, prepared by alkali metal reduction (396–398), show ESR spectra very similar to those of the radicals L^-, but the smaller proton hyperfine couplings, reduced for both the complexed and uncomplexed rings (L = $RC_6H_4COC_6H_4R'$, R or R' = H, o-Me, p-Me, p-Cl, etc.) (397–398), suggest delocalization over the entire ligand L.

The one-electron oxidation potential of the chpt derivatives $[Cr(CO)_3\text{-}(\eta^6\text{-}C_7H_6RR')]$ (R = H, R' = H, exo- or $endo$-Ph, or exo-CN; R = R' = OMe) ($E_{1/2} = 0.65$–1.03 V) depends on the position of the substituents at the methylene carbon atom. For the chromium complexes (R = H, R' = CN; R = R' = OMe) (388) and $[Mo(CO)_3(\eta^6\text{-chpt})]$ (84), one-electron reduction is also observed.

IX

η^7- AND η^8-BONDED COMPLEXES

The irreversible one-electron reduction of the cycloheptatrienyl complexes $[M(CO)_3(\eta^7\text{-}C_7H_6R)]^+$ (M = Cr, R = H or Me) (388) leads to reductive ring coupling [M = Cr (*131, 399*) or W (*400*)], presumably via the neutral radical. Chromium(II) (*131*), zinc dust, or anions such as [OCN]$^-$ (*399, 400*) give [Cr$_2$(CO)$_6$(η^6,η'^6-C$_{14}$H$_{14}$)] and [W(CO)$_3$(η^6-C$_{14}$H$_{14}$)], in which one W(CO)$_3$ group has been removed from the dicycloheptatrienyl ligand; electrochemical reduction provides a route to [W$_2$(CO)$_6$(η^6,η'^6-C$_{14}$H$_{14}$)] (*401*). Sodium amalgam and paramagnetic [MoCl(dppe)(η-C$_7$H$_7$)]$^+$ give the neutral complex which is reversibly reoxidized in butyronitrile ($E^0 = -0.47$ V vs Ag/AgCl) (*402*).

Redox-related lanthanide and actinide complexes of cot have been described, usually without electrochemical studies. Potassium reduces [Ce(η-cot)$_2$] to mono- and dianions which are isolable as salts of the cation [K(monoglyme)$_2$]$^+$ (*403*). CV of [U(η-cot)$_2$] shows that the dication is formed via two one-electron transfers; the first oxidation leads to a geometric or solvation change so that the second occurs at a more negative potential. The dication subsequently reacts with [U(η-cot)$_2$] to give [U$_2$(η-cot)$_4$]$^{2+}$ (*404*).

Lithium naphthalide and uranocene in thf give the monoanion [U(η-cot)$_2$]$^-$, isolated as a solvated lithium salt (*405*); neutral (*406*) and anionic (*407*) neptunium and plutonium analogs have also been prepared.

X

SANDWICH COMPOUNDS

For the purposes of this article, a sandwich compound is regarded as one in which a metal is π bonded to two planar, delocalized rings. Such compounds almost invariably undergo electron-transfer reactions which are, however, rarely uncomplicated. The enhanced reactivity of the more reduced and oxidized sandwich complexes is finding increased synthetic application.

A. *Titanium*

Iodine and [TiCp(η-cot)] yield the air-stable monocation as an iodide or triiodide salt (*408*).

B. *Vanadium*

Vanadocene, [VCp$_2$], is oxidized and reduced in one-electron steps to highly reactive ions. The reversible formation of [VCp$_2$]$^-$ at the very negative E^0 of -2.74 V in thf can be observed only under vacuum electrochemical conditions, and the quasi-reversible oxidation to $[VCp_2]^+$ is apparently accompanied by solvation to $[V(solvent)_2Cp_2]^+$ (299).

Although $[V(\eta\text{-arene})_2]$ has not been studied electrochemically, both the cation and the anion are well authenticated. Indeed, $[V(\eta\text{-arene})_2]^+$ is the precursor to the neutral compound, being formed in the reaction between VCl_4 and the arene under Friedel–Crafts conditions (409).

The addition of lithium naphthalide to $[V(\eta\text{-arene})_2]$ leads to the disappearance of the ESR spectrum typical of the neutral vanadium radical. The postulated (410, 411) one-electron reduction was confirmed by NMR spectroscopy which showed that the blue solution generated when $[V(\eta\text{-}C_6H_6)_2]$ reacts with potassium in 1,2-dimethoxyethane contains $[V(\eta\text{-}C_6H_6)_2]^-$ (412).

In contrast to $[V(\eta\text{-arene})_2]$, the isoelectronic mixed-sandwich $[VCp(\eta\text{-}C_7H_7)]$ is not reduced prior to electrolytic discharge $(-2.3$ V) in MeCN. However, it is easily oxidized $(E^0 = 0.19$ V) by coulometry (413) or with iodine (414) to give a stable monocation.

C. Chromium, Molybdenum, and Tungsten

The electrochemistry of bis(arene)chromium compounds has probably been studied more extensively than that of any other organometallic π-complex apart from ferrocene. Both $[Cr(\eta\text{-}C_6H_6)_2]$ and $[Cr(\eta\text{-}C_6H_6)_2]^+$ are isolable and have a formal potential of about -0.8 V. The redox process has been shown to be highly reversible by polarography (415), CV (416), and rotating disk voltametry (417), and its potential is largely independent of solvent effects, leading to proposals that the cation should be used as an internal reference (418–420).

Early work showed predictable but small shifts in potential on alkyl or aryl substitution (421–424). A good correlation between $E_{1/2}$ and meta substituent constants led to the conclusion that the substituent effect is transferred to chromium by an inductive mechanism (417). However, a second series of derivatives showed good correlations with both the σ_m and σ_p Hammett constants, a puzzling result since the former is an inductive parameter wheras the latter is sensitive to both inductive and resonance effects (425). The negative E^0 values of $[Cr(\eta\text{-arene})_2]$, which can be varied over a 1-V range by ring substitution (417, 425, 426), suggest the possible use of such complexes as mild, selective, one-electron reducing agents.

The oxidation potentials of bis(arene)chromium compounds with more extensive π systems, such as naphthalene, pyrene, and phenanthrene, show little variation (85, 416) despite the fact that the uncoordinated ligands show very large changes in E^0. However, this is consistent with the accepted bonding scheme for $[Cr(\eta\text{-}C_6H_6)_2]$ (427, 428), in which the HOMO is composed of the $3d_{z^2}$ orbital on chromium and the σ framework of the benzene rings. Thus, changes in the π-energy levels of the condensed hydrocarbon have little effect on the redox potential of the complex.

Further oxidation of $[Cr(\eta\text{-arene})_2]^+$ leads to free chromium(III) ions (426), and unsubstituted derivatives do not reduce at potentials to -2.7 V (416). However, the benzophenone complex $[Cr(\eta\text{-}C_6H_6)(\eta^6\text{-}C_6H_5COPh)]$ gives a monoanion at -1.91 V and a dianion at -2.01 V. The reductions are typical of free aromatic ketones, and the ultimate electrolysis products involve first the protonation of the carbonyl group, and, second, either the cleavage of that group or the carbonyl-containing ring to give $[Cr(\eta\text{-}C_6H_6)_2]$ (429).

The reduction of $[Cr(\eta^6\text{-naphthalene})_2]^+$ by lithium naphthalide was thought to give $[Cr(\eta^6\text{-naphthalene})_2]^-$ (410), but this is doubtful since the formal potential of the naphthalene–naphthalide couple is at least several hundred millivolts positive of that of the complex (416). A similar reaction between $[Cr(\eta\text{-}C_6H_6)_2]^+$ and alkali metals in thf or 1,2-dimethoxyethane yielded the radical $[Cr(\eta\text{-}C_6H_6)(\eta\text{-}C_6H_5)]$ (36); ESR spectroscopy showed an intramolecular, interannular hydrogen exchange reaction (Scheme 32) with a rate of $\sim 10^7$ sec^{-1} (430).

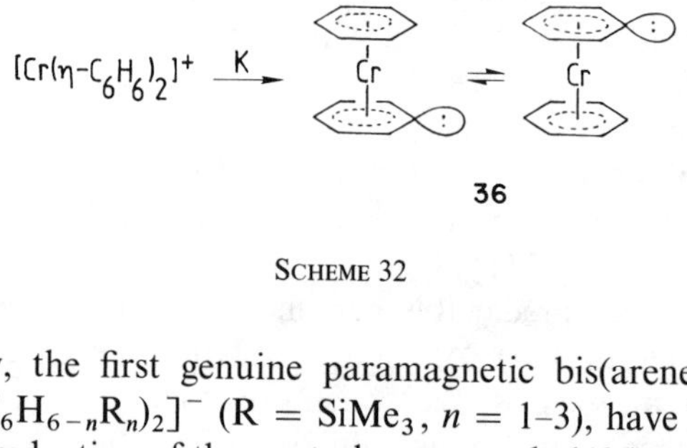

36

SCHEME 32

Very recently, the first genuine paramagnetic bis(arene)metal anions, namely $[Cr(\eta\text{-}C_6H_{6-n}R_n)_2]^-$ (R = SiMe$_3$, n = 1–3), have been generated by alkali metal reduction of the neutral compounds (431).

Chromocene, $[CrCp_2]$, is easily oxidized to an isolable monocation (432) $[E^0 \approx 0.6$ V (299, 433–435)], but gives an unstable anion; the reduction to $[CrCp_2]^-$ is reversible only under stringent, high-vacuum conditions (299, 435).

The mixed-sandwich compound $[CrCp(\eta\text{-}C_6H_6)]$ is oxidized ($E^0 = -0.19$ V in CH$_2$Cl$_2$) to an unstable monocation which loses benzene and forms $[CrCp_2]^+$. Reduction, apparently in a one-electron step ($E^0 = -2.16$

V in thf), gives unknown products (436). Related species, such as [CrCp-(η^7-azulene)] and [Cr(η^5-azulene)(η^7-azulene)], undergo facile oxidation to isolable monocations (437, 438).

The potential of the couple [CrCp(η-C$_7$H$_7$)]–[CrCp(η-C$_7$H$_7$)]$^+$ (84), estimated at ~ -0.5 V vs sce, is in the same range as those for the oxidations of [CrCp$_2$] and [Cr(η-arene)$_2$]. The reduction of [CrCp(η-C$_7$H$_7$)] with potassium in ether solvents yields [CrCp(η-C$_7$H$_7$)]$^-$, which is the most stable unsubstituted chromium sandwich anion yet reported (439); the ESR spectrum is consistent with a ligand-based half-filled orbital. The neutral compound is easier to reduce than [Cr(η-arene)$_2$], apparently because the C$_7$H$_7$ ligand has a higher electron affinity than an arene.

No detailed study of the electrochemistry of molybdenum or tungsten sandwiches seems to have been completed. Nevertheless, [M(η-arene)$_2$] (83, 440) and [MCp(η-C$_7$H$_7$)] (441) (M = Mo or W) are readily oxidized by iodine to isolable monocations.

D. *Manganese*

The complex [Mn(η-C$_5$Me$_5$)$_2$] is oxidized to the monocation ($E^0 = -0.56$ V) and reduced to [Mn(η-C$_5$Me$_5$)$_2$]$^-$ ($E^0 = -2.50$ V in MeCN), which is isolable as a pyrophoric sodium salt using sodium naphthalide in thf. The anion is isoelectronic with ferrocene but does not add electrophiles to the rings and is stable to ring exchange. Needless to say, it is an extremely powerful one-electron reductant (442).

E. *Iron, Ruthenium, and Osmium*

Since the original observation (443) of the oxidation of ferrocene to the ferrocenium ion ($E^0 = 0.31$ V, 0.2 M LiClO$_4$ in MeCN) (444), the formal potentials of well over 100 derivatives have been determined. Numerous studies assessing the effect of substitution on E^0 have been reported; tabulated data and a discussion of the significance of the observed trends are available elsewhere (445, 446). As expected, donor substituents render the compound more easily oxidized so that, for example, [Fe(η-C$_5$Me$_5$)$_2$] has an E^0 value of -0.12 V in MeCN (447).

On the assumption that E^0 for the couple [FeCp$_2$]–[FeCp$_2$]$^+$ is essentially solvent independent, ferrocene has been proposed (448) as an internal reference potential standard, designed to eliminate problems associated with liquid junction potentials. While this proposal seems useful, it should be remembered that it is based only on an assumption; the solvent independence of the ferrocene E^0 value has not yet been definitively demonstrated.

Attention has also been drawn to the fact that the ferrocene–ferrocenium couple is not perfectly reversible in either the chemical or electrochemical sense, despite popular belief. The values of the heterogeneous electron-transfer rates, k_s, for ferrocene in MeCN and dmf are 0.044 and 0.033 cm sec^{-1}, respectively (at platinum, 0.1 M [NEt$_4$][ClO$_4$] as supporting electrolyte), making the couple [FeCp$_2$]–[FeCp$_2$]$^+$ far from Nernstian in behavior. Thus, CV peak separations in MeCN should be between 70 and 80 mV at scan rates, v, in the range 0.1 to 1.0 V sec^{-1}, not the 60 mV commonly assumed. A peak separation of 100 mV was reported for ferrocene oxidation in dmf ($v = 1$ V sec^{-1}) (449).

NMR spectroscopy has also been used to measure electron exchange rates between ferrocenes and the corresponding ferrocenium ions. The rates do not vary with solvent dielectric constant in the manner predicted by Marcus theory. Rather, they are more dramatically dependent on substituent effects, with [Fe(η-C$_5$Me$_5$)$_2$] undergoing a 10-fold faster electron exchange than ferrocene (450).

The further oxidation of [Fe(η-C$_5$Me$_5$)$_2$]$^+$ by electrolysis in an acidic AlCl$_3$-n-butylpyridinium chloride melt gives [Fe(η-C$_5$Me$_5$)$_2$]$^{2+}$, which is stable under argon; the E^0 value for the cation–dication couple is ~ 1.6 V. The ferrocenium ion is also oxidizable in this medium, albeit irreversibly (451).

Substituted ferrocenes also form monoanions at very negative potentials; electron addition is genuinely associated with the ferrocene nucleus rather than with an electroactive substituent. The E^0 value for [Fe(η-C$_5$H$_4$Ph)$_2$] is -2.62 V, and [FeCp$_2$] itself shows a quasi-reversible reduction at -2.93 V in dmf (452), with a peak separation of ~ 250 mV at $-37°$C ($v = 1$ V sec^{-1}). Exhaustive electrolytic reduction of ferrocene derivatives yields solutions containing the substituted cyclopentadienide anions; the latter may be used in the syntheses of other cyclopentadienylmetal complexes (453). Ferrocenes are also finding use as mediators in electron-transfer reactions, especially at electrode surfaces (454–456).

The oxidation of ruthenocene, [RuCp$_2$], is not straightforward. On a mercury electrode it is a relatively reversible one-electron step, forming [Hg{RuCp$_2$}$_2$]$^{2+}$ (457) which can be isolated by CPE or via [Hg(CN)$_2$] oxidation in HBF$_4$ solution (458, 459). By contrast, an irreversible two-electron oxidation occurs at platinum at about 0.7 V. Although the process has been studied by a variety of electrochemical techniques (*444, 457, 459–461*), the nature of the chemical transformations involved is still unclear; the two-electron product can be reduced to [$RuCp_2$], but there may well be ligand gain and loss accompanying what is overall a chemically reversible system. The complex [$RuClCp_2$]$^+$ has been isolated by reacting ferric chloride with ruthenocene (*459*).

In basic molten salts, the oxidation of [$RuCp_2$] at carbon is highly irreversible, but in a neutral melt (1:1 $AlCl_3$:*n*-butylpyridinium chloride) it behaves more like a quasi-reversible one-electron process. Under acid conditions (excess $AlCl_3$), waves are seen at 0.76 and 0.93 V, suggesting stabilization of the ruthenocene dication (*462*).

Osmocene, [$OsCp_2$], apparently undergoes two separate one-electron oxidations at platinum ($E_{1/2} \approx 0.7$ and 1.4 V). Despite agreement that the first step is irreversible (*444, 459, 461*), a salt of composition [$OsCp_2$]-[BF_4], but otherwise poorly characterized, has been isolated by exhaustive electrolytic oxidation (*459*).

The complexes [$Fe(\eta\text{-}C_5R_5)(\eta\text{-}C_6R'_6)$]^Z form a three- ($Z = +1$ to -1) or possibly four-membered ($Z = +2$ to -1) redox series. Early polarographic studies showed that [$FeCp(\eta\text{-arene})$]$^+$ is reduced in two one-electron steps (*84, 463, 464*). The first is fully reversible (e.g., arene = C_6H_6, $E^0 \approx -1.25$ V in thf) (*465*), but the second is so only in strictly aprotic solvents (*465, 466*) (e.g., arene = C_6H_6, $E^0 = -2.13$ V in thf).

Exhaustive reduction at the second wave in the presence of CO_2, H^+, or RI (R = Me or Ph) gives cyclohexadienyls such as [$FeCp(\eta^5\text{-}C_6H_6Ph)$] via electrophilic addition to the monoanion. Interestingly, whereas [$FeCp(\eta^5\text{-}C_6H_7)$] is oxidized in an irreversible two-electron step, substitution at the methylene carbon atom leads to the formation of [$FeCp(\eta^5\text{-}C_6H_6R)$]$^+$, isoelectronic with [$FeCp_2$]$^+$ (*465*).

The reductive route to the cyclohexadienyl compounds is said (*466*) to be more regiospecific than that involving direct hydride addition to [$FeCp(\eta\text{-arene})$]$^+$. For example, [$FeCp(\eta^6\text{-biphenyl})$]$^+$ and sodium amalgam give *endo*-[$FeCp(\eta^5\text{-}C_6H_6Ph)$] whereas $LiAlH_4$ in thf yields an isomeric mixture. It is interesting to note, then, that at $-60°C$ $NaBH_4$ and $LiAlH_4$ also react via initial one-electron transfer. Thus, the ESR spectra of the radicals [$FeCp(\eta\text{-}C_6H_5X)$] (X = H or F) are readily identifiable in the reaction with [$FeCp(\eta\text{-}C_6H_5X)$]$^+$. One-electron reduction is also detected in the reaction of $LiAlH_4$ with [$FeCp(\eta\text{-}C_4Me_4S)$]$^+$, [$FeCp(\eta\text{-}C_6Et_6)$]$^+$, and [$Fe(\eta\text{-}C_6Me_6)_2$]$^{2+}$. The last affords [$Fe(\eta^4\text{-}C_6Me_6H_2)(\eta\text{-}C_6Me_6)$] (*467*) rather than [$Fe(\eta^5\text{-}C_6Me_6H)_2$], which would be expected on the basis of qualitative rules for the addition of nucleophiles to organometallic cations (*468*).

In CH_2Cl_2 at 203 K, the reaction of $SbCl_5$ with $[FeCp(\eta\text{-arene})]^+$ (arene = naphthalene or C_6H_5X, X = OMe, aryl, etc.) generates radical species formulated as $[FeCp(\eta\text{-arene})]^{2+}$ on the basis of ESR spectroscopy (469). However, electrochemical data in support of this formulation have not yet been reported.

On electrolysis, $[Fe(\eta\text{-}C_5H_4COR)(\eta\text{-}1,3,5\text{-}C_6H_3R'_3)]^+$ are regiospecifically reduced at the carbonyl group whereas the reaction with $NaBH_4$ leads to attack at both the carbonyl and the arene ring. At pH 0, reduction at -1.0 V gives the alcohols $[Fe\{\eta\text{-}C_5H_4CH(OH)R\}(\eta\text{-}C_6H_6)]^+$ (R = Me or Ph), but at pH 13–14, reduction at -1.2 V affords the pinacol (37). The electroreduction of 38 is also stereospecific, giving the endo alcohol 39 (470).

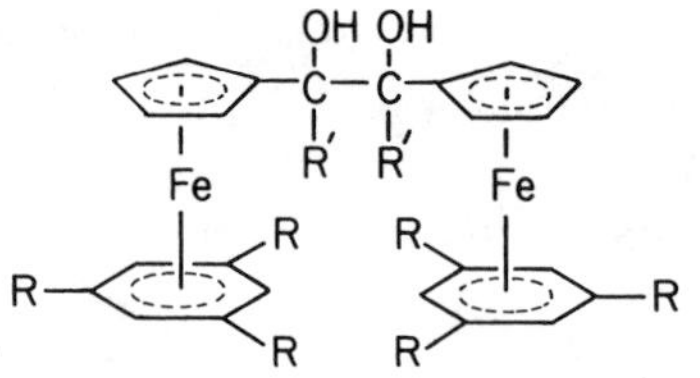

37; R = Me, R′ = Me or PH

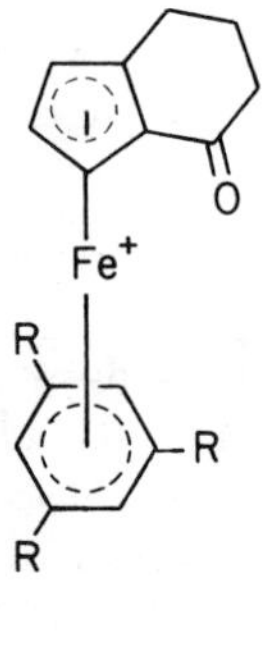

38 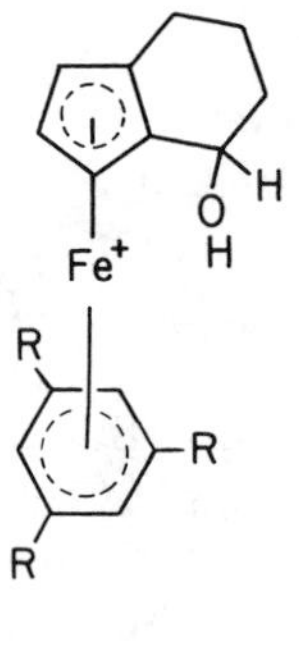   **39**

The neutral radicals $[Fe(\eta\text{-}C_5R_5)(\eta\text{-}C_6R'_6)]$ are generally, and simply, prepared by sodium amalgam reduction of the monocations (471, 472). The highly substituted complexes (R = H, R′ = Me; R = Me, R′ = Me or Et) (472, 473) and the naphthalene analog (R = H) (471) are thermally stable solids. By contrast, $[FeCp(\eta\text{-}C_6Me_{6-n}H_n)]$ (n = 0–5) (473, 474), $[Fe(\eta\text{-}C_5Me_5)(\eta\text{-}C_6H_6)]$ (472), and $[FeCp(\eta^6\text{-tetralin})]$ (471) dimerize in the solid state and in nonpolar solvents to give dicyclohexadienyls (**40**) via intra-molecular electron transfer from the metal atom to the arene ring (*473*).

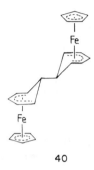

40

The dimers **40** are oxidatively cleaved to $[Fe(\eta\text{-}C_5R_5)(\eta\text{-}C_6R'_6)]^+$, either electrolytically (*475*) or by chemical means (*474*).

The X-ray structure of $[FeCp(\eta\text{-}C_6Me_6)]$ (*472*) confirms the 19-electron configuration of the metal atom; the two rings are planar and parallel, although the C_5 ligand is considerably further from the iron atom than in the monocation. Detailed ESR studies (*476–478*) confirm the d^7 formulation for $[Fe(\eta\text{-}C_5R_5)(\eta\text{-}C_6R'_6)]$ and show the single electron to occupy a metal-based d_{xz}, d_{yz} orbital (*476*). The spectra also reveal isomerism in the α-methyl- and β-fluoronaphthalene complexes in which the metal may bond to the substituted or unsubstituted arene ring (*477*).

The radicals $[Fe(\eta\text{-}C_5R_5)(\eta\text{-}C_6R'_6)]$ have very negative oxidation potentials (e.g., R = R′ = Me, $E^0 = -1.87$ V) leading to their description as electron reservoirs. Indeed, their ionization potentials (R = H, R′ = Me or Et; R = R′ = Me) are similar to that of potassium metal (*473*). Interestingly, $[Fe(\eta\text{-}C_5H_4R)(\eta\text{-}C_6Me_6)]$ (R = H or CO_2H) catalyze the reduction of nitrate ion to ammonia (*479*).

The action of oxygen on $[Fe(\eta\text{-}C_5R_5)(\eta\text{-}C_6R'_6)]$ is very arene dependent. In strictly aprotic solvents, $[FeCp(\eta\text{-}C_6H_6)]$ gives the peroxy complex **41** (*480*) whereas $[Fe(\eta\text{-}C_5R_5)(\eta\text{-}C_6Et_6)]$ (R = Me or Et) (*472*) and $[FeCp(\eta^6\text{-}$naphthalene)] (*471*) yield superoxide salts $[Fe(\eta\text{-}C_5R_5)(\eta\text{-arene})][O_2]$. The

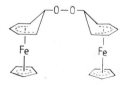

41

most interesting reaction, however, is that with $[Fe(\eta\text{-}C_5R_5)(\eta\text{-}C_6Me_6)]$, which leads to C—H bond activation. After one-electron transfer, an arene methyl proton is abstracted by $[O_2]^-$ to give the cyclohexadienyl $[Fe(\eta\text{-}C_5R_5)(\eta^5\text{-}C_6Me_5CH_2)]$ (481, 482). The activation of an N—H bond of $[FeCp(\eta^6\text{-}C_6Me_5NH_2)]^+$ is similarly achieved, with sodium amalgam reduction of the cation and subsequent treatment with oxygen giving $[FeCp(\eta^5\text{-}C_6Me_5NH)]$ (483).

The exocyclic methylene (482, 484) and NH groups (483) are very reactive toward nucleophiles. Thus, $[FeCp(\eta^5\text{-}C_6Me_5CH_2)]$ and halides, RX, yield $[FeCp(\eta^6\text{-}C_6Me_5CH_2R)]^+$ [R = Me, COPh, SiMe$_3$, PPh$_2$, Fe(CO)$_2$Cp, etc.]. With metal carbonyls, zwitterionic metallates such as $[FeCp\{\eta^6\text{-}C_6Me_5CH_2Fe(CO)_4\}]$ result, and the organometallic cations $[Fe(CO)_3\text{-}Cp]^+$ and $[Fe(CO)_2(\eta\text{-}C_2H_4)Cp]^+$ give **42** and **43**, respectively.

42   43

The complex $[FeCp(\eta^5\text{-}C_6Me_5CH_2)]$ can be prepared directly from $[FeCp(\eta\text{-}C_6Me_6)]^+$ and $t\text{-}BuOK$. This route is also applicable to other arenes but the extensive organic chemistry ensuing (485–488) is not a matter for this article.

The cyclohexadienyl $[FeCp(\eta^5\text{-}C_6Me_5CH_2)]$ is, like $[FeCp(\eta^5\text{-}C_6H_6R)]$ (see above), oxidized to a paramagnetic monocation by iodine (489), although other authors state that the reaction yields $[Fe_2Cp_2(\eta^6,\eta'^6\text{-}C_6MeCH_2CH_2C_6Me_5)]^{2+}$ (482). The monocation, and the anion $[FeCp(\eta^5\text{-}C_6Me_5CH_2)]^-$ generated by sodium reduction, show ESR spectra consistent with d^7 metal centers and, therefore, with electron transfer associated with the cyclohexadienyl ligand (489). The formation of a ligand-based radical cation might well yield the dimeric species described above.

The ruthenium complex $[RuCp(\eta\text{-}C_6Me_6)]^+$ is not reduced before -2.7 V in MeCN and is irreversibly oxidized (490).

The dications $[Fe(\eta\text{-}C_6Me_nH_{6-n})]^{2+}$ ($n = 3$ or 6) are reduced in two stages. The first is reversible (e.g., $n = 6$, $E^0 = -0.48$ V in aqueous acetone), whereas the second is not (*491*), probably because the redox product reacts with the solvent; $[Fe(\eta\text{-}C_6Me_6)_2]$ has been synthesized independently (*492*).

By contrast, $[Ru(\eta\text{-arene})_2]^{2+}$ appear to undergo a single two-electron process (*493, 494*), which is only partially chemically reversible in MeCN (e.g., arene = C_6Me_6, $E_{1/2} = -1.02$ V). However, the neutral complex $[Ru(C_6Me_6)_2]$ may be isolated via sodium reduction of the dication in liquid ammonia (*495*). It does not have a 20-electron configuration and should be formulated as $[Ru(\eta^4\text{-}C_6Me_6)(\eta^6\text{-}C_6Me_6)]$ (*496*).

If one of the C_6Me_6 ligands of $[Ru(\eta\text{-}C_6Me_6)]^{2+}$ is replaced by a cyclophane, the E^0 value is more positive (e.g., -0.50 V for **44**) and the reduction

44

is more reversible. Complexes with cyclophanes having a boat shape are the most readily reduced implying that the major factor determining the E^0 value of $[Ru(\eta\text{-arene})_2]^{2+}$ is the ease with which one arene ring can undergo distortion to the η^4 form (*493, 494*).

F. Cobalt and Rhodium

The cobaltocene–cobaltocenium couple, $[CoCp_2]\text{-}[CoCp_2]^+$, has an E^0 value of ~ -0.90 V in nonaqueous solvents (*497–499*) and approximates to a Nernstian couple much better than $[FeCp_2]\text{-}[FeCp_2]^+$; its k_s value of 0.86 cm sec^{-1} (*500*) is about 25 times higher than that of ferrocene (*449*).

The E^0 values of substituted cobaltocenium ions show the expected variations (*415, 447, 501, 502*). For example, the bis(indenyl) complex $[Co(\eta^5\text{-}C_9H_7)_2]^+$ ($E^0 = -0.53$ V) (*415*) and the borinate analog $[CoCp(\eta^5\text{-}C_5H_5BPh)]^+$ ($E^0 = -0.44$ V) (*503*) have more positive reduction potentials than $[CoCp_2]^+$, whereas that of $[Co(\eta\text{-}C_5Me_5)_2]^+$ is much more negative ($E^0 = -1.48$ V). In spite of the last value, $[Co(\eta\text{-}C_5Me_5)_2]$ is isolable in good yield from the alkali metal reduction of the cation (*447, 501*), and may prove to be a powerful and useful one-electron reductant. ESR spectroscopy shows it to have a $^2E_{1g}$ ground state (*447*).

Cobaltocene is reduced to a monoanion ($E^0 = -1.9$ V in MeCN) (*491*, *504*) which is subject to electrophilic attack. Protonation by weak acids such as water or phenol occurs directly at the cyclopentadienyl ligand with no evidence for initial formation of metal hydrides (*505*). In dmf, the anion adds other electrophiles to give the substituted cyclopentadiene compounds $[Co(\eta^4\text{-}C_5H_5R)Cp]$. Hydride abstraction by $[CPh_3]^+$ then provides a route to substituted cobaltocenium salts (*502*, *506*).

The mixed-sandwich compounds $[Co(\eta\text{-}C_5R_5)(\eta\text{-}C_6R'_6)]^{2+}$ ($R = R' =$ H or Me) are reduced in two one-electron steps in propylene carbonate (*507*), and the oxidation of the cyclobutadiene complexes $[Co(\eta\text{-}C_4R_4)Cp]$ (*371*, *508*) has been described in Section VII,F. Although electrochemical studies have not been made, the three-membered redox series $[Co(\eta\text{-}C_6Me_6)_2]^Z$ ($Z = 0$, 1, and 2) has been characterized. The mono- and dications are prepared from $CoCl_2$, C_6Me_6, and $AlCl_3$ in the presence and absence, respectively, of aluminum powder (*509*); the neutral species is formed on reduction of the monocation by sodium in liquid ammonia (*510*). The X-ray structure of $[Co(\eta\text{-}C_6Me_6)_2][PF_6]$ shows both rings to be η^6 bonded (*511*), which contrasts with $[Rh(\eta^4\text{-}C_6Me_6)(\eta\text{-}C_6Me_6)]^+$; the cobalt cation has two unpaired electrons in an antibonding e_{1g} orbital.

The reduction of $[RhCp_2]^+$ ($E^0 = -1.4$ V in MeCN) (*512*) is considerably more difficult than that of $[CoCp_2]^+$, and the formation of $[RhCp_2]$ is complicated by dimerization to **45**. The latter, isolated in low yield from the reaction of $[RhCp_2]^+$ with alkali metals (*513*), is better synthesized by electrolytic reduction in CH_2Cl_2 (*512*).

45

The couple $[RhCp_2]^+$–$[RhCp_2]$ is electrochemically highly reversible ($k_s = 0.73$ cm sec^{-1}). Additional reduction to $[RhCp_2]^-$ is possible ($E^0 = -2.2$ V), but the anion has a lifetime of but a few seconds in dmf at

$-50°C$. Thus, both $[RhCp_2]$ and $[RhCp_2]^-$ are considerably less stable than their cobalt analogs.

G. *Nickel*

Nickelocene, $[NiCp_2]$, is the only metallocene to support a four-membered electron-transfer series, $[NiCp_2]^Z$ ($Z = -1$ to $+2$); ironically all four compounds are air sensitive and reactive. The best characterized couple is that relating the neutral and monocationic species ($E^0 = 0.1$ V) (*514*). Further oxidation ($E^0 = 0.74$ V) gives the dication whose reactivity (*498, 515, 516*) renders $[NiCp_2]^+$–$[NiCp_2]^{2+}$ chemically reversible only at low temperatures (*515*) or in rigorously purified solvent (*516*). The dication has also been generated by anodic oxidation of nickelocene in molten salts (*517*).

The oxidation of $[Ni(\eta\text{-}C_5Me_5)_2]$ is much more facile ($E^0 = -0.7$ and 0.35 V) (*447, 518*), and both the mono- and dications are isolable. The former reacts with radical sources and the latter with nucleophiles to give $[Ni(\eta^4\text{-}C_5Me_5R)(\eta\text{-}C_5Me_5)]^+$ [R = Ph, CN, or C(CN)Me$_2$] (*518*).

At $-60°C$ in dmf, $[NiCp_2]^-$ is stable enough to form a reversible couple with nickelocene ($E^0 = -1.6$ V). The electron-transfer step is fairly slow, suggesting that structural distortion accompanies reduction. Such a distortion would not be surprising in that $[NiCp_2]^-$ formally has a 21-electron metal (*299, 435*).

Bulk reduction of $[NiCp_2]$ in MeCN or thf yields some $[Ni(\eta^3\text{-}C_5H_7)Cp]$ and uncharacterized products (*299, 498*). Sodium naphthalide in thf gives a mixture of remarkable polynuclear, cyclopentadienylnickel compounds containing between two and six metal atoms. The octahedral cluster $[\{NiCp\}_6]$ and its monocation are the first to be structurally characterized (*519*).

XI

ADDENDUM

Recent important papers are summarized in this section.

A review (*520*) on inorganic electrochemistry contains sections on organometallic compounds. The oxidatively induced isomerizations of *cis*-$[Mo(CO)_2(CNCy)_4]$ and *cis,trans*-$[Mo(CO)_2(CNR)_2(PR_3)_2]$, to the *trans*- and *trans,trans* cations, respectively, compare with those of related complexes described in section II,C,6 (*521*).

^{60}Co γ-irradiation of single crystals of $[Mn_2(CO)_{10}]$ showed (522) that the reduced product is likely to be carbonyl-bridged $[Mn_2(CO)_9]^-$ rather than $[Mn_2(CO)_{10}]^-$ (Section II,D). Electrolytic reduction (523) of $[Mn(CO)_3$-$(NCMe)_{3-n}L_n]^+$ (L = phosphine, n = 0–2) provides a route to high yields of $[MnH(CO)_3L_2]$ (Section II,D).

Lithium reduction of $[Co(\eta\text{-}C_2H_4)(PMe_3)_3]$ gives $Li[Co(\eta\text{-}C_2H_4)$-$(PMe_3)_3]$, a hexane-soluble complex with a lithium-bridged trimeric structure (524) (Section IV).

The allyl complexes $[MoX(CO)_2(L\text{–}L)(\eta^3\text{-}C_3H_4R)]$ (X = Cl or CO_2CF_3, L–L = bipy, dppe, etc.; R = H or Me) are oxidized (525) to 17-electron cations (Section V).

X-Ray structural studies (526) have verified the reversibility of the electron-transfer series $[\overline{Nb(CH_2C_6H_4C_6H_4CH_2})Cp_2]^Z$ (Z = $-1, 0$, and 1), and show the CH_2—Nb—CH_2 angle to narrow from 106.3 to 83.0, and then to 80°, on reduction (Section VII,B).

The radical $[Mo(CO)_3\{HB(pz)_3\}]$ (pz = pyrazolyl), isoelectronic with $[Mo(CO)_3Cp]$ (Section VII,C), has been isolated and structurally characterized (527); it is implicated in the syntheses of carbyne and aroyl complexes from $[Mo(CO)_3\{HB(pz)_3\}]^-$ and arenediazonium salts (528).

^{60}Co γ-irradiation of $[FeX(CO)_2Cp]$ (X = Cl or I), and $[MI(CO)_3Cp]$ (M = Mo or W) gives paramagnetic species formulated as the corresponding monoanions (529). Electrochemical reduction of $[Fe(MPh_3)(CO)_2Cp]$ (M = Si, Ge, or Sn) affords $[Fe(MPh_3)(CO)_2Cp]^-$, characterised by ESR spectroscopy and presumably stabilized by the phenyl substituents on M (530); the corresponding halides, and $SnCl_3$ or $GeCl_3$ derivatives, do not give stable 19-electron radical anions (Section VII,E).

Redox-catalyzed migratory insertion of CO into $[FeMe(CO)(PPh_3)Cp]$ rapidly gives a high yield of $[Fe(COMe)(CO)(PPh_3)Cp]$ (531). Synthesis of formic acid esters via oxidation of $[FeH(CO)_2Cp]$ by cerium(IV) or copper(II) in alcohols, ROH, also proceeds via initial one-electron oxidation; the paramagnetic formyl $[Fe(CHO)(CO)(ROH)Cp]^+$ is implicated as an intermediate (532)[2] (Section VII,E).

Sophisticated electrochemical studies have shown that the reduction of $[Co(\eta^4\text{-}1,5\text{-cot})Cp]$ to $[Co(\eta^4\text{-}1,3\text{-cot})Cp]^-$ involves electron transfer before isomerization; the rate of isomerization is $2 \pm 1 \times 10^3$ sec^{-1} (533) (Section VII,F).

The 17-electron cations $[Cr(CO)_3(\eta^6\text{-arene})]^+$ are stabilized in CF_3CO_2H (534), and readily undergo substitution (535). The species formed on electro-

---

[2] This work has since been retracted [A. Cameron, V. H. Smith, and M. C. Baird, *Organometallics* 3, 338 (1984)].

chemical two-electron reduction of $[Cr(CO)_3(\eta^6\text{-naphthalene})]$ has been fully characterized as the benzocyclohexadienyl complex $[Cr(CO)_3(\eta^5\text{-}C_{10}H_9)]^-$ (536). Two-electron reduction of complexes such as $[Cr_2(CO)_6\text{-}(\eta^6,\eta'^6\text{-biphenyl})]$ is chemically reversible but involves an ECE mechanism (537) (Section VIII).

Brief mention has been made (538) of the potassium reduction of $[Ti(\eta\text{-toluene})_2]$ to the green monoanion $[Ti(\eta\text{-toluene})_2]^-$ (Section X,A).

Ferrocene is reduced to $[FeCp_2]^-$ in dme. The process is chemically reversible at $-45°C$, but large CV peak separations may imply structural distortion in the anion (539). By contrast to ruthenocene, $[Ru(\eta\text{-}C_5Me_5)_2]$ gives a stable monocation (540) (Section X,E).

The cation $[FeCp(\eta^6\text{-arene})]^+$ undergoes reductively catalyzed arene decomplexation to give $[Fe\{P(OMe)_3\}_3Cp]^+$ (541) (Section X,E).

REFERENCES

1. R. E. Dessy and L. A. Bares, *Acc. Chem. Res.* **5**, 415 (1972).
2. L. I. Denisovich and S. P. Gubin, *Russ. Chem. Rev. (Engl. Transl.)* **46**, 27 (1977).
3. D. de Montauzon, R. Poilblanc, P. Lemoine, and M. Gross, *Electrochim. Acta* **23**, 1247 (1978).
4. S. P. Gubin, *Pure Appl. Chem.* **23**, 463 (1970).
5. J. A. McCleverty, *in* "Reactions of Molecules at Electrodes" (N. Hush, ed.), p. 460. Wiley (Interscience), New York, 1971.
6. C. T. -W. Chu, F. Y. -K. Lo, and L. F. Dahl, *J. Am. Chem. Soc.* **104**, 3409 (1982).
7. J. K. Kochi, *Acc. Chem. Res.* **7**, 351 (1974).
8. J. K. Kochi, "Organometallic Mechanisms and Catalysis." Academic Press, New York, 1978.
9. A. M. Bond, B. S. Grabaric, and J. J. Jackowski, *Inorg. Chem.* **17**, 2153 (1978).
10. A. M. Bond, R. Colton, and M. E. McDonald, *Inorg. Chem.* **17**, 2842 (1978).
11. J. Moraczewski and W. E. Geiger, *J. Am. Chem. Soc.* **103**, 4779 (1981).
12. J. W. Hershberger and J. K. Kochi, *J. Chem. Soc., Chem. Commun.*, p. 212 (1982); J. W. Hershberger, C. Amatore, and J. K. Kochi, *J. Organomet. Chem.* **250**, 345 (1983); J. W. Hershberger, R. J. Klingler, and J. K. Kochi, *J. Am. Chem. Soc.* **105**, 61 (1983).
13. M. I. Bruce, J. G. Matisons, B. K. Nicholson, and M. L. Williams, *J. Organomet. Chem.* **236**, C57 (1982).
14. A. F. Hepp and M. S. Wrighton, *J. Am. Chem. Soc.* **103**, 1258 (1981).
15. D. P. Summers, J. C. Luong, and M. S. Wrighton, *J. Am. Chem. Soc.* **103**, 5238 (1981).
16. H. Lehmkuhl, *Synthesis*, p. 377 (1973).
17. J. Grobe, M. Keil, B. Schneider, and H. Zimmermann, *Z. Naturforsch. B.: Anorg. Chem., Org. Chem.* **35**, 428 (1980).
18. J. J. Habeeb and D. G. Tuck, *J. Organomet. Chem.* **139**, C17 (1977).
19. J. Grobe, B. H. Schneider, and H. Zimmermann, *Z. Anorg. Allg. Chem.* **481**, 107 (1981).
20. J. Grobe and B. H. Schneider, *Z. Naturforsch. B.: Anorg. Chem., Org. Chem.* **36**, 8 (1981).
21. G. Schiavon, G. Bontempelli, and B. Coran, *J. Chem. Soc., Dalton Trans.*, p. 1074 (1981).

22. C. Gosden and D. Pletcher, *J. Organomet. Chem.* **186**, 401 (1980).
23. F. M. Dayrit and J. Schwartz, *J. Am. Chem. Soc.* **103**, 4466 (1981).
24. D. Ballivet-Tkatchenko, M. Riveccie, and N. El Murr, *J. Am. Chem. Soc.* **101**, 2763 (1979).
25. A. J. Bard and L. R. Faulkner, "Electrochemical Methods." Wiley, New York, 1980.
26. D. T. Sawyer and J. L. Roberts, Jr., "Experimental Electrochemistry for Chemists." Wiley, New York, 1974.
27. R. Adams, "Electrochemistry at Solid Electrodes." Dekker, New York, 1969.
28. A. M. Bond, "Modern Polarographic Methods in Analytical Chemistry." Dekker, New York, 1980.
29. D. D. McDonald, "Transient Techniques in Electrochemistry." Plenum, New York, 1977.
30. W. E. Geiger, *in* "Inorganic Reactions and Methods" (J. J. Zuckerman, ed.), in press. Verlag Chemie, Weinheim.
31. W. E. Geiger and N. G. Connelly, *Adv. Organomet. Chem.* **24**, in press (1985).
32. D. G. Brown, *Prog. Inorg. Chem.* **18**, 177 (1973).
33. A. A. Vlcek, *Electrochim. Acta* **13**, 1063 (1968).
34. D. de Montauzon and R. Poilblanc, *J. Organomet. Chem.* **104**, 99 (1976).
35. C. J. Pickett and D. Pletcher, *J. Organomet. Chem.* **102**, 327 (1975).
36. J. Chatt, C. T. Kan, G. J. Leigh, C. J. Pickett, and D. R. Stanley, *J. Chem. Soc., Dalton Trans.*, p. 2032 (1980).
37. A. M. Bond and R. Colton, *Inorg. Chem.* **15**, 2036 (1976).
38. A. M. Bond, J. W. Bixler, E. Mocellin, S. Datta, E. J. James, and S. S. Wreford, *Inorg. Chem.* **19**, 1760 (1980).
39. R. J. Klingler, J. C. Huffman, and J. K. Kochi, *J. Am. Chem. Soc.* **102**, 208 (1980).
40. C. J. Pickett and D. Pletcher, *J. Chem. Soc., Dalton Trans.*, p. 879 (1975).
41. J. Grobe and H. Zimmermann, *Z. Naturforsch. B.: Anorg. Chem., Org. Chem.* **36**, 301 (1981).
42. C. J. Pickett and D. Pletcher, *J. Chem. Soc., Dalton Trans.*, p. 636 (1976).
43. H. L. Chum, D. Koran, and R. A. Osteryoung, *J. Organometal. Chem.* **140**, 349 (1977).
44. C. J. Pickett and D. Pletcher, *J. Chem. Soc., Dalton Trans.* p. 749 (1976).
45. P. Lemoine and M. Gross, *C. R. Hebd. Seances Acad. Sci.* **280**, 797 (1975).
46. R. J. Klinger, J. C. Huffman, and J. K. Kochi, *Inorg. Chem.* **20**, 34 (1981).
47. C. P. Casey, L. D. Albin, M. C. Saeman, and D. H. Evans, *J. Organomet. Chem.* **155**, C37 (1978).
48. O. S. Mills and A. D. Redhouse, *J. Chem. Soc. A.* p. 642 (1968).
49. P. J. Krusic, U. Klabunde, C. P. Casey, and T. F. Block, *J. Am. Chem. Soc.* **98**, 2015 (1976).
50. M. K. Lloyd, J. A. McCleverty, D. G. Orchard, J. A. Connor, M. B. Hall, I. H. Hillier, E. M. Jones, and G. K. McEwen, *J. Chem. Soc., Dalton Trans.*, p. 1743 (1973).
51. M. F. Lappert, R. W. McCabe, J. J. MacQuitty, P. L. Pye, and P. I. Riley, *J. Chem. Soc., Dalton Trans.*, p. 90 (1980).
52. R. D. Rieke, H. Kojima, and K. Öfele, *Angew. Chem., Int. Ed. Engl.* **19**, 538 (1980).
53. R. D. Rieke, H. Kojima, and K. Öfele, *J. Am. Chem. Soc.* **98**, 6735 (1976).
54. E. O. Fischer, M. Schluge, and J. O. Besenhard, *Angew. Chem., Int. Ed. Engl.* **15**, 683 (1976).
55. E. O. Fischer, D. Wittmann, D. Himmelreich, and D. Neugebauer, *Angew. Chem., Int. Ed. Engl.* **21**, 444 (1982).
56. W. P. Fehlhammer and F. Degel. *Angew. Chem., Int. Ed. Engl.* **18**, 75 (1979).
57. H. Behrens and D. Herrmann, *Z. Naturforsch. B.: Anorg. Chem., Org. Chem.* **21**, 1236 (1966).
58. J. A. Connor, E. M. Jones, G. A. McEwen, M. K. Lloyd, and J. A. McCleverty, *J. Chem. Soc., Dalton Trans.*, p. 1246 (1972).

59. B. E. Bursten, *J. Am. Chem. Soc.* **104**, 1299 (1982).
60. M. K. Lloyd and J. A. McCleverty, *J. Organomet. Chem.* **61**, 261 (1973).
61. P. M. Treichel and G. J. Essenmacher, *Inorg. Chem.* **15**, 146 (1976).
62. G. J. Essenmacher and P. M. Treichel, *Inorg. Chem.* **16**, 800 (1977).
63. P. M. Treichel, D. W. Firsich, and G. J. Essenmacher, *Inorg. Chem.* **18**, 2405 (1979).
64. D. A. Bohling, J. F. Evans, and K. R. Mann, *Inorg. Chem.* **21**, 3546 (1982).
65. W. S. Mialki, D. E. Wigley, T. E. Wood, and R. A. Walton, *Inorg. Chem.* **21**, 480 (1982).
66. D. D. Klendworth, W. W. Welters, III, and R. A. Walton, *Organometallics* **1**, 336 (1982).
67. C. M. Giandomenico, L. H. Hanau, and S. J. Lippard, *Organometallics* **1**, 142 (1982).
68. W. S. Mialki, R. E. Wild, and R. A. Walton, *Inorg. Chem.* **20**, 1380 (1981).
69. C. Caravana, C. M. Giandomenico, and S. J. Lippard, *Inorg. Chem.* **21**, 1860 (1982).
70. A. Bell, D. D. Klendworth, R. E. Wild, and R. A. Walton, *Inorg. Chem.* **20**, 4456 (1981).
71. A. J. L. Pombeiro and R. L. Richards, *J. Organomet. Chem.* **179**, 459 (1979).
72. D. E. Wigley and R. A. Walton, *Organometallics* **1**, 1322 (1982).
73. J. Chatt, A. J. Pearman, and R. L. Richards, *J. Chem. Soc., Dalton Trans.*, p. 2139 (1977).
74. G. J. Leigh and C. J. Pickett, *J. Chem. Soc., Dalton Trans.*, p. 1797 (1977).
75. J. Chatt, G. J. Leigh, H. Neukomm, C. J. Pickett, and D. R. Stanley, *J. Chem. Soc., Dalton Trans.*, p. 121 (1980).
76. Y. Mizobe, R. Ono, Y. Uchida, M. Hidai, M. Tezuka, S. Moue, and A. Tsuchiya, *J. Organomet. Chem.* **204**, 377 (1981).
77. C. J. Pickett, J. E. Tolhurst, A. Copenhaver, T. A. George, and R. K. Lester, *J. Chem. Soc., Chem. Commun.*, p. 1071 (1982).
78. W. Hussain, G. J. Leigh, and C. J. Pickett, *J. Chem. Soc., Chem. Commun.*, p. 747 (1982).
79. C. J. Pickett and G. J. Leigh, *J. Chem. Soc., Chem. Commun.*, p. 1033 (1981).
80. H. M. Colquhoun and K. Henrick, *Inorg. Chem.* **20**, 4074 (1981).
81. H. M. Colquhoun, *J. Chem. Res. (S)*, p. 275 (1981).
82. H. M. Colquhoun and T. J. King, *J. Chem. Soc., Chem. Commun.*, p. 879 (1980).
83. W. E. Silverthorn, *Inorg. Chem.* **18**, 1835 (1979).
84. R. E. Dessy, F. E. Stary, R. B. King, and M. Waldrop, *J. Am. Chem. Soc.* **88**, 471 (1966).
85. H. tom Dieck, K.-D. Franz, and F. Hohmann, *Chem. Ber.* **108**, 163 (1975).
86. R. E. Dessy, R. B. King, and M. Waldrop, *J. Am. Chem. Soc.* **88**, 5112 (1966).
87. Y. Kaizu and H. Kobayashi, *Bull. Chem. Soc. Jpn.* **45**, 470 (1972).
88. J. A. Connor, E. J. James, C. Overton, and N. El Murr, *J. Organomet. Chem.* **218**, C31 (1981).
89. R. E. Dessy, J. C. Charkoudian, T. P. Abeles, and A. L. Rheingold, *J. Am. Chem. Soc.* **92**, 3947 (1970).
90. H. tom Dieck and E. Kühl, *Z. Naturforsch. B.: Anorg. Chem., Org. Chem.* **37**, 324 (1982).
91. K. H. Pannell and R. Inglesias, *Inorg. Chim. Acta* **33**, L161 (1979).
92. W. Kaim, *Inorg. Chim. Acta* **53**, L151 (1981).
93. N. G. Connelly, *J. Chem. Soc., Dalton Trans.*, p. 2183 (1973).
94. P. K. Ashford, P. K. Baker, N. G. Connelly, R. L. Kelly, and V. A. Woodley, *J. Chem. Soc., Dalton Trans.*, p. 477 (1982).
95. N. G. Connelly, Z. Demidowicz, and R. L. Kelly, *J. Chem. Soc., Dalton Trans.*, p. 2335 (1975).
96. J. C. Kotz and C. L. Nivert, *J. Organomet. Chem.* **52**, 387 (1973).
97. J. C. Kotz, C. L. Nivert, J. M. Lieber, and R. C. Reed, *J. Organomet. Chem.* **91**, 87 (1975).
98. J. C. Kotz, C. L. Nivert, J. M. Lieber, and R. C. Reed, *J. Organomet. Chem.* **84**, 255 (1975).
99. C. Elschenbroich and F. Stohler, *Angew. Chem., Int. Ed. Engl.* **14**, 174 (1975).
100. A. M. Bond, D. J. Darensbourg, E. Mocellin, and B. J. Stewart, *J. Am. Chem. Soc.* **103**, 6827 (1981).
101. A. M. Bond, R. Colton, and J. J. Jackowski, *Inorg. Chem.* **14**, 274 (1975).

102. F. L. Wimmer, M. R. Snow, and A. M. Bond, *Inorg. Chem.* **13**, 1617 (1974).
103. C. M. Elson, *Inorg. Chem.* **15**, 469 (1976).
104. P. F. Crossing and M. R. Snow, *J. Chem. Soc. A.*, p. 610 (1971).
105. A. M. Bond, R. Colton, and J. J. Jackowski, *Inorg. Chem.* **14**, 2526 (1975).
106. J. Lewis and R. Whyman, *J. Chem. Soc.*, p. 5486 (1965).
107. S. Datta, T. J. McNeese, and S. S. Wreford, *Inorg. Chem.* **16**, 2661 (1977).
108. D. M. P. Mingos, *J. Organomet. Chem.* **179**, C29 (1979).
109. S. Clamp, N. G. Connelly, G. E. Taylor, and T. S. Louttit, *J. Chem. Soc., Dalton Trans.*, p. 2162 (1980).
110. M. R. Snow and F. L. Wimmer, *Aust. J. Chem.* **29**, 2349 (1976).
111. J. A. Connor and P. I. Riley, *J. Chem. Soc., Dalton Trans.*, p. 1231 (1979).
112. J. A. Connor and P. I. Riley, *J. Organomet. Chem.* **174**, 173 (1979).
113. J. A. Connor and P. I. Riley, *J. Chem. Soc., Dalton Trans.*, p. 1318 (1979).
114. D. Weir and J. K. S. Wan, *J. Organomet. Chem.* **220**, 323 (1981).
115. B. A. L. Crichton, J. R. Dilworth, C. J. Pickett, and J. Chatt, *J. Chem. Soc., Dalton Trans.*, p. 892 (1981).
116. A. M. Bond, J. A. Bowden, and R. Colton, *Inorg. Chem.* **13**, 602 (1974).
117. A. M. Bond and R. Colton, *Inorg. Chem.* **15**, 446 (1976).
118. H. Behrens and H. Zizlsperger, *Z. Naturforsch. B.: Anorg. Chem., Org. Chem.* **16**, 349 (1961).
119. M. C. R. Symons, S. W. Bratt, and J. L. Wyatt, *J. Chem. Soc., Dalton Trans.*, p. 991 (1982).
120. A. M. Bond, R. Colton, and J. J. Jackowski, *Inorg. Chem.* **17**, 105 (1978).
121. M. A. Fox, K. A. Campbell, and E. P. Kyba, *Inorg. Chem.* **20**, 4163 (1981).
122. S. W. Bratt and M. C. R. Symons, *J. Chem. Soc., Dalton Trans.*, p. 1314 (1977).
123. P. J. Krusic, H. Stoklosa, L. E. Manzer, and P. Meakin, *J. Am. Chem. Soc.* **97**, 667 (1975).
124. A. Hudson, M. F. Lappert, and B. K. Nicholson, *J. Chem. Soc., Dalton Trans.*, p. 551 (1977).
125. P. Lemoine, A. Giraudeau, and M. Gross, *Electrochim. Acta* **21**, 1 (1976).
126. P. Lemoine and M. Gross, *J. Organomet. Chem.* **133**, 193 (1977).
127. L. I. Denisovich, A. A. Ioganson, S. P. Gubin, N. E. Kolobova, and K. N. Anisimov, *Bull. Acad. Sci. USSR, Div. Chem. Sci. Engl. Transl.* **18**, 218 (1969).
128. R. Colton, J. Dalziel, W. P. Griffith, and G. Wilkinson, *J. Chem. Soc.*, p. 71 (1960).
129. M. C. R. Symons, J. Wyatt, B. M. Peake, J. Simpson, and B. H. Robinson, *J. Chem. Soc., Dalton Trans.*, p. 2037 (1982).
130. S. P. Church, M. Poliakoff, J. A. Timney, and J. J. Turner, *J. Am. Chem. Soc.* **103**, 7515 (1981).
131. J. A. Armstead, D. J. Cox, and R. Davis, *J. Organomet. Chem.* **236**, 213 (1982).
132. S. B. McCullen, H. W. Walker, and T. L. Brown, *J. Am. Chem. Soc.* **104**, 4007 (1982).
133. D. R. Kidd, C. P. Cheng, and T. L. Brown, *J. Am. Chem. Soc.* **100**, 4103 (1978).
134. R. L. Harlow, P. J. Krusic, R. J. McKinney, and S. S. Wreford, *Organometallics* **1**, 1506 (1982).
135. R. E. Dessy, P. M. Weissman, and R. L. Pohl, *J. Am. Chem. Soc.* **88**, 5117 (1966).
136. O. P. Anderson, S. A. Fieldhouse, C. E. Forbes, and M. C. R. Symons, *J. Chem. Soc., Dalton Trans.*, p. 1329 (1976).
137. A. M. Bond, R. Colton, and M. J. McCormick, *Inorg. Chem.* **16**, 155 (1977).
138. A. M. Bond, B. S. Grabaric, and Z. Grabaric, *Inorg. Chem.* **17**, 1013 (1978).
139. R. H. Reimann and E. Singleton, *J. Chem. Soc., Dalton Trans.*, p. 2658 (1973).
140. F. Bombin, G. A. Carriedo, J. A. Miguel, and V. Riera, *J. Chem. Soc., Dalton Trans.*, p. 2049 (1981).
141. C. Eaborn, N. Farrell, J. L. Murphy, and A. Pidcock, *J. Chem. Soc., Dalton Trans.*, p. 58 (1976).

142. R. Seeber, G. A. Mazzocchin, E. Roncari, and U. Mazzi, *Transition Met. Chem.* (*Weinheim, Ger.*) **6**, 123 (1981).
143. U. Mazzi, E. Roncari, R. Seeber, and G. A. Mazzocchin, *Inorg. Chim. Acta* **41**, 95 (1980).
144. G. J. Leigh, R. H. Morris, C. J. Pickett, D. R. Stanley, and J. Chatt, *J. Chem. Soc., Dalton Trans.*, p. 800 (1981).
145. A. J. L. Pombeiro, C. J. Pickett, and R. L. Richards, *J. Organomet. Chem.* **224**, 285 (1982).
146. M. R. Snow and M. H. B. Stiddard, *J. Chem. Soc. A.*, p. 777 (1966).
147. P. M. Treichel and H. J. Mueh, *Inorg. Chem.* **16**, 1167 (1977).
148. P. M. Treichel and D. W. Firsich, *J. Organomet. Chem.* **172**, 223 (1979).
149. P. M. Treichel and J. P. Williams, *J. Organomet. Chem.* **135**, 39 (1977).
150. P. M. Treichel and H. J. Mueh, *J. Organomet. Chem.* **122**, 229 (1976).
151. P. M. Treichel, G. E. Dirreen, and H. J. Mueh, *J. Organomet. Chem.* **44**, 339 (1972).
152. A. C. Sarapu and R. F. Fenske, *Inorg. Chem.* **14**, 247 (1975).
153. J. R. Morton, K. F. Preston, and S. J. Strach, *J. Phys. Chem.* **84**, 2478 (1980).
154. N. El Murr and A. Chaloyard, *Inorg. Chem.* **21**, 2206 (1982).
155. A. M. Bond, P. A. Dawson, B. M. Peake, B. H. Robinson, and J. Simpson, *Inorg. Chem.* **16**, 2199 (1977).
156. A. A. Vlcek, *Nature (London)* **177**, 1043 (1956).
157. P. J. Krusic, J. San Filippo, Jr., B. Hutchinson, R. L. Hance, and L. M. Daniels, *J. Am. Chem. Soc.* **103**, 2129 (1981).
158. P. A. Dawson, B. M. Peake, B. H. Robinson, and J. Simpson, *Inorg. Chem.* **19**, 465 (1980).
159. P. K. Baker, N. G. Connelly, B. M. R. Jones, J. P. Maher, and K. R. Somers, *J. Chem. Soc., Dalton Trans.*, p. 579 (1980).
160. S. W. Blanch, A. M. Bond, and R. Colton, *Inorg. Chem.* **20**, 755 (1981).
161. G. Bellachioma, G. Cardaci, and G. Reichenbach, *J. Organomet. Chem.* **221**, 291 (1981).
162. P. K. Baker, K. Broadley, and N. G. Connelly, *J. Chem. Soc., Dalton Trans.*, p. 471 (1982).
163. A. L. Balch, *J. Am. Chem. Soc.* **95**, 2723 (1973).
164. N. G. Connelly, I. Manners, J. R. C. Protheroe, and M. W. Whiteley, *J. Chem. Soc., Dalton Trans.*, in press (1984).
165. G. Zotti, S. Zecchin, and G. Pilloni, *J. Organomet. Chem.* **181**, 375 (1978).
166. M. F. Lappert, J. J. MacQuitty, and P. L. Pye, *J. Chem. Soc., Dalton Trans.*, p. 1583 (1981).
167. J. D. Buhr and H. Taube, *Inorg. Chem.* **18**, 2208 (1979).
168. J. Hanzlik, G. Albertin, E. Bordignon, and A. A. Orio, *J. Organomet. Chem.* **224**, 49 (1982).
169. S. M. Murgia, G. Paliani, and G. Cardaci, *Z. Naturforsch. B.: Anorg. Chem., Org. Chem.* **27**, 134 (1972).
170. G. Piazza and G. Paliani, *Z. Phys. Chem.* (*Wiesbaden*) **71**, 91 (1970).
171. C. Couture, J. R. Morton, K. F. Preston, and S. J. Strach, *J. Magn. Reson.* **41**, 88 (1980).
172. R. E. Dessy, J. C. Charkoudian, and A. L. Rheingold, *J. Am. Chem. Soc.* **94**, 738 (1972).
173. A. A. Vlcek, *Collect. Czech. Chem. Commun.* **24**, 1748 (1959).
174. R. E. Dessy, R. L. Pohl, and R. B. King, *J. Am. Chem. Soc.* **88**, 5121 (1966).
175. O. P. Anderson, S. A. Fieldhouse, C. E. Forbes, and M. C. R. Symons, *J. Organomet. Chem.* **110**, 247 (1976).
176. M. Absi-Halabi and T. L. Brown, *J. Am. Chem. Soc.* **99**, 2982 (1977).
177. P. J. Krusic, P. J. Fagan, and J. San Filippo, Jr., *J. Am. Chem. Soc.* **99**, 250 (1977).
178. G. Schiavon, S. Zecchin, G. Zotti, and G. Pilloni, *Inorg. Chim. Acta* **20**, L1 (1976).
179. S. Zecchin, G. Schiavon, G. Zotti, and G. Pilloni, *Inorg. Chim. Acta* **34**, L267 (1979).
180. S. Zecchin, G. Zotti, and G. Pilloni, *Inorg. Chim. Acta* **33**, L117 (1979).
181. R. Hammer and H.-F. Klein, *Z. Naturforsch. B.: Anorg. Chem., Org. Chem.* **32**, 138 (1977).
182. E. L. Muetterties, J. R. Bleeke, Z.-Y. Yang, and V. W. Day, *J. Am. Chem. Soc.* **104**, 2940 (1982).

183. E. L. Muetterties and P. L. Watson, *J. Am. Chem. Soc.* **100**, 6978 (1978).
184. G. Pilloni, G. Zotti, and M. Martelli, *Inorg. Chem.* **21**, 1283 (1982).
185. G. Pilloni, G. Zotti, and M. Martelli, *Inorg. Chim. Acta* **13**, 213 (1975).
186. J. A. Sofranko, R. Eisenberg, and J. A. Kampmeier, *J. Am. Chem. Soc.* **101**, 1042 (1979).
187. J. A. Sofranko, R. Eisenberg, and J. A. Kampmeier, *J. Am. Chem. Soc.* **102**, 1163 (1980).
188. G. Pilloni, E. Vecchi, and M. Martelli, *J. Electroanal. Chem. Interfacial Electrochem.* **45**, 483 (1973).
189. G. Pilloni, G. Schiavon, G. Zotti, and S. Zecchin, *J. Organomet. Chem.* **134**, 305 (1977).
190. G. Schiavon, S. Zecchin, G. Pilloni, and M. Martelli, *Inorg. Chim. Acta* **14**, L4 (1975).
191. D. Olson and W. Keim, *Inorg. Chem.* **8**, 2028 (1969).
192. B.-K. Teo, A. P. Ginsberg, and J. C. Calabrese, *J. Am. Chem. Soc.* **98**, 3027 (1976).
193. G. Pilloni, S. Valcher, and M. Martelli, *J. Electroanal. Chem. Interfacial Electrochem.* **40**, 63 (1972).
194. D. de Montauzon and R. Poilblanc, *J. Organomet. Chem.* **93**, 397 (1975).
195. G. Schiavon, S. Zecchin, G. Pilloni, and M. Martelli, *J. Inorg. Nucl. Chem.* **39**, 115 (1977).
196. G. Schiavon, S. Zecchin, G. Pilloni, and M. Martelli, *J. Organomet. Chem.* **121**, 261 (1976).
197. J. Vecernik, J. Masek, and A. A. Vlcek, *J. Chem. Soc., Chem. Commun.*, p. 736 (1975).
198. T. A. B. M. Bolsman and J. A. Van Doorn, *J. Organomet. Chem.* **178**, 381 (1979).
199. J. Vecernik, J. Masek, and A. A. Vlcek, *Inorg. Chim. Acta* **16**, 143 (1976).
200. S. Zecchin, G. Schiavon, G. Pilloni, and M. Martelli, *J. Organomet. Chem.* **110**, C45 (1976).
201. M. Martelli, G. Schiavon, S. Zecchin, and G. Pilloni, *Inorg. Chim. Acta* **15**, 217 (1975).
202. A. L. Balch, *J. Am. Chem. Soc.* **98**, 285 (1976).
203. J. M. Pratt and P. R. Silverman, *J. Chem. Soc. A.*, p. 1286 (1967).
204. F. A. Cotton, T. G. Dunne, and J. S. Wood, *Inorg. Chem.* **3**, 1495 (1964).
205. M. E. Kimball and W. C. Kaska, *Inorg. Nucl. Chem. Lett.* **7**, 119 (1971).
206. J. W. Dart, M. K. Lloyd, R. Mason, J. A. McCleverty, and J. Williams, *J. Chem. Soc., Dalton Trans.*, p. 1747 (1973).
207. R. Pribl, Jr., J. Masek, and A. A. Vlcek, *Inorg. Chim. Acta* **5**, 57 (1971).
208. J. Masek, *Inorg. Chim. Acta, Rev.* **3**, 99 (1969).
209. M. Di Vaira, C. A. Ghilardi, and L. Sacconi, *Inorg. Chem.* **15**, 1555 (1976).
210. R. Seeber, G. Albertin, and G.-A. Mazzocchin, *J. Chem. Soc., Dalton Trans.*, p. 2561 (1982).
211. M. Martelli, G. Pilloni, G. Zotti, and S. Daolio, *Inorg. Chim. Acta* **11**, 155 (1974).
212. G. A. Bowmaker, P. D. W. Boyd, G. K. Campbell, J. M. Hope, and R. L. Martin, *Inorg. Chem.* **21**, 1152 (1982).
213. L. Sacconi, P. Dapporto, and P. Stoppioni, *Inorg. Chem.* **15**, 325 (1976).
214. A. Gleizes, M. Dartiguenave, Y. Dartiguenave, J. Galy, and H.-F. Klein, *J. Am. Chem. Soc.* **99**, 5187 (1977).
215. H.-F. Klein, H. H. Karsch, and W. Buchner, *Chem. Ber.* **107**, 537 (1974).
216. T. T. Tsou and J. K. Kochi, *J. Am. Chem. Soc.* **101**, 6319 (1979).
217. M. Troupel, Y. Rollin, C. Chevrot, F. Pfluger, and J.-F. Fauvarque, *J. Chem. Res. (S)*, p. 50 (1979).
218. M. Troupel, Y. Rollin, S. Sibille, J.-F. Fauverque, and J. Perichon, *J. Chem. Res. (S)*, p. 24 (1980).
219. M. Troupel, Y. Rollin, S. Sibille, J.-F. Fauvarque, and J. Perichon, *J. Chem. Res. (S)*, p. 26 (1980).
220. G. Kreisel and W. Seidel, *Z. Anorg. Allg. Chem.* **478**, 106 (1981).
221. W. Seidel and G. Kreisel, *Z. Chem.* **16**, 115 (1976).
222. W. Seidel and G. Kreisel, *Z. Anorg. Allg. Chem.* **426**, 150 (1976).
223. W. Seidel, P. Scholz, and G. Kreisel, *Z. Anorg. Allg. Chem.* **458**, 263 (1979).
224. W. Seidel and I. Bürger, *Z. Anorg. Allg. Chem.* **426**, 155 (1976).

225. K. Schmiedeknecht, *J. Organomet. Chem.* **133**, 187 (1977).
226. W. Seidel and I. Bürger, *J. Organomet. Chem.* **171**, C45 (1979).
227. W. Seidel and I. Bürger, *J. Organomet. Chem.* **177**, C19 (1979); R. Kirmse and W. Seidel, *Z. Anorg. Allg. Chem.* **490**, 19 (1982); C. Cauletti, R. Zanoni, and W. Seidel, *Z. Anorg. Allg. Chem.* **496**, 143 (1983).
228. M. F. Lappert, C. L. Raston, B. W. Skelton, and A. H. White, *J. Chem. Soc., Chem. Commun.*, p. 485 (1981).
229. A. Rusina, A. A. Vlcek, and K. Schmiedeknecht, *J. Organomet. Chem.* **192**, 367 (1980).
230. W. Lau, J. C. Huffman, and J. K. Kochi, *Organometallics* **1**, 155 (1982).
231. D. J. Cole-Hamilton and G. Wilkinson, *J. Chem. Soc., Dalton Trans.*, p. 1283 (1979).
232. E. O. Sherman and P. R. Schreiner, *J. Chem. Soc., Chem. Commun.*, p. 3 (1976).
233. G. Schiavon, G. Bontempelli, M. De Nobili, and B. Corain, *Inorg. Chim. Acta* **42**, 211 (1980).
234. J. Y. Chen and J. K. Kochi, *J. Am. Chem. Soc.* **99**, 1450 (1977).
235. R. J. Klingler, J. C. Huffman, and J. K. Kochi, *J. Am. Chem. Soc.* **104**, 2147 (1982).
236. P. J. Krusic and J. San Filippo, Jr., *J. Am. Chem. Soc.* **104**, 2645 (1982).
237. M. R. Blake, J. L. Garnett, I. K. Gregor, and S. B. Wild, *J. Chem. Soc., Chem. Commun.*, p. 496 (1979).
238. M. R. Blake, J. L. Garnett, I. K. Gregor, and S. B. Wild, *J. Organomet. Chem.* **178**, C37 (1979).
239. O. P. Anderson and M. C. R. Symons, *Inorg. Chem.* **12**, 1932 (1973).
240. N. El Murr, M. Riveccie, and P. Dixneuf, *J. Chem. Soc., Chem. Commun.*, p. 552 (1978).
241. S. P. Gubin and L. I. Denisovich, *J. Organomet. Chem.* **15**, 471 (1968).
242. P. A. Wegner and M. S. Delaney, *Inorg. Chem.* **15**, 1918 (1976).
243. S. D. Ittel, P. J. Krusic, and P. Meakin, *J. Am. Chem. Soc.* **100**, 3264 (1978).
244. S. D. Ittel, F. A. Van-Catledge, and J. P. Jesson, *J. Am. Chem. Soc.* **101**, 6905 (1979).
245. R. K. Brown, J. M. Williams, A. J. Schultz, G. D. Stucky, S. D. Ittel, and R. L. Harlow, *J. Am. Chem. Soc.* **102**, 981 (1980).
246. R. L. Harlow, R. J. McKinney, and S. D. Ittel, *J. Am. Chem. Soc.* **101**, 7496 (1979).
247. H. W. Vanden Born and W. E. Harris, *J. Electroanal. Chem. Interfacial Electrochem*, **42**, 151 (1973).
248. G. Paliani, S. M. Murgia, and G. Cardaci, *J. Organomet. Chem.* **30**, 221 (1971).
249. G. Cardaci, S. M. Murgia, and G. Paliani, *J. Organomet. Chem.* **77**, 253 (1974).
250. R. J. McKinney, *J. Chem. Soc., Chem. Commun.*, p. 603 (1980).
251. L. S. Hegedus and E. L. Waterman, *J. Am. Chem. Soc.* **96**, 6789 (1974).
252. S. P. Gubin and L. I. Denisovich, *Bull. Acad. Sci. USSR, Div. Chem. Sci. Engl. Transl.* 125 (1966).
253. N. El Murr, M. Riveccie, E. Laviron, and G. Deganello, *Tetrahedron Lett.*, p. 3339 (1976).
254. B. Tulyathan and W. E. Geiger, *J. Electroanal. Chem. Interfacial Electrochem.* **109**, 325 (1980).
255. T. A. Albright, W. E. Geiger, J. Moraczewski, and B. Tulyathan, *J. Am. Chem. Soc.* **103**, 4787 (1981).
256. G. Zotti, R. D. Rieke, and J. S. McKennis, *J. Organomet. Chem.* **228**, 281 (1982).
257. N. G. Connelly and R. L. Kelly, *J. Organomet. Chem.* **120**, C16 (1976).
258. N. G. Connelly, R. L. Kelly, M. D. Kitchen, R. M. Mills, R. F. D. Stansfield, M. W. Whiteley, S. M. Whiting, and P. Woodward, *J. Chem. Soc., Dalton Trans.*, p. 1317 (1981).
259. N. G. Donnelly, R. L. Kelly, and M. W. Whiteley, unpublished results.
260. N. G. Connelly, R. L. Kelly, and M. W. Whiteley, *J. Chem. Soc., Dalton Trans.*, p. 34 (1981).

261. N. G. Connelly, M. Freeman, A. G. Orpen, M. W. Whiteley, and P. Woodward, unpublished results.
262. N. G. Connelly, R. M. Mills, M. W. Whiteley, and P. Woodward, *J. Chem. Soc., Chem. Commun.*, p. 17 (1981).
263. N. G. Connelly, A. R. Lucy, R. M. Mills, J. B. Sheridan, M. W. Whiteley, and P. Woodward, *J. Chem. Soc., Chem. Commun.*, p. 1057 (1982).
264. W. A. Fordyce, K. H. Pool, and G. A. Crosby, *Inorg. Chem.* **21**, 1027 (1982).
265. E. Makrlik, J. Hanzlik, A. Camus, G. Mestroni, and G. Zassinovich, *J. Organomet. Chem.* **142**, 95 (1977).
266. A. N. Nesmeyanov, L. S. Isaeva, and T. A. Peganova, *Bull. Acad. Sci. USSR, Div. Chem. Sci. Engl. Transl.* **24**, 2203 (1975).
267. A. N. Nesmeyanov, L. S. Isaeva, T. A. Peganova, and A. A. Slinkin, *Bull. Acad. Sci. USSR, Div. Chem. Sci. Engl. Transl.* **23**, 148 (1974).
268. N. El Murr, A. Chaloyard, and J. Tirouflet, *J. Chem. Soc., Chem. Commun.*, p. 446 (1980).
269. M. F. Lappert, C. J. Pickett, P. I. Riley, and P. J. W. Yarrow, *J. Chem. Soc., Dalton Trans.*, p. 805 (1981).
270. E. Laviron, J. Besançon, and F. Huq, *J. Organomet. Chem.* **159**, 279 (1978).
271. Y. Mugnier, C. Moise, and E. Laviron, *J. Organomet. Chem.* **204**, 61 (1981).
272. S. Valcher and M. Mastragostino, *J. Electroanal. Chem. Interfacial Electrochem.* **14**, 219 (1967).
273. S. P. Gubin and S. A. Smirnova, *J. Organomet. Chem.* **20**, 229 (1969).
274. R. G. Doisneau and J.-C. Marchon, *J. Electroanal. Chem. Interfacial Electrochem.* **30**, 487 (1971).
275. T. Chivers and E. D. Ibrahim, *Can. J. Chem.* **51**, 815 (1973).
276. V. Kadlec, H. Kadlecova, and O. Strouf, *J. Organomet. Chem.* **82**, 113 (1974).
277. Y. Mugnier, C. Moise, and E. Laviron, *J. Organomet. Chem.* **210**, 69 (1981).
278. N. El Murr and A. Chaloyard, *J. Organomet. Chem.* **212**, C39 (1981).
279. N. El Murr and A. Chaloyard, *J. Organomet. Chem.* **231**, 1 (1982).
280. M. C. R. Symons and S. P. Mishra, *J. Chem. Soc., Dalton Trans.*, p. 2258 (1981).
281. A. Chaloyard, A. Dormond, J. Tirouflet, and N. El Murr, *J. Chem. Soc., Chem. Commun.*, p. 214 (1980).
282. J. A. McCleverty, T. A. James, and E. J. Wharton, *Inorg. Chem.* **8**, 1340 (1969).
283. J. G. Kenworthy, J. Myatt, and P. F. Todd, *J. Chem. Soc. B.*, p. 791 (1970).
284. H. H. Brintzinger, *J. Am. Chem. Soc.* **89**, 6871 (1967).
285. G. Henrici-Olivé and S. Olivé, *Angew. Chem., Int. Ed. Engl.* **7**, 386 (1968).
286. S. P. Gubin and S. A. Smirnova, *J. Organomet. Chem.* **20**, 241 (1969).
287. M. Kira, H. Bock, H. Umino, and H. Sakurai, *J. Organomet. Chem.* **173**, 39 (1979).
288. M. F. Lappert and C. L. Raston, *J. Chem. Soc., Chem. Commun.*, p. 1284 (1980).
289. M. F. Lappert and C. L. Raston, *J. Chem. Soc., Chem. Commun.*, p. 173 (1981).
290. M. J. S. Gynane, J. Jeffery, and M. F. Lappert, *J. Chem. Soc., Chem. Commun.*, p. 34 (1978).
291. N. E. Schore and H. Hope, *J. Am. Chem. Soc.* **102**, 4251 (1980).
292. A. M. McPherson, B. F. Fieselmann, D. L. Lichtenberger, G. L. McPherson, and G. D. Stucky, *J. Am. Chem. Soc.* **101**, 3425 (1979).
293. M. L. H. Green and C. R. Lucas, *J. Chem. Soc., Dalton Trans.*, p. 1000 (1972).
294. D. R. Corbin, W. S. Willis, E. W. Duesler, and G. D. Stucky, *J. Am. Chem. Soc.* **102**, 5969 (1980).
295. N. El Murr, C. Moise, M. Riveccie, and J. Tirouflet, *Inorg. Chim. Acta* **32**, 189 (1979).
296. C. Moise, N. El Murr, M. Riveccie, and J. Tirouflet, *C. R. Hebd. Seances Acad. Sci.* **287**, 329 (1978).
297. N. G. Connelly and M. D. Kitchen, *J. Chem. Soc., Dalton Trans.*, p. 2165 (1976).
298. A. M. Bond, A. T. Casey, and J. R. Thackeray, *Inorg. Chem.* **13**, 84 (1974).

299. J. D. L. Holloway and W. E. Geiger, *J. Am. Chem. Soc.* **101**, 2038 (1979).
300. E. N. Gladyshev, P. Ya. Bayushkin, V. K. Cherkasov, and V. S. Sokolov, *Bull. Acad. Sci. USSR, Div. Chem. Sci. Engl. Transl.* **28**, 1070 (1979).
301. G. A. Razuvaev, P. Ya. Bayushkin, V. K. Cherkasov, E. N. Gladyshev, and A. P. Phokeev, *Inorg. Chim. Acta* **44**, L103 (1980).
302. M. F. Lappert and C. R. C. Milne, *J. Chem. Soc., Chem. Commun.*, p. 925 (1978).
303. A. M. Bond, A. T. Casey, and J. R. Thackeray, *Inorg. Chem.* **12**, 887 (1973).
304. A. M. Bond, A. T. Casey, and J. R. Thackeray, *J. Chem. Soc., Dalton Trans.*, p. 773 (1974).
305. T. Madach and H. Vahrenkamp, *Z. Naturforsch. B.: Anorg. Chem., Org. Chem.* **34**, 573 (1979).
306. L. I. Denisovich, S. P. Gubin, Yu. A. Chapovskii, and N. A. Ustynok, *Bull. Acad. Sci. USSR, Div. Chem. Sci. Engl. Transl.*, p. 891 (1968).
307. L. I. Denisovich, I. V. Polovyanyuk, B. V. Lokshin, and S. P. Gubin, *Bull. Acad. Sci. USSR, Div. Chem. Sci. Engl. Transl.* **20**, 1851 (1971).
308. T. Madach and H. Vahrenkamp, *Z. Naturforsch. B.: Anorg. Chem., Org. Chem.* **33**, 1301 (1978).
309. E. Samuel, M. D. Rausch, T. E. Gismondi, E. A. Mintz, and C. Giannotti, *J. Organomet. Chem.* **172**, 309 (1979).
310. S. R. Allen, M. Green, and N. G. Connelly, unpublished results.
311. S. W. Anderson, C. W. Fong, and M. D. Johnson, *J. Chem. Soc., Chem. Commun.*, p. 163 (1973).
312. M. L. H. Green, J. Knight, and J. A. Segal, *J. Chem. Soc., Dalton Trans.*, p. 2189 (1977).
313. R. B. King and M. B. Bisnette, *Inorg. Chem.* **6**, 469 (1967).
314. K. Tanaka, K. U-Eda, and T. Tanaka, *J. Inorg. Nucl. Chem.* **43**, 2029 (1981).
315. A. Nakamura, *J. Organomet. Chem.* **164**, 183 (1979).
316. J. C. Hayes and N. J. Cooper, *J. Am. Chem. Soc.* **104**, 5570 (1982).
317. W. E. Lindsell, *J. Chem. Soc., Dalton Trans.*, p. 2548 (1975).
318. M. C. R. Symons and M. M. Maguire, *J. Chem. Res. (S)*, p. 330 (1981).
319. Y. S. Yu, R. A. Jacobson, and R. J. Angelici, *Inorg. Chem.* **21**, 3106 (1982).
320. T. A. James and J. A. McCleverty, *J. Chem. Soc. A*, p. 3308 (1970).
321. L. I. Denisovich and S. P. Gubin, *J. Organomet. Chem.* **57**, 87 (1973).
322. T. Würminghausen and D. Sellmann, *J. Organomet. Chem.* **199**, 77 (1980).
323. J. Tirouflet, R. Dabard, and E. Laviron, *Bull. Soc. Chim. Fr.*, p. 1655 (1963).
324. N. J. Gogan, C.-K. Chu, and G. W. Gray, *J. Organomet. Chem.* **51**, 323 (1973).
325. P. M. Treichel, K. P. Wagner, and H. J. Mueh, *J. Organomet. Chem.* **86**, C13 (1975).
326. N. G. Connelly and M. D. Kitchen, *J. Chem. Soc., Dalton Trans.*, p. 931 (1977).
327. P. M. Treichel and H. J. Mueh, *Inorg. Chim. Acta* **22**, 265 (1977).
328. L. I. Denisovich, N. V. Zakurin, S. P. Gubin, and A. G. Ginzburg, *J. Organomet. Chem.* **101**, C43 (1975).
329. J. A. McCleverty, D. G. Orchard, J. A. Connor, E. M. Jones, J. P. Lloyd, and P. D. Rose, *J. Organomet. Chem.* **30**, C75 (1971).
330. B. V. Lokshin, E. B. Nazarova, and A. G. Ginzburg, *J. Organomet. Chem.* **129**, 379 (1977).
331. D. Sellmann, J. Müller, and P. Hofmann, *Angew. Chem., Int. Ed. Engl.* **21**, 691 (1982).
332. P. Hydes, J. A. McCleverty, and D. G. Orchard, *J. Chem. Soc. A.*, p. 3660 (1971).
333. J. A. McCleverty and D. G. Orchard, *J. Chem. Soc. A.*, p. 3315 (1970).
334. D. Miholova and A. A. Vlcek, *Inorg. Chim. Acta* **41**, 119 (1980).
335. K. Jonas and L. Schieferstein, *Angew. Chem., Int. Ed. Engl.* **18**, 549 (1979).
336. J. A. Ferguson and T. J. Meyer, *Inorg. Chem.* **10**, 1025 (1971).
337. E. C. Johnson, T. J. Meyer, and N. Winterton, *Inorg. Chem.* **10**, 1673 (1971).
338. J. N. Braddock and T. J. Meyer, *Inorg. Chem.* **12**, 723 (1973).
339. J. A. Ferguson and T. J. Meyer, *Inorg. Chem.* **11**, 631 (1972).

340. R. J. Haines and A. L. du Preez, *Inorg. Chem.* **11**, 330 (1972).
341. M. E. Grant and J. J. Alexander, *J. Coord. Chem.* **9**, 205 (1979).
342. P. M. Treichel and D. C. Molzahn, *J. Organomet. Chem.* **179**, 275 (1979).
343. P. M. Treichel, D. C. Molzahn, and K. P. Wagner, *J. Organomet. Chem.* **174**, 191 (1979).
344. P. M. Treichel and D. A. Komar, *J. Organomet. Chem.* **206**, 77 (1981).
345. D. L. Reger, D. J. Fauth, and M. D. Dukes, *J. Organomet. Chem.* **170**, 217 (1979).
346. P. M. Treichel and L. D. Rosenhein, *J. Am. Chem. Soc.* **103**, 691 (1981).
347. G. Nagao, K. Tanaka, and T. Tanaka, *Inorg. Chim. Acta* **42**, 43 (1980).
348. J. G. M. van der Linden, A. H. Dix, and E. Pfeiffer, *Inorg. Chim. Acta* **39**, 271 (1980).
349. L. I. Denisovich and S. P. Gubin, *J. Organomet. Chem.* **57**, 109 (1973).
350. L. I. Denisovich, S. P. Gubin, and Yu. A. Chapovskii, *Bull. Acad. Sci. USSR, Div. Chem. Sci. Engl. Transl.*, p. 2271 (1967).
351. D. Miholova and A. A. Vlcek, *Inorg. Chim. Acta* **43**, 43 (1980).
352. D. Miholova and A. A. Vlcek, *J. Organomet. Chem.* **240**, 413 (1982).
353. W. Rogers, J. A. Page, and M. C. Baird, *J. Organomet. Chem.* **156**, C37 (1978).
354. D. A. Slack and M. C. Baird, *J. Am. Chem. Soc.* **98**, 5539 (1976).
355. R. H. Magnuson, S. Zulu, W.-M. T'sai, and W. P. Giering, *J. Am. Chem. Soc.* **102**, 6887 (1980).
356. M. F. Joseph, J. A. Page, and M. C. Baird, *Inorg. Chim. Acta* **64**, L121 (1982).
357. R. J. Klingler and J. K. Kochi, *J. Organomet. Chem.* **202**, 49 (1980).
358. P. S. Waterman and W. P. Giering, *J. Organomet. Chem.* **155**, C47 (1978).
359. C. S. Ilenda, N. E. Schore, and R. G. Bergman, *J. Am. Chem. Soc.* **98**, 255 (1976).
360. M. C. R. Symons and S. W. Bratt, *J. Chem. Soc., Dalton Trans.*, p. 1739 (1979).
361. K. Broadley, N. G. Connelly, and W. E. Geiger, *J. Chem. Soc., Dalton Trans.*, p. 121 (1983).
362. R. J. McKinney, *Inorg. Chem.* **21**, 2051 (1982).
363. N. G. Connelly, M. Freeman, I. Manners, and A. G. Orpen, *J. Chem. Soc., Dalton Trans.*, in press (1984).
364. N. G. Connelly, A. R. Lucy, J. D. Payne, A. M. R. Galas, and W. E. Geiger, *J. Chem. Soc., Dalton Trans.*, p. 1879 (1983).
365. M. Freeman, A. G. Orpen, and N. G. Connelly, unpublished results.
366. N. G. Connelly, A. R. Lucy, and A. M. R. Galas, *J. Chem. Soc., Chem. Commun.*, p. 43 (1981).
367. U. Koelle, *J. Organomet. Chem.* **184**, 379 (1980).
368. E. Pfeiffer, M. W. Kokkes, and K. Vrieze, *Transition Met. Chem. (Weinheim, Ger.)* **4**, 393 (1979).
369. N. El Murr, A. Chaloyard, and W. Kläui, *Inorg. Chem.* **18**, 2629 (1979).
370. W. Kläui and K. Dehnicke, *Chem. Ber.* **111**, 451 (1978).
371. U. Koelle, *Inorg. Chim. Acta* **47**, 13 (1981).
372. J. Moraczewski and W. E. Geiger, *Organometallics* **1**, 1385 (1982).
373. S. P. Gubin and V. S. Khandkarova, *J. Organomet. Chem.* **12**, 523 (1968).
374. H. van Willigen, W. E. Geiger, and M. D. Rausch, *Inorg. Chem.* **16**, 581 (1977).
375. G. Lane and W. E. Geiger, *Organometallics* **1**, 401 (1982).
376. G. Bermudez and D. Pletcher, *J. Organomet. Chem.* **231**, 173 (1982).
377. U. Koelle and H. Werner, *J. Organomet. Chem.* **221**, 367 (1981).
378. E. K. Barefield, D. A. Krost, J. S. Edwards, D. G. Van Derveer, R. L. Trytko, S. P. O'Rear, and A. N. Williamson, *J. Am. Chem. Soc.* **103**, 6219 (1981).
379. K. Broadley, G. A. Lane, N. G. Connelly, and W. E. Geiger, *J. Am. Chem. Soc.* **105**, 2486 (1983).
380. J. J. Maj, A. D. Rae, and L. F. Dahl, *J. Am. Chem. Soc.* **104**, 4278 (1982).
381. G. Paliani, *Z. Naturforsch. B.: Anorg. Chem., Org. Chem.* **25**, 786 (1970).
382. P. L. Watson, *J. Chem. Soc., Chem. Commun.*, p. 652 (1980).

383. Y. Mugnier, A. Dormond, and E. Laviron, *J. Chem. Soc., Chem. Commun.*, p. 257 (1982).
384. R. G. Finke, G. Gaughan, and R. Voegeli, *J. Organomet. Chem.* **229**, 179 (1982).
385. P. Reeb, Y. Mugnier, A. Dormond, and E. Laviron, *J. Organomet. Chem.* **239**, C1 (1982).
386. M. Schneider and E. Weiss, *J. Organomet. Chem.* **114**, C43 (1976).
387. A. Davison and D. L. Reger, *J. Organomet. Chem.* **23**, 491 (1970).
388. M. K. Lloyd, J. A. McCleverty, J. A. Connor, and E. M. Jones, *J. Chem. Soc., Dalton Trans.*, p. 1768 (1973).
389. S. P. Gubin and V. S. Khandkarova, *J. Organomet. Chem.* **22**, 449 (1970).
390. N. G. Connelly and G. A. Johnson, *J. Organomet. Chem.* **77**, 341 (1974).
391. K. L. Amos and N. G. Connelly, *J. Organomet. Chem.* **194**, C57 (1980).
392. A. M. Bond, R. Colton, J. J. Jackowski, *Inorg. Chem.* **18**, 1977 (1979).
393. R. D. Rieke, I. Tucker, S. N. Milligan, D. R. Wright, B. R. Willeford, L. J. Radonovich, and M. W. Eyring, *Organometallics* **1**, 938 (1982).
394. I. A. Suskina, L. I. Denisovich, and S. P. Gubin, *Bull. Acad. Sci. USSR, Div. Chem. Sci. Engl. Transl.* **24**, 402 (1975).
395. R. D. Rieke, J. S. Arney, W. E. Rich, B. R. Willeford, Jr., and B. S. Poliner, *J. Am. Chem. Soc.* **97**, 5951 (1975).
396. A. Ceccon, A. Romanin, and A. Venzo, *Transition Met. Chem. (Weinheim, Ger.)* **1**, 25 (1976).
397. N. J. Gogan, I. L. Dickinson, J. Doull, and J. R. Patterson, *J. Organomet. Chem.* **212**, 71 (1981).
398. A. Ceccon, C. Corvaja, G. Giacometti, and A. Venzo, *J. Chem. Soc., Perkin Trans. 2*, p. 283 (1978).
399. J. D. Munro and P. L. Pauson, *J. Chem. Soc.*, p. 3484 (1961).
400. G. Hoch, R. Panter, and M. L. Ziegler, *Z. Naturforsch. B.: Anorg. Chem., Org. Chem.* **31**, 294 (1975).
401. R. Panter and M. L. Ziegler, *Z. Anorg. Allg. Chem.* **453**, 14 (1979).
402. M. L. H. Green and R. B. A. Pardy, *J. Organomet. Chem.* **117**, C13 (1976).
403. A. Greco, S. Cesca, and G. Bertolini, *J. Organomet. Chem.* **113**, 321 (1976).
404. J. A. Butcher, Jr., R. M. Pagni, and J. Q. Chambers, *J. Organomet. Chem.* **199**, 223 (1980).
405. F. Billiau, G. Folcher, H. Marquet-Ellis, P. Rigny, and E. Saito, *J. Am. Chem. Soc.* **103**, 5603 (1981).
406. D. G. Karraker, J. A. Stone, E. R. Jones, Jr., and N. Edelstein, *J. Am. Chem. Soc.* **92**, 4841 (1970).
407. D. G. Karraker and J. A. Stone, *J. Am. Chem. Soc.* **96**, 6885 (1974).
408. J. Knol, A. Westerhof, H. O. van Oven, and H. J. de Liefde Meijer, *J. Organomet. Chem.* **96**, 257 (1975).
409. F. Calderazzo, *Inorg. Chem.* **3**, 810 (1964).
410. G. Henrici-Olivé and S. Olivé, *J. Am. Chem. Soc.* **92**, 4831 (1970).
411. G. Henrici-Olivé and S. Olivé, *J. Organomet. Chem.* **9**, 325 (1967).
412. Ch. Elschenbroich and F. Gerson, *J. Am. Chem. Soc.* **97**, 3556 (1976).
413. W. M. Gulick and D. H. Geske, *Inorg. Chem.* **6**, 1320 (1967).
414. J. Müller, P. Göser, and P. Laubereau, *J. Organomet. Chem.* **14**, P7 (1968).
415. H.-S. Hsiung and G. H. Brown, *J. Electrochem. Soc.* **110**, 1085 (1963).
416. N. Ito, T. Saji, K. Suga, and S. Aoyagui, *J. Organomet. Chem.* **229**, 43 (1982).
417. L. P. Yureva, S. M. Peregudova, L. N. Nekrasov, A. P. Korotkov, N. N. Zaitseva, N. V. Zakurin, and A. Yu. Vasilkov, *J. Organomet. Chem.* **219**, 43 (1981).
418. O. Duschek and V. Gutmann, *Z. Anorg. Allg. Chem.* **394**, 243 (1972).
419. V. Gutmann and G. Peychal-Heiling, *Monatsh. Chem.* **100**, 1423 (1969).
420. V. Gutmann and R. Schmid, *Monatsh. Chem.* **100**, 2113 (1969).

421. A. Rusina and H.-P. Schröer, *Collect. Czech. Chem. Commun.* **31**, 2600 (1966).
422. H. P. Schröer and A. A. Vlcek, *Z. Anorg. Allg. Chem.* **334**, 205 (1964).
423. C. Furliani, *Ric. Sci.* **36**, 989 (1966).
424. C. Furliani and G. Sartori, *Ric. Sci.* **28**, 973 (1958).
425. P. M. Treichel, G. P. Essenmacher, H. F. Efner, and K. J. Klabunde, *Inorg. Chim. Acta* **48**, 41 (1981).
426. H. Brunner and H. Koch, *Chem. Ber.* **115**, 65 (1982).
427. S. E. Anderson, Jr. and R. S. Drago, *Inorg. Chem.* **11**, 1564 (1972).
428. R. Prins, *J. Chem. Phys.* **50**, 4804 (1969).
429. S. Valcher and G. Casalbore, *J. Electroanal. Chem. Interfacial Electrochem.* **50**, 359 (1974).
430. Ch. Elschenbroich, F. Gerson, and J. Heinzer, *Z. Naturforsch. B.: Anorg. Chem., Org. Chem.* **27**, 312 (1972).
431. Ch. Elschenbroich and J. Koch, *J. Organomet. Chem.* **229**, 139 (1982).
432. E. O. Fischer and K. Ulm, *Chem. Ber.* **95**, 692 (1962).
433. J. D. L. Holloway, F. C. Senftleber, and W. E. Geiger, *Anal. Chem.* **50**, 1010 (1978).
434. U. Koelle, *J. Organomet. Chem.* **157**, 327 (1978).
435. J. D. L. Holloway, W. L. Bowden, and W. E. Geiger, *J. Am. Chem. Soc.*, **99**, 7089 (1977).
436. U. Koelle, W. Holzinger, and J. Muller, *Z. Naturforsch. B.: Anorg. Chem., Org. Chem.* **34**, 759 (1979).
437. E. O. Fischer and S. Breitschaft, *Chem. Ber.* **96**, 2451 (1963).
438. E. O. Fischer and J. Müller, *J. Organomet. Chem.* **1**, 464 (1964).
439. Ch. Elschenbroich, F. Gerson, and F. Stohler, *J. Am. Chem. Soc.* **95**, 6956 (1973).
440. E. O. Fischer, F. Scherer, and H. O. Stahl, *Chem. Ber.* **93**, 2065 (1960).
441. H. W. Wehner, E. O. Fischer, and J. Müller, *Chem. Ber.* **103**, 2258 (1970).
442. J. C. Smart and J. L. Robbins, *J. Am. Chem. Soc.* **100**, 3936 (1978).
443. J. A. Page and G. Wilkinson, *J. Am. Chem. Soc.* **74**, 6149 (1952).
444. T. Kuwana, D. E. Bublitz, and G. Hoh, *J. Am. Chem. Soc.* **82**, 5811 (1960).
445. C. K. Mann and K. K. Barnes, "Electrochemical Reactions in Nonaqueous Solvents," pp. 422–425. Dekker, New York, 1970.
446. M. D. Morris, *in* "Electroanalytical Chemistry" (A. J. Bard, ed.), Vol. 7, pp. 149–150. Dekker, New York, 1974.
447. J. L. Robbins, N. Edelstein, B. Spencer, and J. C. Smart, *J. Am. Chem. Soc.* **104**, 1882 (1982).
448. R. R. Gagné, C. A. Koval, and G. C. Lisensky, *Inorg. Chem.* **19**, 2854 (1980).
449. J. W. Diggle and A. J. Parker, *Electrochim. Acta* **18**, 975 (1973).
450. E. S. Yang, M.-S. Chan, and A. C. Wahl, *J. Phys. Chem.* **84**, 3094 (1980).
451. R. J. Gale, P. Singh, and R. Job, *J. Organomet. Chem.* **199**, C44 (1980).
452. Y. Mugnier, C. Moise, J. Tirouflet, and E. Laviron, *J. Organomet. Chem.* **186**, C49 (1980).
453. N. El Murr, A. Chaloyard, and E. Laviron, *Nouv. J. Chim.* **2**, 15 (1978).
454. K. W. Willman, R. D. Rocklin, R. Nowak, K.-N. Kuo, F. A. Schultz, and R. W. Murray, *J. Am. Chem. Soc.* **102**, 7629 (1980).
455. A. B. Bocarsly, E. G. Walton, and M. S. Wrighton, *J. Am. Chem. Soc.* **102**, 3390 (1980).
456. P. Daum, J. R. Lenhard, D. Rolison, and R. W. Murray, *J. Am. Chem. Soc.* **102**, 4649 (1980).
457. S. P. Gubin, S. A. Smirnova, L. I. Denisovich, and A. A. Lubovich, *J. Organomet. Chem.* **30**, 243 (1971).
458. D. N. Hendrickson, Y. S. Sohn, W. H. Morrison, and H. B. Gray, *Inorg. Chem.* **11**, 808 (1972).
459. L. I. Denisovich, N. V. Zakurin, A. A. Bezrukova, and S. P. Gubin, *J. Organomet. Chem.* **81**, 207 (1974).

460. D. E. Walker, R. N. Adams, and A. L. Julliard, 136th Meet. Am. Chem. Soc., Atlantic City, New Jersey, Sept. 1959. Quoted in ref. 444.
461. D. E. Bublitz, G. Hoh, and T. Kuwana, *Chem. Ind. (London)* **78**, 635 (1959).
462. R. J. Gale and R. Job, *Inorg. Chem.* **20**, 42 (1981).
463. A. N. Nesmeyanov, L. I. Denisovich, S. P. Gubin, N. A. Volkenau, E. I. Sirotkina, and I. N. Bolesova, *J. Organomet. Chem.* **20**, 169 (1969).
464. D. Astruc and R. Dabard, *Bull. Soc. Chim. Fr.*, p. 228 (1976).
465. N. El Murr, *J. Chem. Soc., Chem. Commun.*, p. 251 (1981).
466. A. N. Nesmeyanov, N. A., Volkenau, P. V. Petrovskii, L. S. Kotova, V. A. Petrakova, and L. I. Denisovich, *J. Organomet. Chem.* **210**, 103 (1981).
467. P. Michaud, D. Astruc, and J. H. Ammeter, *J. Am. Chem. Soc.* **104**, 3755 (1982).
468. S. G. Davies, M. L. H. Green, and D. M. P. Mingos, *Tetrahedron* **34**, 3047 (1978).
469. S. P. Solodovnikov, A. N. Nesmeyanov, N. A. Volkenau, and L. S. Kotova, *J. Organomet. Chem.* **201**, C45 (1980).
470. E. Roman, D. Astruc, and A. Darchen, *J. Organomet. Chem.* **219**, 221 (1981).
471. A. N. Nesmeyanov, S. P. Solodovnikov, N. A. Volkenau, L. S. Kotova, and N. N. Sinitsyna, *J. Organomet. Chem.* **148**, C5 (1978).
472. D. Astruc, J.-R. Hamon, G. Althoff, E. Roman, P. Batail, P. Michaud, J.-P. Mariot, F. Varret, and D. Cozak, *J. Am. Chem. Soc.* **101**, 5445 (1979).
473. J.-R. Hamon, D. Astruc, and P. Michaud, *J. Am. Chem. Soc.* **103**, 758 (1981).
474. A. N. Nesmeyanov, N. A. Volkenau, and V. A. Petrakova, *Bull. Acad. Sci. USSR, Div. Chem. Sci. Engl. Transl.* **23**, 2083 (1974).
475. C. Moinet, E. Roman, and D. Astruc, *J. Organomet. Chem.* **128**, C45 (1977).
476. M. V. Rajasekharan, S. Giezynski, J. H. Ammeter, N. Oswald, P. Michaud, J.-R. Hamon, and D. Astruc, *J. Am. Chem. Soc.* **104**, 2400 (1982).
477. S. P. Solodovnikov, A. N. Nesmeyanov, N. A. Volkenau, and L. S. Kotova, *J. Organomet. Chem.* **201**, 447 (1980).
478. S. P. Solodovnikov, A. N. Nesmeyanov, N. A. Volkenau, N. N. Sinitsyna, and L. S. Kotova, *J. Organomet. Chem.* **182**, 239 (1979).
479. A. Buet, A. Darchen, and C. Moinet, *J. Chem. Soc., Chem. Commun.*, p. 447 (1979).
480. N. A. Volkenau and V. A. Petrakova, *J. Organomet. Chem.* **233**, C7 (1982).
481. D. Astruc, E. Roman, J.-R. Hamon, and P. Batail, *J. Am. Chem. Soc.* **101**, 2240 (1979).
482. D. Astruc, J.-R. Hamon, E. Roman, and P. Michaud, *J. Am. Chem. Soc.* **103**, 7502 (1981).
483. P. Michaud and D. Astruc, *J. Chem. Soc., Chem. Commun.*, p. 416 (1982).
484. J.-R. Hamon, D. Astruc, E. Roman, P. Batail, and J. J. Mayerle, *J. Am. Chem. Soc.* **103**, 2431 (1981).
485. C. Moinet and E. Raoult, *J. Organomet. Chem.* **231**, 245 (1982).
486. J. W. Johnson and P. M. Treichel, *J. Am. Chem. Soc.* **99**, 1427 (1977).
487. J. F. Helling and W. A. Hendrickson, *J. Organomet. Chem.* **168**, 87 (1979).
488. R. G. Sutherland, B. R. Steele, K. J. Demchuk, and C. C. Lee, *J. Organomet. Chem.* **181**, 411 (1979).
489. S. P. Solodovnikov, N. A. Volkenau, and L. S. Kotova, *J. Organomet. Chem.* **231**, 45 (1982).
490. I. W. Robertson, T. A. Stephenson, and D. A. Tocher, *J. Organomet. Chem.* **228**, 171 (1982).
491. D. M. Braitsch and R. Kumarappan, *J. Organomet. Chem.* **84**, C37 (1975).
492. E. O. Fischer and F. Röhrscheid, *Z. Naturforsch. B.: Anorg. Chem., Org. Chem.* **17**, 483 (1962).
493. E. D. Laganis, R. H. Voegeli, R. T. Swann, R. G. Finke, H. Hopf, and V. Boekelheide, *Organometallics* **1**, 1415 (1982).

494. R. G. Finke, R. H. Voegeli, E. D. Laganis, and V. Boekelheide, *Organometallics* **2**, 347 (1983).
495. E. O. Fischer and Ch. Elschenbroich, *Chem. Ber.* **103**, 162 (1970).
496. G. Huttner and S. Lange, *Acta Crystallogr. Sect. B* **28**, 2049 (1972).
497. A. A. Vlcek, *Collect. Czech. Chem. Commun.* **30**, 952 (1965).
498. S. P. Gubin, S. A. Smirnova, and L. I. Denisovich, *J. Organomet. Chem.* **30**, 257 (1971).
499. W. E. Geiger, Jr., *J. Am. Chem. Soc.* **96**, 2632 (1974).
500. W. E. Geiger, Jr. and D. E. Smith, *J. Electroanal. Chem. Interfacial Electrochem.* **50**, 31 (1974).
501. U. Koelle and F. Khouzami, *Angew. Chem., Int. Ed. Engl.* **19**, 640 (1980).
502. N. El Murr and E. Laviron, *Can. J. Chem.* **54**, 3350 (1976).
503. U. Koelle, *J. Organomet. Chem.* **152**, 225 (1978).
504. N. El Murr, R. Dabard, and E. Laviron, *J. Organomet. Chem.* **47**, C13 (1973).
505. W. E. Geiger, Jr., W. L. Bowden, and N. El Murr, *Inorg. Chem.* **18**, 2358 (1979).
506. N. El Murr and E. Laviron, *Tetrahedron Lett.*, p. 875 (1975).
507. U. Koelle and F. Khouzami, *Chem. Ber.* **114**, 2929 (1981).
508. M. Rosenblum, B. North, D. Wells, and W. P. Giering, *J. Am. Chem. Soc.* **94**, 1239 (1972).
509. E. O. Fischer and H. H. Lindner, *J. Organomet. Chem.* **1**, 307 (1964).
510. E. O. Fischer and H. H. Lindner, *J. Organomet. Chem.* **2**, 222 (1964).
511. M. R. Thompson, C. S. Day, V. W. Day, R. I. Mink, and E. L. Muetterties, *J. Am. Chem. Soc.* **102**, 2979 (1980).
512. N. El Murr, J. E. Sheats, W. E. Geiger, Jr., and J. D. L. Holloway, *Inorg. Chem.* **18**, 1443 (1979).
513. E. O. Fischer and H. Wawersik, *J. Organomet. Chem.* **5**, 559 (1966).
514. J. Tirouflet, E. Laviron, R. Dabard, and J. Komenda, *Bull. Soc. Chim. Fr.*, p. 857 (1963).
515. R. J. Wilson, L. F. Warren, and M. F. Hawthorne, *J. Am. Chem. Soc.* **91**, 758 (1969).
516. R. P. Van Duyne and C. N. Reilley, *Anal. Chem.* **44**, 158 (1972).
517. R. J. Gale and R. Job, *Inorg. Chem.* **20**, 40 (1981).
518. U. Koelle, F. Khouzami, and H. Lueken, *Chem. Ber.* **115**, 1178 (1982).
519. M. S. Paquette and L. F. Dahl, *J. Am. Chem. Soc.* **102**, 6621 (1980).
520. C. J. Pickett, "Electrochemistry," *Chem. Soc. Spec. Per. Rep.* **8**, (1983).
521. K. A. Conner and R. A. Walton, *Organometallics* **2**, 169 (1983).
522. T. Lionel, J. R. Morton, and K. F. Preston, *Inorg. Chem.* **22**, 145 (1983).
523. B. A. Narayanan, C. Amatore, and J. K. Kochi, *J. Chem. Soc., Chem. Commun.*, p. 397 (1983).
524. H. F. Klein, H. Witty, and V. Schubert, *J. Chem. Soc., Chem. Commun.*, p. 231 (1983).
525. B. J. Brisdon, K. A. Conner, and R. A. Walton, *Organometallics* **2**, 1159 (1983).
526. L. M. Engelhardt, W-P. Leung, C. L. Raston, and A. H. White, *J. Chem. Soc., Chem. Commun.*, p. 386 (1983).
527. K.-B Shiu, M. D. Curtis, and J. C. Huffman, *Organometallics* **2**, 936 (1983).
528. T. Desmond, F. J. Lalor, G. Ferguson, and M. Parvez, *J. Chem. Soc., Chem. Commun.*, p. 457 (1983).
529. M. C. R. Symons, S. W. Bratt, and J. L. Wyatt, *J. Chem. Soc., Dalton Trans.*, p. 1377 (1983).
530. D. Miholova and A. A. Vlcek, *Inorg. Chim. Acta*, **73**, 249 (1983).
531. R. H. Magnuson, R. Meirowitz, S. J. Zulu, and W. P. Giering, *Organometallics* **2**, 460 (1983).
532. A. Cameron, V. H. Smith, and M. C. Baird, *Organometallics* **2**, 465 (1983).
533. M. Grzeszczuk, D. E. Smith, and W. E. Geiger, *J. Am. Chem. Soc.* **105**, 1772 (1983).
534. S. N. Milligan, I. Tucker, and R. D. Rieke, *Inorg. Chem.* **22**, 987 (1983).

535. M. G. Peterleitner, M. V. Tolstaya, V. V. Krivykh, L. I. Denisovich, and M. I. Rybinskaya, *J. Organomet. Chem.* **254**, 313 (1983).
536. W. P. Henry and R. D. Rieke, *J. Am. Chem. Soc.* **105**, 6314 (1983).
537. S. N. Milligan and R. D. Rieke, *Organometallics* **2**, 171 (1983).
538. P. N. Hawker and P. L. Timms, *J. Chem. Soc., Dalton Trans.*, p. 1123 (1983).
539. N. Ito, T. Saji, and S. Aoyagui, *J. Organomet. Chem.* **247**, 301 (1983).
540. U. Koelle and A. Salzer, *J. Organomet. Chem.* **243**, C27 (1983).
541. A. Darchen, *J. Chem. Soc., Chem. Commun.*, p. 768 (1983).

ADVANCES IN ORGANOMETALLIC CHEMISTRY, VOL. 23

Redistribution Reactions of Transition Metal Organometallic Complexes

PHILIP E. GARROU

Dow Chemical USA
Central Research—New England Laboratory
Wayland, Massachusetts

I

INTRODUCTION

We define a redistribution reaction in a transition metal system as a reaction in which an exchange takes place of the type shown in Eq. (1).

$$MA + M'B \rightleftharpoons MB + M'A \tag{1}$$

The metals M and M' and the ligands A and B may be the same or different. Such reactions have also been described as disproportionations, symmetrizations, metathesis reactions, scrambling reactions, and transfer reactions. In transition metal systems such reactions are intermolecular in nature and result in transfer of ligands from one metal center to another.

In contrast to the extensive accumulation of kinetic and thermodynamic data on the redistribution of main group complexes (1, 2) there are far fewer similar studies on transition metal species. Those reports that do exist are generally qualitative in nature.

For transition metal complexes four types of redistributions can be envisaged as shown in Eqs. (2)–(5).

Type I—Metal centers the same and substituents different:

$$LMA_n + LMB_n \;\rightleftharpoons\; LMA_{n-1}B + LMB_{n-1}A \ldots \tag{2}$$

Type II—Metal centers different and substituents the same:

$$LMA_n + LM'A^*_n \;\rightleftharpoons\; LMA_{n-1}A^* + LM'A^*_{n-1}A \ldots \tag{3}$$

Type III—Both metal centers and substituents different:

$$LMA_n + LM'B_n \;\rightleftharpoons\; LMA_{n-1}B + LM'B_{n-1}A + LMB_n + LM'A_n \ldots \tag{4}$$

Type IV—Both metal centers and substituents the same:

$$MA_n + M^*A^*_n \;\rightleftharpoons\; MA_{n-1}A^* + M^*A^*_{n-1}A \ldots \tag{5}$$

Ligand transfers of types I–III give perspective as to the chemical "inertness" of certain bonds. As we shall see later (Section II), systems that fail to achieve a statistically random equilibrium position do so because there is a net change in ΔH, which is to say a more stable system is being formed. In addition, when two complexes are mixed and it is known that a kinetic pathway exists for exchange, if they do not exchange, it is evident that the ligand–metal bond strength of one or both of the reactants is stronger than the ligand–metal bonds of the exchange products would be.

Type IV reactions are the most significant to the basic understanding of metal complexes in solution, and in particular the catalytic activity of certain metal species. To transpose nomenclature from the olefin metathesis field such reactions are "nonproductive," i.e., after the transfer the same species exist in solution as before the transfer. Such reactions are the most difficult to follow and can only be studied by isotopic labeling of the substituents. I believe that to fully understand the molecular dynamics of transition metal complexes and/or catalytic systems, it is not sufficient to study and understand intramolecular rearrangements or intermolecular exchanges of added ligands (substitution reactions) if such intermolecular exchanges between complexes are competing at a comparable rate.

The relative importance of such redistribution processes has been neglected in mechanistic studies until recently when examination of a number of catalytically active species such as $MCl(CO)(PR_3)_2$ (M = Rh, Ir) (3) and $MCl_2(PR_3)_2$ (M = Pt, Pd) (4) revealed intermolecular scrambling of all ligands. Other elegant studies have offered evidence to show that certain H–H (5), R–H (6), and R–CHO (7) eliminations are bimolecular processes and can be envisioned as occuring during ligand transfer reactions between metal sites.

It is generally accepted that coordinate unsaturation is a prerequisite for catalytic activity and that exchange reactions occur via a S_E2 (cyclic) intermediate such as **1**. It is, therefore, not unusual that such coordinately

$$L_nM \diamond ML_n$$

with X groups bridging

1

saturated species have been neglected in mechanistic proposals. We wish to point out that such species *could* (a) enhance dissociation of L to produce a catalytically active metal center and/or (b) be the actual reactive intermediates when the X groups are the functionalities undergoing reaction. We shall see in several instances that the "reaction order" obtained by kinetic studies can be misleading since the rate determining step may occur before or after the bimolecular redistribution intermediates are involved in the process. It is not our contention that redistribution reactions are involved in all solution chemistry or are the key to all catalytic reactions. It is hoped, however, that investigators will be more aware of such species and their potential involvement in transition metal reaction mechanisms.

The examination of the literature that follows is not meant to be all inclusive but rather representative of the functionalities that are capable of transfer between transition metal centers. Emphasis is placed on such reactions in the VIB, VIIB, and VIII subgroups.

II

THE THERMODYNAMICS OF REDISTRIBUTION REACTIONS

We have noted that redistribution reactions are characterized by the fact that bonds change in relative position but not in type. If the bond energies do not alter appreciably (no enthalpy change), the reactions may be regarded as thermoneutral. For such thermoneutral reactions, ΔG will depend on the entropy term only, and purely statistical redistributions should result. When nonrandom reactions are observed, they are indicative of nonzero enthalpies, and thus changes in the bond strengths during the scrambling reactions.

If there is no enthalpy change, the equilibrium position will be governed by the $T\Delta S$ term, and K_1 for a two-site exchange shown in Eqs. (6)–(7) will

$$MX_2 + MY_2 \rightleftharpoons 2MXY \tag{6}$$

$$K_1 = \frac{[MX_2][MY_2]}{[MXY]^2} \tag{7}$$

be 0.25. For nonrandom redistributions the K values will deviate from purely statistical (random) values. The enthalpy change for such reactions can be derived from the equilibrium constant using Eqs. (8)–(10). The error in such

$$\Delta G° = -RT \ln K \tag{8}$$

$$\ln\left\{\frac{KT_2}{KT_1}\right\} = \frac{-\Delta H°}{R}\left[\frac{1}{T_2} - \frac{1}{T_1}\right] \tag{9}$$

$$\Delta H° = \frac{-R[\ln K(T_2) - \ln K(T_1)]}{(1/T_2) - (1/T_1)} \tag{10}$$

a $\Delta H°$ calculation arises mainly from the error associated with determining the equilibrium constant. Lee (8) has shown that a reaction with a $\Delta H°$ error of 0.5–1.0 kcal/mol occurs for $T_1 = 30°C$, $T_2 = 80°C$ with a 4 and 8% error in the measurement of K at those temperatures.

Moedritzer (2) has pointed out that a very rough measure of $\Delta H°$ can be obtained from measurement of K at just one temperature since

$$\Delta G° = \Delta H° - T\Delta S° \tag{11}$$

$$\Delta G° = \frac{-RT}{n} \ln K \tag{12}$$

For the completely random case $\Delta H°$ is 0 thus

$$\Delta G°_{random} = -T(\Delta S°)_{random} = \frac{-RT}{n} \ln K_{random} \tag{13}$$

The difference $\Delta G°_{observed} - \Delta G°_{random}$ is the driving force of the reaction, thus

$$\Delta G°_{difference} = \frac{-RT}{n} \ln K\left\{\frac{K_{OBS}}{K_{random}}\right\} = \Delta H° \tag{14}$$

Equation (14) assumes that the difference between the measured and statistical $\Delta G°$ is due solely to the $\Delta H°$ contribution and that $\Delta S°$ does not differ appreciably between statistical and nonstatistical exchange. It has been pointed out (8) that such an approximation will be very poor when

$\Delta G°_{observed} - \Delta G°_{random}$ is less than 1–2 kcal/mol since differences in molecular solvation and aggregation in solution could lead to $\Delta S°$ values >0.

When the systems under study contain numerous exchangable substituents, the experimental difficulties inherent in obtaining good thermodynamic parameters become more apparent due to the complexity of the resulting equilibria. For example, for three substituents exchanging on two equivalent metal atoms at least two equilibrium constants need to be fully defined as shown in Eqs. (15)–(19). Complete analysis of such a system

$$MX_3 + MY_3 \rightleftharpoons MX_2Y + MXY_2 \tag{15}$$

$$2MX_2Y \rightleftharpoons MX_3 + MXY_2 \tag{16}$$

$$K_1 = \frac{[MX_3][MXY_2]}{[MX_2Y]^2} \tag{17}$$

$$2MXY_2 \rightleftharpoons MY_3 + MX_2Y \tag{18}$$

$$K_2 = \frac{[MX_2Y][MY_3]}{[MXY_2]^2} \tag{19}$$

therefore cannot be achieved by simply mixing MX_3 and MY_3! Approach of this equilibrium from at least two directions is necessary in order to calculate the minimum two equilibrium constants.

For a single metal atom M exhibiting x exchangable functionalities there are $x + 1$ possible products of formula MA_yB_{x-y} where $y = 0, 1, 2, \ldots x$. For such a reaction there will be $x - 1$ equilibrium constants.

The lack of complete thermodynamic data for the majority of studies examined in this article point out, I feel, the complexity of the equilibria involved and the unavailability of intermediates to approach such equilibria from more than one direction. This is unfortunate since thermodynamic data for such scrambling reactions could lead to very valuable bond strength data for vast classes of organometallic complexes and would shed light on the intricacies of numerous catalytic processes and point toward rational creation of many others.

III

EXPERIMENTAL TECHNIQUES

Much of the early work on redistribution reactions of main group complexes was handicapped by the lack of modern instrumental techniques. Determination of the equilibrium position was usually obtained by distillative separation of products. This method obviously suffered due to the small boiling point differences usually encountered in such complexes and the possibility of rapid rearrangement during the distillation. Such a separation technique is applicable only when the forward and reverse rate constants of the reaction are small. If the half-life of the reaction is on the time scale of the separation and the forward and reverse rate constants are comparable, then obviously such separations cannot be used to determine the equilibrium. Physical separations of organometallic complexes would usually necessitate fractional crystallizations of closely related complexes and thus would be both difficult and laborious. The advent of modern instrumental methods has, however, allowed quantitative examination of such systems *in situ*.

Any analytical method used to examine redistribution reactions must allow for rapid, quantitative, and precise determination of all the reaction products present in the mixture (*1*). In order to measure quantitatively an equilibrium mixture by spectroscopy each component of the equilibrium must contain at least one characteristic signal and the concentration of the component must be obtainable from the spectrum.

A. *Ultraviolet–Visible Spectroscopy*

In the ultraviolet and visible region (*9*) (50,000–12,500 cm^{-1} frequency range) spectral patterns are caused by the absorption of incident radiation, and one observes transitions between electronic energy levels. The frequencies absorbed are characteristic of the substrate, and as long as the absorption follows Beer's law, the intensity of the absorption can be used to measure the concentration of the species. Wavelengths of maximum absorption (λ_{max}) and extinction coefficients (ε_{max}) of all components of the redistribution must be known. Although in general the bands one observes are broad and overlapping, identical functional groups in different molecules will not necessarily absorb at exactly the same wavelength. In addition if two substances A and B have absorption bands that overlap, there will be some wavelength at which the molar absorptivities of the two species are equal. Holding the sum of the concentrations of these species in solution constant will cause invariant absorbance at this wavelength as the ratio of A to B is varied. This is referred to as the isosbestic point. The existence of such points in a system provides information on the number of species present in solution.

The relationship of the extinction coefficient ε to concentration is given by Eq. (20), where A is the absorbance, I_0/I is the ratio of the intensity of the

$$A = \log I_0/I = \varepsilon c b \qquad (20)$$

incident light to the transmitted light, and b is the cell path length. Such calculations allow measurement of the equilibrium constant and, when data are taken at several temperatures, evaluation of thermodynamic parameters. Determination of the change in concentrations of the various species with time also allows evaluation of the rate constants for the reaction.

B. *Infrared Spectroscopy*

In the infrared region (*10*) (4000–667 cm^{-1}) spectral patterns are caused by the absorption of incident radiation, and one observes transitions between vibrational levels of the molecules. There are two fundamental vibrations for a given molecules, stretching and bending. When incident light of the same frequency as a vibration is absorbed by a molecule, the amplitude of the vibration is increased, producing the spectrum. Measurement of rates and equilibrium constants is more difficult by this method since Beer's Law often does not hold for IR work and because maintaining a constant temperature during the study is difficult due to irradiation of the sample by the radiation source.

C. *Nuclear Magnetic Resonance Spectroscopy*

If a molecule is placed in an external magnetic field, its magnetic moment can be aligned with or against the external field. Alignment with the field is more stable and energy must be absorbed to "flip" to the less stable alignment against the field. In the NMR experiment, one maintains a constant radiation frequency and varies the strength of the magnetic field. At some value of field strength the energy required to flip the spin state coincides with the energy of the radiation, absorption occurs, and a signal (resonance) is observed (*11*). The energy of the resonance one observes depends on the electronic environment of the nucleus. Since electrons shield the nucleus, the magnitude of the field seen by the nucleus differs from the applied field. The chemical shift of the resonance (difference in the shielding constants of the sample and a reference compound) is therefore a very sensitive measure of chemical environment, and as such, NMR is a powerful tool in the examination of redistribution reactions.

When considering the study of redistribution reactions, one must remember that NMR spectra are extremely sensitive to the rates of the processes that cause the exchanges. For very slow exchange, the lifetime at each site is long with reference to the interaction with the rf field; thus the spectrum will consist of two sharp lines as it would in the absence of exchange. For moderate exchange, the timescale of interaction with the rf field is of the order of the site exchange timescale, and one observes an increase in line width. For very fast exchange, one observes only a single resonance representing an average position (chemical shift) of the two exchange sites. This possibility can and should be eliminated by cooling the sample and thus slowing down the exchange processes.

When using NMR to measure the rate and equilibrium position of re-distribution reactions, one must be aware of the accuracy of signal integra-tion. One normally uses the integrated intensity of a ^1H NMR signal as a measure of the concentration of the species under examination. When the concentrations of the species present are similar, an error of $\pm 5\%$ in their ratio is normally assumed. Thus when a K of 0.01 or 100 is being examined, an error in K of 15% would be possible. It must also be noted that when one is observing a nucleus other than ^1H, one usually decouples the ^1H spin–spin coupling by applying an rf field of specific frequency while the other nucleus is being observed. This results in what is known as Nuclear Overhauser Enhancement (NOE) (12). The effect of this is to cause differences in signal intensities for equal concentrations of nuclei having different functional groups. One must measure the NOE of the system under study and com-pensate for its effect on the integrated intensity of the signals under observa-tion in order to assess properly the concentrations of the various species. In addition, the accuracy of a normal variable temperature probe is $\pm 2°$ and thus errors of up to 10% in calculated rate constants can be expected.

Thus, although as a quantitative tool for examination of transfer reactions NMR is unsurpassed, one must realize its limitations and understand that the accuracy obtained is not on par with that obtained by standard kinetic techniques.

D. *Isotopic Labeling*

A technique that has been used to a lesser extent than spectroscopy but proves invaluable when one undertakes the examination of type II and IV transfers is isotopic labeling (13). The most obvious application would be for carbonyl exchanges using ^{13}CO, but other ^{13}C- or ^2H-labeled systems are potentially examinable and economically feasible.

E. *Mass Spectroscopy*

Mass spectroscopy can be used to detect organometallic redistribution products provided a unique mass spectrum is obtainable for each com-ponent (14). Once can obtain thermodynamic data from mass spectra if the sensitivity of the spectrometer to each of the species is known and if the intensity of the peak is known to be proportional to the partial pressure of the compound.

F. Other Techniques

Obviously there are numerous other techniques that will be applicable to specific reaction mixtures. One must only remember that each component of the reaction must reveal at least one characteristic signal, and the concentration of the component must be obtainable.

In Section IV we shall see specific examples in which the above-mentioned techniques have been used to gather details about specific redistribution reactions.

IV

TRANSFERS BY LIGAND TYPE

A. η^3-Allyl

Nesmeyanov et al. (15, 16) examined the reaction of $[\eta^3\text{-}C_3H_5PdCl]_2$ with $Fe_2(CO)_9$ or $Fe(CO)_5$ and isolated only $(\eta^3\text{-}C_3H_5)FeCl(CO)_3$ and unreacted π-allyl palladium dimer. In a similar fashion, they observed that the reaction of $[\eta^3\text{-}C_3H_5NiX]_2$ (X = halide) with $Fe_2(CO)_9$ resulted in the formation of $(\eta^3\text{-}C_3H_5)FeX(CO)_3$ and proposed the mechanism shown in Scheme 1.

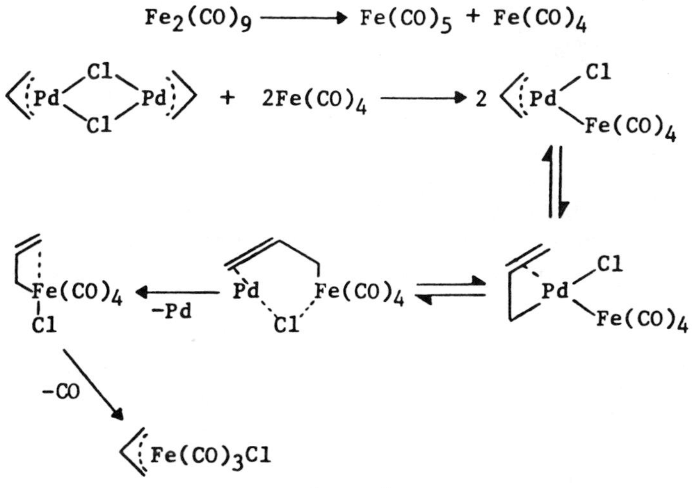

$$Fe_2(CO)_9 \longrightarrow Fe(CO)_5 + Fe(CO)_4$$

SCHEME 1

Heck (17) found that π-allyl ligands are readily transferred from π-allyl palladium halides to a tetracarbonyl cobalt anion. Thus $[(\eta^3\text{-}C_3H_5)PdCl]_2$ and $NaCo(CO)_4$ reacted at room temperature to give a 50% yield of $(\eta^3\text{-}C_3H_5)Co(CO)_3$, which was isolated as its mono triphenylphosphine derivative. He noted that since palladium chloride reacts directly with olefins to produce π-allyl complexes but cobalt derivatives do not, such ligand exchange reactions have synthetic value. The proposed mechanism shown in Scheme 2 is similar to Nesmeyanov's where a palladium–cobalt intermediate undergoes a 1,2 shift of the allyl group possibly via a σ-carbon bonded intermediate.

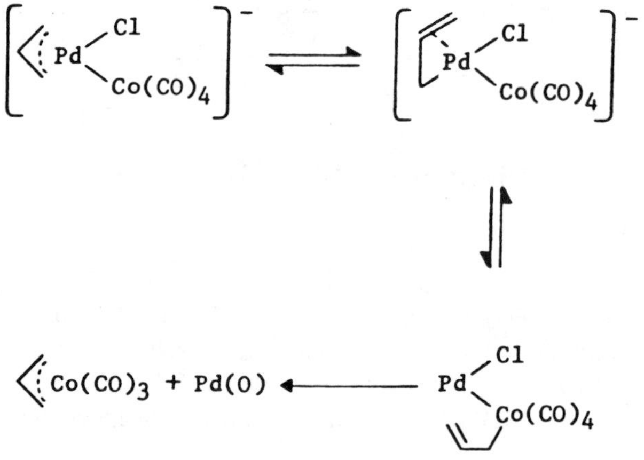

SCHEME 2

A PMR study of $[\eta^3\text{-}C_3H_5Pd(SCN)]_2$ reveals that exchange processes are occurring (18). It was found that $[\eta^3\text{-}MeC_3H_4PdCl]_2$ reacted with $[\eta^3\text{-}MeC_3H_4PdI]_2$ to produce the halide scrambling product $(\eta^3\text{-}MeC_3H_4Pd)_2ICl$. Since kinetics revealed a second order dependence on dimer concentration, Brown proposed intermediate 2 as the structure by which both π-allyl and halide scrambling could occur. Such an intermediate can dissociate into two dimers in any of three equivalent ways.

It is also possible to envision a bridging intermediate for transfer of π-allyl groups. A model structure comes from the studies of Werner and Kuhn (19, 20), who found that the reaction of η^3-allyl-η^5-cyclopentadienyl palladium with tertiary phosphines led to dinuclear complexes having the general structure 3. X-Ray analysis verified the structure. The platinum analog could also be prepared.

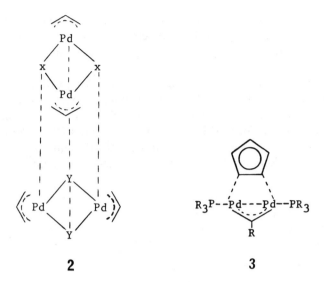

2   **3**

B. *Cyclopentadienyl*

Reaction of nickelocene with iron pentacarbonyl was shown to yield **4** as well as the cyclopentadienyl (Cp) compounds $Cp_2Ni_2(CO)_2$ and

4

$Cp_2Fe_2(CO)_4$, but ferrocene did not react with nickel tetracarbonyl to give Cp–nickel complexes (*21*). They proposed that bridging intermediate **4** was involved in the Cp transfer.

Nesmeyanov's laboratory (*22*) found that η^3-allyl palladium cyclopentadienyl reacted with iron pentacarbonyl to produce $[CpFe(CO)_2]_2$ and Cp_2Fe. The fate of the palladium was not revealed. In a similar fashion, $Cp(\eta^3\text{-}C_3H_5)Pd$ reacted with $FeCl_2$ in THF to produce $[(\eta^3\text{-}C_3H_5)PdCl]_2$ and Cp_2Fe. No conversions or yields were given.

Several groups (*23–25*) have reported that nickelocene reacts with bis-phosphinenickel dihalides to produce $CpNiX(PR_3)$ in 70–90 % yield. Since the reactions were not examined in solution, it is not known whether an equilibrium exists and $CpNiX(PR_3)$ precipitates due to lower solubility or whether the mixed species is simply the thermodynamically more stable material.

NMR examination of the reaction of $MeCpCoMe_2(PR_3)$ with $CpCo(PPh_3)_2$ in THF revealed signals due to $MeCpCo(PPh_3)_2$ and $CpCoMe_2(PPh_3)$ (26). After 48 hours, equilibrium had been reached with a $67:33$ ratio of $MeCpCoMe_2(PPh_3)$ to $CpCoMe_2(PPh_3)$. When the analogous reaction was run with PMe_3 as shown in Eq. (21), the equilibrium ratio of **5** to **7** was $60:40$. Complex **7** was shown to be inert to phosphine dissociation

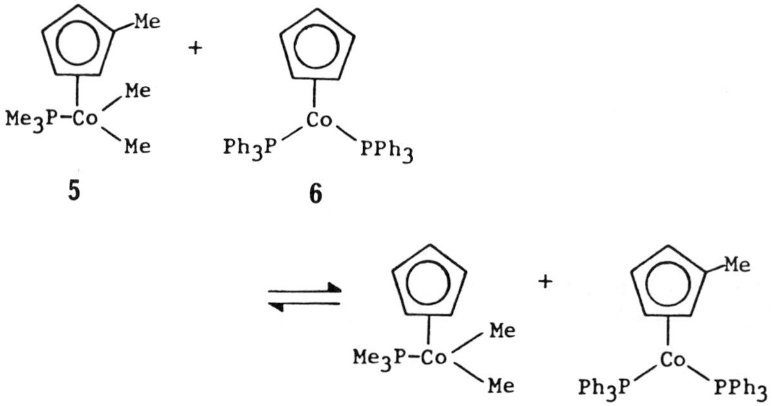

$$(21)$$

under these conditions. Since methyl exchange should have produced $CpCoMe_2(PPh_3)$ and the mixed phosphine complex $MeCpCo(PMe_3)(PPh_3)$, this indicates Cp–ligand exchange between the complexes.

No reactions of $CpCoMe_2(PPh_3)$ with $RhCl(PPh_3)_3$ or $IrCl(PPh_3)_3$ were observed, but reaction with $Fe_2(CO)_9$ or $Fe(CO)_5$ produced $CpCo(CO)_2$ and $Fe(PPh_3)(CO)_4$.

Reaction of $TiCl_4$ with $[CpFe(CO)_2]_2$ resulted in the isolation of 5% $CpTiCl_3$ (27). The reaction of $Fe(CO)_5$ with Cp_2TiCl_2 resulted in chromatographic isolation of $\sim 10\%$ Cp_2Fe. In an analogous fashion reaction of $(\eta^2\text{-}C_8H_{12})PdBr_2$ with $CpFeBr(CO)_2$ resulted in the isolation of 54% $CpPd(\eta^2\text{-}C_8H_{12})^+FeBr_4^-$ (27).

The nature of the transition state for Cp–ligand transfer is not known, but it could be similar to the bridging Cp complexes discovered by Werner and Kraus (28). The reduction of $CpPd(MeCO_2)L$ with NaH was shown to produce **9**, which was characterized by X-ray crystallographic analysis.

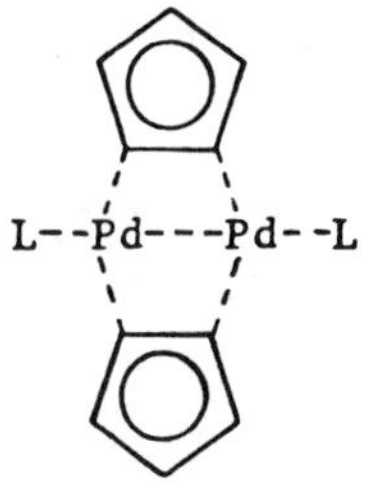

$$L-\overset{|}{\underset{|}{Pd}}---\overset{|}{\underset{|}{Pd}}--L$$

9

C. Halide

Halide redistribution is the best known and most often observed type of transfer between organometallic complexes. Thus halide transfer is often observed during the transfer of other ligands such as Cp, R, H, and CO. Studies concerned with multiple exchanges are covered under other headings. The following studies have been concerned mainly with halide exchange.

On dissolving equimolar quantities of $PtBr_4(PEt_3)(Py)$ and $PtCl_4$-$(PEt_3)(Py)$ in $CDCl_3$ containing Br_2, the ^{31}P NMR spectrum after 1 hour revealed resonances due only to the tetrachloride and tetrabromide (29). Carrying out the same experiment in the absence of Br_2 results in a ^{31}P-NMR spectrum revealing $PtBr_xCl_{4-x}(PEt_3)(Py)$ ($x = 0-4$). After 24 hours a statistical distribution of all five species was obtained. The authors concluded that this supports a mechanism of Pt(II) catalysis of the exchange via Cl–Pt(II)–Br–Pt(IV)–Br type bridged intermediates.

A $^{31}P\{^1H\}$ NMR spectrum of equimolar quantities of $RhCOBr(PPh_3)_2$ and $IrCl(CO)(PPh_3)_2$ at 30°C in CH_2Cl_2 or benzene exhibited four resonances of equal intensity due to $RhCl(CO)(PPh_3)_2$, $RhBr(CO)(PPh_3)_2$, $IrCl(CO)(PPh_3)_2$, and $IrBr(CO)(PPh_3)_2$, indicating that random redistribution of the halides had occurred (3). The redistribution was complete in ≤ 180 seconds. Similar random halide exchanges were observed for numerous rhodium–rhodium and rhodium–iridium systems. The halide exchange can be effectively slowed down at $-75°$, at which temperature after 15 minutes no exchange was observed between $RhBr(CO)(PPh_2Et)_2$ and $RhCl(CO)$-$(PPh_2Et)_2$. An associative mechanism involving an intermediate such as **10**

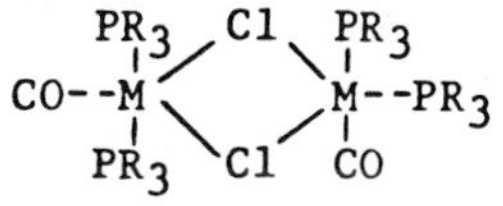

$$CO-\overset{PR_3}{\underset{PR_3}{M}}\overset{Cl}{\underset{Cl}{<}}\overset{PR_3}{\underset{CO}{M}}--PR_3$$

10

was suggested since ionic dissociation was considered less likely for exchanges in benzene.

Shaw *et al.* (*30*) have noticed that a solution containing *trans*-IrCl(CO)(PMe$_2$Ph)$_2$ and RhCl$_3$(CO)(PMe$_2$Ph)$_2$ redistributes to an equimolar mixture of *trans*-RhCl(CO)(PMe$_2$Ph)$_2$ and IrCl$_3$(CO)(PMe$_2$Ph)$_2$ in less than 5 minutes at 22°C. Conversion is very low ($t_{1/2}$ = hours) below -30°C. In a similar fashion equimolar mixtures of *trans*-IrCl(CO)(PEt$_2$Ph)$_2$ and RhCl$_3$(CO)(PEt$_2$Ph)$_2$ reacted more slowly than the PMe$_2$Ph complexes. Solutions of *trans*-IrCl(CO)(PEt$_2$Ph)$_2$ and IrCl$_3$(CO)(PEt$_2$Ph)$_2$ after 15 minutes at 22°C revealed conversion to *trans*-IrCl(CO)(PEt$_2$Ph)$_2$ and IrCl$_3$(CO)(PMe$_2$Ph)$_2$ with only a trace of the mixed phosphine species and a small percentage of the starting complexes. Over a few hours, however, the concentrations of the mixed species IrCl(CO)(PMe$_2$Ph)(PEt$_2$Ph) and IrCl$_3$(CO)(PMe$_2$Ph)(PEt$_2$Ph) gradually increased. In general they noted that phosphine exchange with Rh(I) species is faster than with Ir(I). No phosphine exchange is observed between Rh(III) species. They suggest these halide transfers occur via halo bridged intermediates such as **11**.

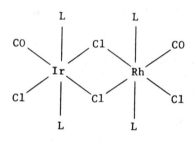

11

During their extensive studies on the reaction and structural chemistry of palladium complexes of bisdiphenylphosphinomethane [bis(dppm)], Hunt and Balch examined the halogen scrambling reactions shown in Eqs. (22)–(23) in detail (*31*).

$$PdX_2(dppm) + PdY_2(dppm) \rightleftharpoons 2PdXY(dppm) \qquad (22)$$

$$Pd_2X_2(dppm) + Pd_2Y_2(dppm)_2 \rightleftharpoons 2Pd_2XY(dppm)_2 \qquad (23)$$

In both cases the equilibria between the three complexes are rapidly established (5 minutes) in nonpolar and noncoordinating solvents. The equilibrium values shown in Table I were obtained by ^{31}P or ^{1}H NMR in CDCl$_3$ at 24°C. It was assumed that scrambling occurs via bridged intermediates. A ^{31}P{^{1}H} NMR of equimolar quantities of PtI(Ph)(PPh$_3$)$_2$ and

TABLE I

EQUILIBRIUM POSITION AND THERMODYNAMIC PARAMETERS FOR
$PdX_2(dppm) + PdY_2(dppm)$ SCRAMBLING REACTIONS

Complex	X	Y	K	ΔH (kcal/mol)	ΔS (cal/mol · deg)
$PdLX_2$	Cl	Br	3.3 ± 0.3		
	Br	I	3.1 ± 0.1		
	Cl	I	2.1 ± 0.3		
Pd_2XYL_2	Cl	I	6.4 ± 0.3	-1.2 ± 0.5	3.0 ± 1

$PdBr(Ph)(PPh_3)_2$ in $CDCl_3$ or C_6H_6 at 30°C consists of four resonances of equal intensity indicating that random redistribution of halide had occurred in ≤ 5 minutes (32). At $-75°C$ no sign of exchange is observed within 15 minutes. Halide exchange is envisioned to occur through a bimolecular chloride bridged intermediate similar to that proposed for the Rh–Ir systems (3).

D. Carbon Monoxide

Like halide transfer, CO transfer is often observed during the transfer of other ligands. Such studies concerned with multiple exchanges are covered under both headings.

When cobalt naphthenate and an excess of $Fe(CO)_5$ were heated under 50 atm of CO, $Co_2(CO)_8$ was obtained in 43% yield. In the absence of $Fe(CO)_5$, no $Co_2(CO)_8$ was produced (33). In a similar fashion of $RhCl_3$ with $Fe(CO)_5$ at 100°C for 12 hours resulted in 65% yield of $Rh_6(CO)_{16}$. Shorter reaction times resulted in the isolation of $[RhCl(CO)_2]_2$. Powell and Shaw (34) found that equimolar quantities of $Rh_2Cl_2(CO)_4$ and $Rh_2Cl_2(C_2H_4)_4$ reacted to give $Rh_2Cl_2(CO)_2(C_2H_4)_2$ [Eq. (24)]. They assumed the equilib-

$$Rh_2Cl_2(CO)_4 + Rh_2Cl_2(C_2H_4)_4 \rightleftharpoons Rh_2Cl_2(CO)_2(C_2H_4)_2 \qquad (24)$$

rium to lay far to the right since it is the only complex identified in solution by IR spectroscopy.

To determine whether CO exchange occurs between Vaska type complexes $[MX(CO)(PR_3)_2, M = Rh, Ir]$, ^{13}C enriched $RhCl(CO)(PPh_3)_2$ was allowed to react at room temperature with several Rh(I) and Ir(I) complexes (3). In ≤ 60 seconds the intensity of the ^{13}CO IR absorption due to $RhCl(CO)(PPh_3)_2$ and the ^{13}CO absorption due to the other reactant species were of equal intensity indicating random redistribution had occurred. Values for $\nu(^{12}CO)$ and $\nu(^{13}CO)$ are shown in Table II.

TABLE II

INFRARED DATA FOR ^{13}CO-LABELED $MX(CO)L_2$ (M = Rh, Ir; X = Cl, NCS) SCRAMBLING REACTIONS

	$v(^{12}CO)(cm^{-1})$	$v(^{13}CO)(cm^{-1})$ (OBS)	$v(^{13}CO)(cm^{-1})$ (THEO)
RhCl(CO)(PPh$_3$)$_2$	1962	1917	1918
Rh(acac)(CO)(PPh$_3$)	1983	1936	1939
Rh(NCS)(CO)(PPh$_3$)$_2$	1983	1936	1939
IrCl(CO)(PPh$_3$)$_2$	1953	1907	1909
Ir(NCS)(CO)(PPh$_3$)$_2$	1974	1927	1930

The interaction of ^{13}C enriched RhCl(CO)(PPh$_3$)$_2$ with RhCl$_3$(CO)-(PPh$_3$)$_2$ for 4 hours gave an IR spectrum that indicated that CO exchange had not occurred. The lack of exchange between the Rh(I) and Rh(III) complexes suggests that coordinative unsaturation is a prerequisite for fast carbonyl exchange under these conditions. Carbon monoxide exchange could potentially occur by associative or dissociative mechanisms. A dissociative process should result in the precipitation of insoluble [RhCl-(PPh$_3$)$_2$]$_2$ with the evolution of CO; however, this dimer was not observed even in a refluxing acetone solution under a rapid flow of nitrogen. Exchange was thus envisioned to occur through an intermediate such as 12.

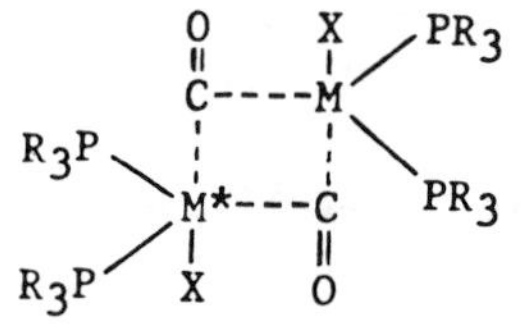

12

A common problem associated with the use of metal carbonyls as synthetic reagents is the drastic conditions required to induce CO dissociation. Catalytic amounts of a variety of transition metal complexes have been found to induce stepwise CO substitution (35). Thus Fe(CO)$_5$ reacts with 2,6-dimethylphenyl isocyanide only in the presence of catalytic CoCl$_2$ to give Fe(RNC)(CO)$_4$. The data are consistent with the mechanisms presented in Scheme 3.

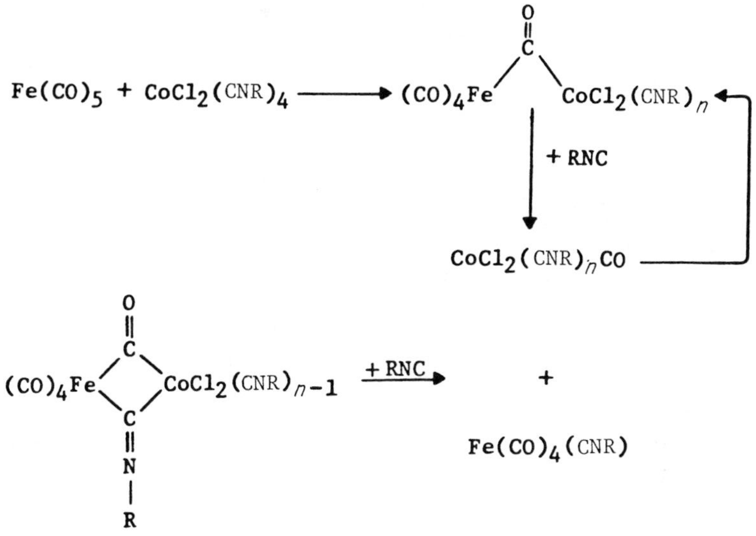

$$Fe(CO)_5 + CoCl_2(CNR)_4 \longrightarrow (CO)_4Fe \overset{\overset{\displaystyle O}{\underset{\displaystyle \|}{C}}}{\diagdown} CoCl_2(CNR)_n$$

$+ \text{RNC} \downarrow$

$$CoCl_2(CNR)_nCO$$

$$(CO)_4Fe \begin{array}{c} \overset{\displaystyle O}{\underset{\displaystyle \|}{C}} \\ \diagup \diagdown \\ \underset{\displaystyle \underset{\displaystyle \|}{\underset{\displaystyle N}{\underset{\displaystyle |}{R}}}}{C} \end{array} CoCl_2(CNR)_{n-1} \xrightarrow{+ \text{RNC}} +$$

$$Fe(CO)_4(CNR)$$

SCHEME 3

The structures $M(CO)_6$ (M = Cr, Mo, W), $M_3(CO)_{12}$ (M = Fe, Ru, Os), and $Ir_4(CO)_{12}$ also reveal CO replacement in the presence of catalytic $CoCl_2$. The reaction between $Fe(CO)_5$ and RNC is also catalyzed by NiX_2, $RuCl_2$-$(PPh_3)_3$, and $RhCl(PPh_3)_3$. Transfer of CO from $M(CO)_6$ (M = Mo, W) at room temperature can also be induced by reaction with $Rh(acac)(PPh_3)_2$ or $RhCl(PPh_3)_3$; the isolated products are $M(CO)_5(PPh_3)$, cis- and trans-$M(CO)_4(PPh_3)_2$, and $Rh(acac)(CO)(PPh_3)$ or $RhCl(CO)(PPh_3)_2$ (36).

E. Tetraphenylcyclobutadiene

The reaction of a variety of η^4-Ph_4C_4 complexes with a variety of Ni, Co, and Mo, Cp, or CO complexes results in the isolation of mixed complexes such as $CpPd(\pi$-$Ph_4C_4)^+$ and $CpCo(\pi$-$Ph_4C_4)$, as shown in Table III (37–40). In the $Fe(CO)_5$ and $Ni(CO)_4$ reactions both Ph_4C_4 and halide are being transferred. All of these studies involved product isolation, and thus nothing is known about potential equilibria that might exist in solution.

TABLE III

TETRAPHENYLCYCLOBUTADIENE TRANSFER REACTIONS

Reactant	Cp reagent	Product	Yield (%)
$(\pi\text{-}Ph_4C_4)PdBr_2$	$[CpFe(CO)_2]_2$	$CpPd(\pi\text{-}Ph_4C_4)^+$	42
$(\pi\text{-}Ph_4C_4)PdBr_2$	$CpFe(CO)_2Br$	$CpPd(\pi\text{-}Ph_4C_4)^+$	85
$(\pi\text{-}Ph_4C_4)NiBr_2$	$CpFe(CO)_2Br$	$CpPd(\pi\text{-}Ph_4C_4)^+$	77
$(\pi\text{-}Ph_4C_4)CoBr(CO)_2$	$CpFe(CO)_2Br$	$CpCo(\pi\text{-}Ph_4C_4)$	> 50
$(\pi\text{-}Ph_4C_4)PdX_2$	$[CpMo(CO)_3]_2$	$(\pi\text{-}Ph_4C_4)MoX(CO)Cp$	18
$(\pi\text{-}Ph_4C_4)PdBr_2$	$Fe(CO)_5$	$(\pi\text{-}Ph_4C_4)Fe(CO)_3$	88
$(\pi\text{-}Ph_4C_4)PdBr_2$	$Ni(CO)_4$	$(\pi\text{-}Ph_4C_4)NiBr_2$	47
$(\pi\text{-}Ph_4C_4)PdBr_2$	Cp_2Co	$CpCo(\pi\text{-}Ph_4C_4)$	12

F. Phosphine

During an attempt to prepare $PdCl_2(PPh_3)_2$ by redistribution of $PdCl_2$ with $Pd(PPh_3)_4$, a rapid reaction occurred [Eq. (25)] to give products resulting from cleavage of a phosphorus–aryl bond (41). In order to evaluate

$$6Pd(PPh_3)_4 + 7PdCl_2 \longrightarrow 3PdCl_2(PPh_3)_2 +$$
$$4\text{ }trans\text{-}Pd(Ph)Cl(PPh_3)_2 + 2Pd_3Cl_2(PPh_3)_3(PPh_2)_2 \qquad (25)$$

trans P–P coupling by $^{31}P\{^1H\}$-NMR, the redistribution of $PdI_2(PMe_3)_2$ and $PdI_2(PEt_3)_2$ was carried out (42). After 24 hours an AB spectrum due to $PdI_2(PMe_3)(PEt_3)$ was observed.

Whereas $Co_2(CO)_8$ reacts with 1 mol of PPh_3 to give $Co_2(CO)_7(PPh_3)$, when $Co_2(CO)_8$ is reacted with PR_3 (R = n-Bu or cyclohexyl) one cannot isolate the monosubstituted dimer (43). Investigation by IR spectroscopy identified the equilibrium in solution [Eq. (26)]. Metal complexes of tri-

$$2Co_2(CO)_7L \rightleftharpoons Co_2(CO)_8 + Co_2(CO)_6L_2 \qquad (26)$$

fluorophosphine have been prepared by reaction of $Ni(PF_3)_4$ with various transition metal derivatives (44). Thus reaction of $[CpMCl_2]_2$ (M = Rh, Ir) with excess $Ni(PF_3)_4$ in refluxing toluene results in the isolation of 60% $CpM(PF_3)_2$. $MnBr(CO)_5$ reacts with $Ni(PF_3)_4$ to give a mixture of Mn_2-$(CO)_{10-x}(PF_3)_x$ derivatives. The rhenium analog behaves similarly. Reaction of $CpFeI(CO)_2$ with $Ni(PF_3)_4$ gives a 77% yield of $CpFeI(CO)(PF_3)$. $CpMoCl(CO)_3$ reacts similarly to give $Cp_2Mo_2(CO)_5PF_3$ and $[CpMo-(CO)_2PF_3]_2$.

Strohmeier et al. (45) examined phosphine exchange between complexes of the type $IrX(CO)(L)_2$. When L = $P(Cy)_3$ and L′ = PPh_3 or $P(OPh)_3$, no exchange was observed by IR spectroscopy after 4 days, whereas when $L = PPh_3$ and $L' = P(OPh)_3$ quantitative conversion to $IrCl(CO)LL'$ occurred in less than 5 minutes.

Reaction of $[RhCl(CO)(PPh_3)]_2$ with PR_3 gave $RhCl(CO)(PPh_3)(PR_3)$ (46). ^{1}H-NMR of such complexes indicated that phosphine exchange was occurring, and the solution equilibria in Eq. (27) were proposed. It appeared that mixed species were isolated due to their lower solubility.

$$2\,RhCl(CO)(PPh_3)(PR_3) \quad \rightleftharpoons \quad RhCl(CO)(PPh_3)_2 + RhCl(CO)(PR_3)_2 \qquad (27)$$

Reaction of $RhCl(CO)(PPh_2Me)_2$ with $RhClCO[P(m\text{-}CH_3C_6H_4)_3]_2$, $RhCl(CO)(PPh_2Et)_2$, $RhCl(CO)(PPhEt_2)_2$, or $RhCl(CO)(PEt_3)_2$ at 30°C in benzene or methylene chloride results in an ABX ^{31}P{^{1}H} NMR spectrum due to $RhClCO(PPh_2Me)(L)$, in addition to resonances due to the starting complexes (3). All the organophosphine redistributions were random as evidenced by an $RhCl(CO)_2(L_2):RhCl(CO)LL':RhCl(CO)(L')_2$ ratio of $1:2:1$ when equimolar quantities of the complexes were mixed. The reaction of $IrCl(CO)(P(OPh)_3)_2$ with $RhCl(CO)(PPhEt_2)_2$ revealed nonrandom exchange. Accurate calculation of the position of the equilibrium from observation of room temperature ^{31}P{^{1}H} NMR spectra of mixed phosphine–phosphite complexes necessitates knowledge of the T_1 relaxation time and NOE of such complexes. The greater distance of protons on phosphite ligands from the phosphorus nuclei should result in longer T_1 times and smaller NOE enhancements (12). This is observed since the phosphite resonances of $IrCl(CO)(PPhEt_2)[P(OPh)_3]$ and $RhCl(CO)$-$(PPh_2Et_2)[P(OPh)_3]$ are of somewhat lower intensity than the phosphine resonances. NMR does qualitatively show that formation of mixed phosphine–phosphite products is strongly favored.

Two mechanisms are possible. Dissociation of PR_3 from $MCl(CO)(PR_3)_2$ (M = Rh, Ir) would result in a 14-electron three-coordinate intermediate which could then react rapidly with PR'_3, or dissociation could occur from a dimeric carbonyl or halide bridged intermediate (as proposed for the halide and carbonyl exchanges). Hard evidence for neither mechanism was presented.

Nixon and Swain (47) found that ligand exchange reactions occur immediately when equimolar solutions of $[RhCl(PF_3)_2]_2$ and $[RhCl(CO)_2]_2$ are mixed at room temperature. IR and mass spectrometry data are given to support the presence of $Rh_2Cl_2(PF_3)_3CO$ and $Rh_2Cl_2(PF_3)(CO)_3$ in solution. In contrast to the data of Shaw (34) the position of this equilibrium does not appear random since only trace amounts of $[RhCl(CO)_2]_2$ and $[RhCl(PF_3)_2]_2$ can be seen.

$Rh_2[P(OMe)_3]_8$, which has a bicapped trigonal antiprismatic structure at room temperature, undergoes a specific equatorial intermolecular ligand exchange rather than any intramolecular type of exchange (48).

An interesting example of both phosphine and halide exchange is observed when a 1:1 mixture of $RuCl_2(PPh_3)_3$ and $cis\text{-}RuCl_2(PF_3)_2(PPh_3)_2$ is refluxed in acetone (49). Three geometric isomers of **13** are isolated.

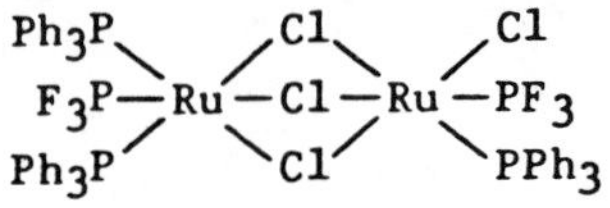

$$
\begin{array}{ccc}
Ph_3P & Cl & Cl \\
F_3P\!-\!Ru\!-\!Cl\!-\!Ru\!-\!PF_3 \\
Ph_3P & Cl & PPh_3
\end{array}
$$

13

The $^{31}P\{^1H\}$ NMR spectrum of equimolar mixtures of $Pt_2Cl_4(PBu_3)_2$ and $Pd_2Cl_4(PBu_3)_2$ revealed immediate formation of the mixed metal complex $PtPdCl_4(PBu_3)_2$, with an equilibrium constant for the reaction [Eq. (28)] of 5.3. At temperatures above $-10°C$, the resonances due to L bound

$$Pt_2Cl_4L_2 + Pd_2Cl_4L_2 \rightleftharpoons 2PtPdCl_4L_2 \qquad (28)$$

to Pd broaden, whereas the analogous broadening due to L bound to Pt is not observed until $+10°C$. Line broadening kinetics suggests both reactions to be first order in substrate, indicating that the exchange occurs via attack of dimer on dimer rather than by dissociation. Since no phosphine scrambling is observed when $Pt_2Cl_4(PBu_3)_2$ is mixed with $Pd_2Cl_4(PPr_3)_2$, they propose **14**, an intermediate similar to Brown's (18). The mixed complexes could not be isolated from solution.

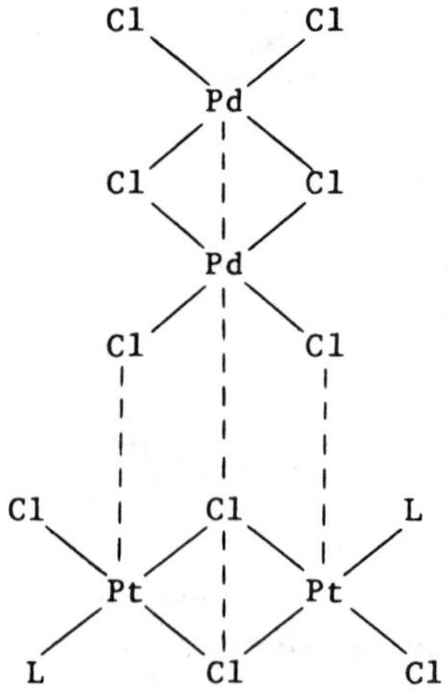

14

When a pentane solution of $Rh_2Co_2(CO)_{12}$ is mixed with an equimolar amount of $Rh_2Co_2(CO)_8(PF_3)_4$, an immediate intermolecular ligand exchange of CO and PF_3 occurs and $Rh_2Co_2(CO)_{10}(PF_3)_2$ is produced (*39*). This is the first reported intermolecular exchange between two compounds having tetranuclear clusters of metal atoms.

During their studies on the metal catalyzed cis–trans isomerizations of complexes of the type MX_2L_2, Nelson *et al.* (*4*) observed the phosphine exchanges shown in Eq. (29). The ratios of mixed to symmetrical species for

$$PdX_2(PR_3)_2 + PdX_2(PR'_3)_2 \rightleftharpoons 2PdX_2(PR_3)(PR'_3) \qquad (29)$$

a variety of PR_3 ligands are given in Table IV. For L = $PPh(OMe)_2$, L' = $PPhBz_2$, L = $P(OPh)_3$, and L' = $P(OBz)_3$ only the symmetric complexes are observed in solution. For most complexes with $AsEt_3$ and for all complexes with piperidine, disproportionation of the mixed-ligand complex

TABLE IV
EQUILIBRIUM POSITION FOR A SERIES OF
$PdCl_2L_2 + PdCl_2(L')_2$ EXCHANGE REACTIONS

$PdCl_2L_2$	+ $PdCl_2(L')_2$	$\dfrac{PdCl_2(L)(L')}{PdCl_2L_2 + PdCl_2(L')_2}$
Me_2PPh	$BzOPPh_2$	2.5
Me_2PPh	$(BzO)_2PPh$	∞
Me_2PPh	$(BzO)_3P$	∞
Me_2PPh	$BzPPh_2$	0.12
Me_2PPh	Bz_2PPh	0.16
Me_2PPh	Bz_3P	0.21
$(MeO)_3P$	Ph_3P	2.5
$(MeO)_3P$	$(PhO)_3P$	∞
$(MeO)_3P$	$(BzO)_3P$	1.4
$(MeO)_3P$	Et_3As	9.5
$(MeO)_2PPh$	$(BzO)PPh_2$	1.8
$(MeO)_2PPh$	$(BzO)_2PPh$	1.3
$(MeO)_2PPh$	$(BzO)_3P$	2.0
$(MeO)_2PPh$	$BzPPh_2$	4.7
$(MeO)_2PPh$	Bz_2PPh	0
$(MeO)_2PPh$	Et_3As	0.45
PPh_3	$(BzO)_3P$	2.6
PPh_3	$(Bz)_3P$	0.5
PPh_3	$(PhO)_3P$	2.0
$(PhO)_3P$	$(BzO)_3P$	0
$(MeO)PPh_2$	Et_3As	0.70
PPh_3	$(EtO)_3P$	4.7
$(PhO)_3P$	$(MeO)PPh_2$	0.13

to symmetric complexes is reported to be more rapid and complete than for most $R_3P-R'_3P$ mixed ligand complexes.

The mixed piperidine complexes are an excellent example of solubility determining an isolated reaction product. Although the literature first reported (50, 51) that $PdCl_2(PR_3)(L)$ (L = piperidine) could be isolated and had a trans configuration, this study by $^{31}P\{^1H\}$ clearly reveals that these complexes completely disproportionate to $PdCl_2(PR_3)_2$ and $PdCl_2(L)_2$ in solution.

Addition of a solution of L [L = PPh_3, $PMePh_2$, $P(i\text{-}Pr)_3$, $P(Cy)_3$, $AsPh_3$, $SbPh_3$] to a solution of $[PtCl_2L']_2$ (L' = PEt_3) resulted in the isolation of trans-$PtLL'Cl_2$ (52). trans-$PtCl_2(PEt_3)(L)$ (L = PPh_3, $PMePh_2$, $AsPh_3$) when suspended in dichloromethane in the presence of catalytic L isomerizes to the cis isomer in 2–3 hours with no observable disproportionation. However, reduction with $NaBH_4$ at 0°C followed by addition of HCl led to scrambled products, the extent of scrambling dependent on L and L'.

When trans-$NiR_2(PMe_3)_2$ (53) [R = 2,6-$(OMe)_2C_6H_3$] and trans-$NiR_2(PMe_2Ph)_2$ are treated together in benzene for 3 hours at 81°C, trans-$NiR_2(PMe_3)(PMe_2Ph)$ is observed in 13.5% yield by NMR. A similar intermolecular exchange between trans-$NiRCl(PMe_3)_2$ and trans-$NiRCl(PMe_2Ph)_2$ took place rapidly at 65°C, attaining an equilibrium of K = 4.0 ± 0.2 in less than 0.5 hours. Exchange between NiR_2L_2 complexes was postulated to involve phosphine dissociation to a three-coordinate intermediate whereas exchange between $NiRClL_2$ was postulated to involve R_3P dissociation from a halide bridged dimer.

G. Hydride

Attempted reduction of $PdCl_2(PR_3)_2$ (R = Cy, i-Pr) with borohydride reagents failed to give hydride derivatives. However, reaction with $NiH(BH_4)(PR_3)_2$ resulted in the scrambling shown in Eq. (30) (54). When PR_3 = PR'_3 the Pd and Ni hydrides could be isolated by fractional recrystallizations.

$$PdCl_2(PR_3)_2 + NiH(BH_4)(PR'_3)_2 \longrightarrow$$

$$MHCl(PR_3)_2 + MHCl(PR'_3)_2 + MHCl(PR_3)(PR'_3) \qquad (30)$$

Addition of the chloro bridged dimer $Pt_2Cl_4(PPr_3)_2$ to $IrH_5(PPr_3)_2$ in toluene resulted in evolution of hydrogen and the appearance in the ^1H NMR spectrum of a mixture of two new Ir–H and one Pt–H resonance (55). $\{^1H\}$ ^{31}P NMR showed that all hydride ligands are mutually coupled, and the $^{31}P\{^1H\}$ spectrum revealed that the phosphorus nucleus of the tertiary phosphine ligand coordinated to the platinum is coupled to two hydride ligands and to two other equivalent phosphorus nuclei. From the combined NMR data, structure **15** was assigned as a mixed metal hydride complex. On standing for one-half hour at room temperature **15** disproportionates to **16** and several uncharacterized iridium hydride complexes. Structure **15** is therfore the isolated intermediate that results in hydride transfer to Pt.

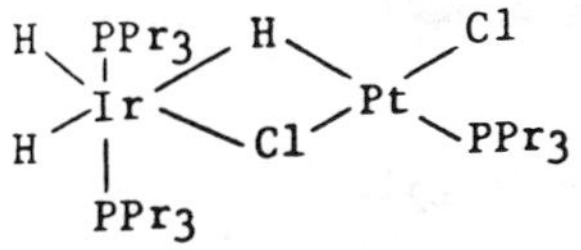

15 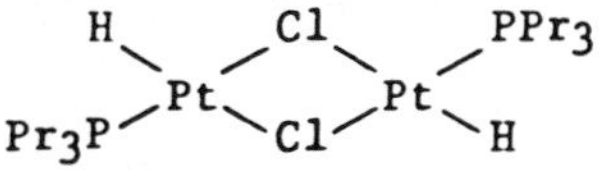   **16**

In a similar fashion, during their studies on bimetallic hydrido-bridged complexes, Musco et al. isolated **17** from the reaction of [Rh(dppe)-(MeOH)$_2$]$^+$ with IrH$_5$[P(i-Pr)$_3$]$_2$ (56). X-Ray diffraction was used to confirm the structure. Another model for hydride transfer can be observed in the work of Jonas and Wilke, who observed that the reduction of NiCl$_2$(dppe) with Na(HBMe$_3$) resulted in the dinuclear complex **18**, whose structure was solved by X-ray diffraction (57, 58).

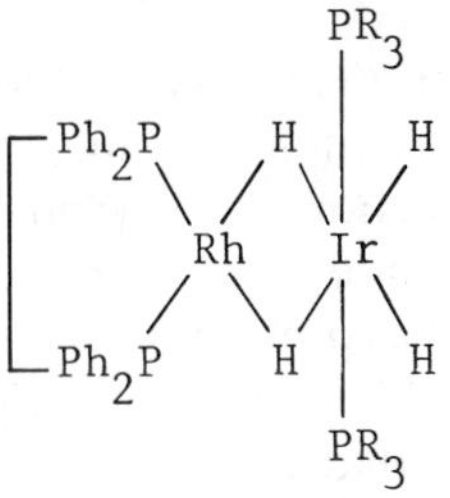

17 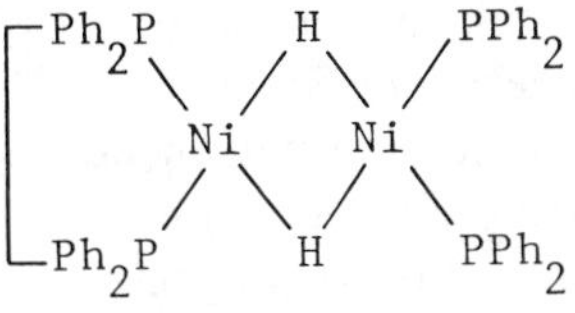   **18**

One of the most significant studies on hydride transfer was carried out by Evans and Norton (5, 59). They found that reaction of OsH$_2$(CO)$_4$ with D$_2$Os(CO)$_4$ gives the crossover product HD. The possibility that HD is produced not in the elimination reaction but by a prior H$_2$–D$_2$ scrambling reaction was ruled out by carrying the reaction out under D$_2$ [Eq. (31)].

$$H_2Os(CO)_4 \xrightarrow{\ D_2\ } H_2 \qquad (31)$$

Although hydrogen elimination is binuclear, the reaction is not kinetically bimolecular, the rate being first order in H$_2$Os(CO)$_4$. The derived rate law and activation parameters suggest a rate determining step involving CO dissociation followed by a fast binuclear elimination of hydrogen and re-coordination of CO, as shown in Scheme 4. In this mechanism coordinated

$$Os(CO)_4H_2 \xrightarrow{\;k_{1/2}\;} Os(CO)_3H_2 + CO$$

$$Os(CO)_3H_2 + Os(CO)_4H_2 \xrightarrow{\;fast\;} H_2Os_2(CO)_7 + H_2$$

$$H_2Os_2(CO)_7 + CO \xrightarrow{\;fast\;} H_2Os_2(CO)_8$$

Scheme 4

hydride successfully competes with CO for a vacant coordination site on $H_2Os(CO)_3$. When the reaction is carried out under ^{13}CO and stopped after one half-life, negligible label is incorporated into recovered $H_2Os(CO)_4$, whereas considerable label is found in the dinuclear product consistent with the proposed mechanism.

H. Alkyls

Alkyl transfers from main-group to transition metals have been known for many years (60, 61). There are now several examples of alkyl transfers between transition metal atoms. Such transfer reactions can be characterized by their electronic character. Transfer from vitamin B_{12}-analog cobalt alkyls to higher oxidation state metals can be thought of as transfer of an essentially nucleophilic alkyl to an electrophilic metal center (62). Among the most interesting of these transfer reactions are those that transfer alkyl groups between similar or identical metal centers. Often these reactions involve the simultaneous transfer of halogen, hydride, or other ligands in the opposite direction. It is becoming more and more apparent that such intermolecular alkyl transfer reactions are intimately involved in C–X (X = C, H, CHO, etc.) coupling reactions.

While examining the oxidative addition reaction of p-FC$_6$H$_4$X with Ni(PEt$_3$)$_4$ to give NiX(C$_6$H$_4$F)(PEt$_3$)$_2$, Parshall (63) noted contamination by NiX$_2$(PEt$_3$)$_2$ and proposed that these by-products were formed by successive redistribution and reductive elimination reactions as shown in Scheme 5. He also postulated that the thermal decomposition of NiX(R)-(PPh$_3$)$_2$ complexes (64) to give R—R could be proceeding in the same manner.

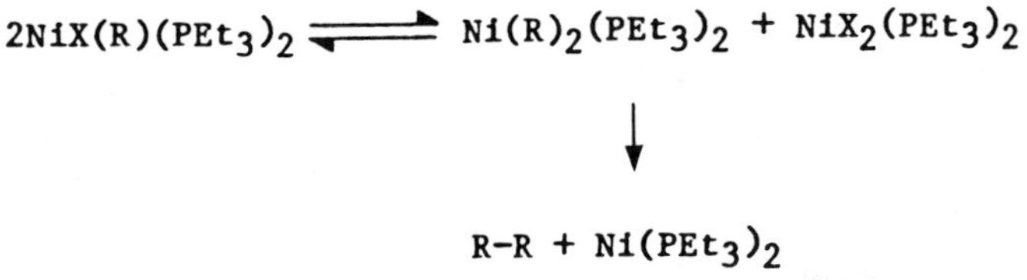

$$2NiX(R)(PEt_3)_2 \rightleftharpoons Ni(R)_2(PEt_3)_2 + NiX_2(PEt_3)_2$$

$$\downarrow$$

$$R\text{--}R + Ni(PEt_3)_2$$

Scheme 5

Homonuclear alkyl transfer occurs rapidly and stereospecifically when equimolar solutions of cis-Pt(Me)$_2$(PEt$_3$)$_2$ and cis-PtCl$_2$(PEt$_3$)$_2$ are mixed at 25°C (65) ^{1}H NMR of the mixture reveals quantitative formation of cis-PtMe(Cl)(PEt$_3$)$_2$. Heteronuclear transfer between platinum and palladium occurs when PtMe$_2$(COD) or PtMeCl(COD) is reacted with PdCl$_2$(PhCN)$_2$.

In a similar fashion reaction of equimolar quantities of PtMe$_2$(COD) with PtI$_2$(COD) in CDCl$_3$ at 30°C results in a symmetrization reaction producing PtMeI(COD) as the final product (66). The half-life for this reaction is ~ 12 hours. PtCl$_2$(PPh$_2$Me)$_2$ reacts with PtMe$_2$(PPh$_2$Me)$_2$ at 30°C to produce PtCl(Me)(PPh$_2$Me) quantitatively in less than 15 minutes (66). However, not all systems led to such transfers. Attempted reaction of PdCl$_2$[P(p-MeC$_6$H$_4$)$_3$]$_2$ with Pt(Me)I[P(p-MeC$_6$H$_4$)$_3$]$_2$ revealed no sign of decomposition or Me transfer to Pd after 72 hours at 30°C. In contrast, reaction of Pt(Me)Br[P(p-MeC$_6$H$_4$)$_3$]$_2$ with Pd(Br)Ph[P(p-MeC$_6$H$_4$)$_3$]$_2$ for 6 hours at 30°C resulted in the observation of toluene but no biphenyl, indicating an intermolecular decomposition pathway. Solutions of the individual components over the same period of time showed no sign of decomposition.

The reaction of cis-PtMe$_2$(PPhMe$_2$)$_2$ and cis-Pt(NO$_3$)$_2$(PMe$_2$Ph)$_2$ in solution gave cis-PtMe(NO$_3$)(PMe$_2$Ph)$_2$ as the only product, indicating that the reaction proceeded with retention of configuration at both platinum centers (67). The product is established by kinetic control since the cis isomer slowly isomerizes to the trans isomer in solution. Structure **19** is the proposed intermediate that is expected to break down and give only cis products.

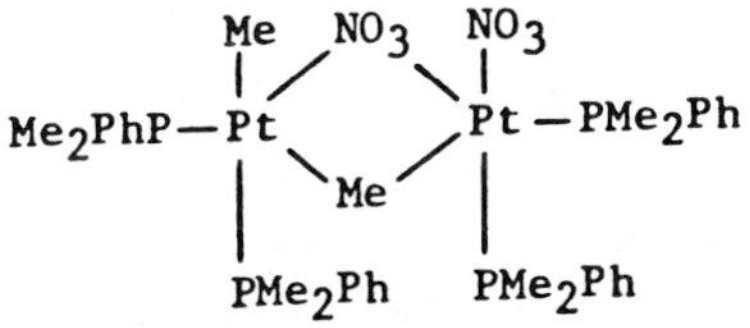

19

The reaction of metal carbonyl anions with methyl metal carbonyl compounds has been studied and grouped into three categories: adduct formation, methyl transfer, and no apparent reaction (68). The formation of acyl anion adducts was observable only in the reaction of $Mn(Me)(CO)_5$ with either $Mn(CO)_5^-$ or $Re(CO)_5^-$ and could be trapped by alkylation with Me_3OBF_4 as shown in Scheme 6.

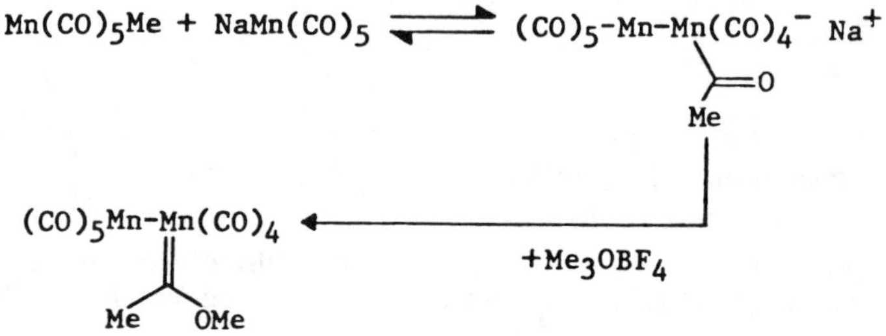

$$Mn(CO)_5Me + NaMn(CO)_5 \;\rightleftharpoons\; (CO)_5\text{-}Mn\text{-}Mn(CO)_4^- \; Na^+$$

(CO)$_5$Mn-Mn(CO)$_4$ ⟵ $+Me_3OBF_4$

Me   OMe

The methyl transfer reactions are envisoned as arising from S_N2 displacement of a less nucleophilic metal carbonyl anion by a more nucleophilic carbonyl anion. $CpFe(CO)_2^-$ is the more basic species (69) and therefore abstracts methyl from the less basic $CpMo(Me)(CO)_3$. $CpMo(CO)_3^-$ and $Mn(CO)_5^-$ are bases of comparable strength, and, therefore, the reaction either of $MnMe(CO)_5$ with $CpMo(CO)_3^-$ or of $CpMo(Me)(CO)_3$ with $Mn(CO)_5^-$ gave mixtures of $CpMo(Me)(CO)_3$ and $Mn(Me)(CO)_5$.

Puddephatt *et al.* examined a series of methyl transfers between Pd(II), Pt(II), Au(I), and Au(III) centers with particular emphasis on determining which methyl metal complexes are the stronger methylating agents (70). The reaction of cis-Pt(Me)$_2$(PMe$_2$Ph)$_2$ with AuX(PMe$_2$Ph) to give $trans$-PtX(Me)(PMe$_2$Ph)$_2$ and AuMe(PMe$_2$Ph) revealed an equilibrium constant dependent on halogen: $K = 0.3$ (X = I), 1.0 (X = Br), and 5.4 (X = Cl). Identical equilibria were reached starting with reagents on either side of the equation.

The complex cis-PtMe$_2$(PMe$_2$Ph)$_2$ reacted with cis-PdCl$_2$(PMe$_2$Ph)$_2$ according to Eq. (32). The reaction was complete after 4 days at 20°C, but

$$cis\text{-PtMe}_2(PMe_2Ph)_2 + cis\text{-PdCl}_2(PMe_2Ph)_2 \longrightarrow$$
$$trans\text{-PtCl(Me)(PMe}_2Ph)_2 + trans\text{-PdCl(Me)(PMe}_2Ph)_2 \qquad (32)$$

the product PdCl(Me)(PPhMe$_2$)$_2$ was not further methylated to PdMe$_2$-(PMe$_2$Ph)$_2$ with time. The methyl exchange was discussed in terms of a bimolecular electrophilic substitution (S_E2) reaction at the methyl group involving an intermediate such as **20.** The reactions all followed overall

$$\begin{array}{ccc}
\text{Me}_2\text{PhP} & \text{Cl} \quad \text{Cl} & \\
| \quad \diagup \quad \backslash \quad | & \\
\text{Me}_2\text{PhP--Pt} & \quad \quad \text{Pt--PMe}_2\text{Ph} & \\
| \quad \backslash \quad \diagup \quad | & \\
\text{Me} \quad \text{Me} \quad \text{PMe}_2\text{Ph} &
\end{array}$$

20

second order kinetics. The driving force for methyl–halogen exchange reactions was presumed to arise from differences in the metal–methyl and metal–halogen bond strengths.

The complex cis-PtMe$_2$(PMe$_2$Ph)$_2$ **(21)** reacts with a fourfold excess of NO$_2$ to give PtMe$_2$(PMe$_2$Ph)$_2$(NO$_3$)$_2$ **(22)** (71). Compound **22** then reacts with **21** to give cis-PtMe(NO$_3$)(PMe$_2$Ph)$_2$ **(23)** and fac-PtMe$_3$(NO$_3$)-(PMe$_2$Ph)$_2$ **(24)**. Then **24** undergoes reductive elimination of ethane to give $trans$-PtMe(NO$_3$)(PMe$_2$Ph)$_2$. Compound **22** may be isolated, depending on solvent, since the reaction to give **22** is faster than the reaction of **21** and **22** to give **23** and **24**.

The rate of reaction of cis-PtMe$_2$L$_2$ (L = PPhMe$_2$) with cis-PtX$_2$L$_2$ followed the order NO$_3 \gg$ NO$_2 >$ SePh $>$ I $>$ NCS $\gg$ C (72). The range in reactivities is large. With X = NO$_3$, the reaction was complete in less than 1 minute, whereas the reaction with X = Cl was only half complete after 15 days at 34°C. The symmetrization of cis-PtCl$_2$L$_2$ with cis-PtMe$_2$L$_2$ to give PtClMeL$_2$ was studied as a function of L. The final product was always the trans isomer, although the cis isomer could be identified at intermediate stages. The rates followed the order PEt$_3$($t_{1/2} \sim$ 1 hour) > PMe$_3$($t_{1/2} \sim$ 2 days) $\gg$ PMe$_2$Ph($t_{1/2} \sim$ 15 days). Reaction of cis-PtMe$_2$(PEt$_3$)$_2$ with cis-PtCl$_2$(PMe$_2$Ph)$_2$ gave a 1:1 mixture of $trans$-PtCl(Me)(PEt$_3$)$_2$ and $trans$-PtCl(Me)(PMe$_2$Ph)$_2$ with a half-life of 8 hours at 20°C. Phosphine exchange proceeded at a slower rate with a half-life of 10 days. In contrast cis-PtMe$_2$(PMe$_2$Ph)$_2$ and cis-PtCl$_2$(PEt$_3$)$_2$ gave no detectable reaction after one week. The dependence of the reaction rate on R follows the order Me > Ph. The authors observed overall second order kinetics, which they felt was inconsistent with a dissociative mechanism.

It was later discovered that the reaction of cis-PtMe$_2$(PPhMe$_2$)$_2$ with cis-PtCl$_2$(PMe$_2$Ph)$_2$ to give cis-PtClMe(PMe$_2$Ph)$_2$ is catalyzed by Pt$_2$Cl$_2$(μ-Cl)$_2$(PMe$_2$Ph)$_2$ (73). The data were rationalized in terms of the catalyst Pt$_2$Cl$_2$(μ-Cl)$_2$(PMe$_2$Ph)$_2$ reacting with cis-PtMe$_2$(PMe$_2$Ph)$_2$ to give cis-PtCl(Me)(PPhMe$_2$)$_2$ and a reactive dimeric methylplatinum intermediate, perhaps with a bridging methyl group. The intermediate may then either react with cis-$PtCl_2(PMe_2Ph)_2$ to give cis-$PtClMe(PPhMe_2)_2$ and regenerate $Pt_2Cl_2(\mu$-$Cl)_2(PMe_2Ph)_2$ or react with more cis-$PtMe_2(PMe_2Ph)_2$ to give inactive $Pt_2Me_2(\mu$-$Cl)_2(PMe_2Ph)_2$ which does not react with cis-$PtCl_2(PMe_2Ph)_2$.

Thermolysis of compounds of the type Cp_2VR occurs with formation of RH, Cp_2V, and a vanadocene homolog with an R group substituting one of the Cp rings (74). It was shown that abstraction of hydrogen from one of the Cp rings is an intermolecular process as is the alkyl group transfer and the RH elimination.

Reaction of Cp_2ZrMe_2 with $CpMoH(CO)_3$ occurs at room temperature with the rapid evolution of methane to give **25** (75). Compound **25** displays

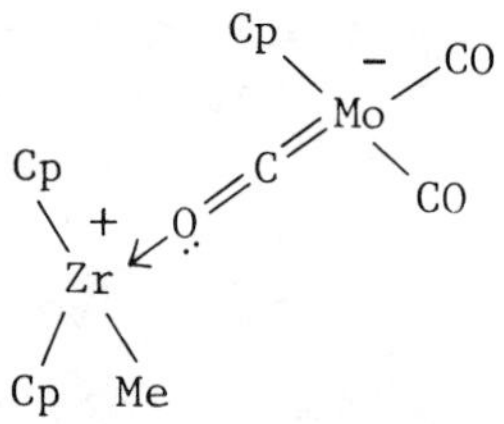

25

an unusually low νCO stretch (1545 cm^{-1}). Preliminary X-ray results suggest a carbonyl–OC bridged structure. The authors favor the depicted intimate ion pair zwitterion structure.

Elimination of methane from cis-$HOsMe(CO)_4$ proceeds by a binuclear mechanism as evidenced by the observation of CD_4 evolution from the thermolysis of $DOsMe(CO)_4$ in the presence of $DOsCD_3(CO)_4$ (63). Norton (59) proposed that such binuclear elimination should occur whenever coordinatively unsaturated alkyls such as $AuMe(PPh_3)$ are reacted with metal hydrides as in the reaction of $HMn(CO)_5$ with $AuMe(PPh_3)$ (76). Methane is also formed from the reaction of $H_2Os(CO)_4$ with $RhI_3(COMe)(CO)$ or $IrCl(COMe)(CO)(AsPh_3)$. Compound **26** is proposed as a general intermediate for such reactions.

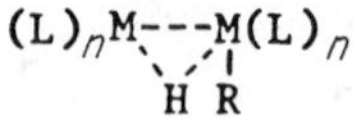

$$(L)_n M\text{-}\text{-}\text{-}M(L)_n$$
$$H \quad R$$

26

$HCo(CO)_4$ is known to react stoichiometrically with $Co(COR)(CO)_4$ to give aldehydes [Eq. (32)]. The reaction is inhibited by CO suggesting the intermediacy of $Co(COR)(CO)_3$. Norton suggested (59) that the reaction follows a bimolecular pathway by analogy to the $H_2Os(CO)_4$ bimolecular eliminations. Again the overall reaction would appear to be first order since

$$RCOCo(CO)_3 + HCo(CO)_4 \longrightarrow RCHO + Co_2(CO)_7 \qquad (33)$$

the rate determining step would be the creation of a vacant coordination site and not the bimolecular reaction.

Bergman et al. (77) found that heating a mixture of CO, 27, and (27)-d_6 yields only acetone-d_0 and -d_6. However, heating a mixture of 27 and (27)-d_6 without CO then treating with CO under conditions previously established to preclude crossover during carbonylation produced a statistical mixture of acetone-d_6, -d_3 and -d_0. Scrambling clearly takes place prior to the carbonylation reaction. When (27)-d_6 and 28 were heated at 62°C, a resonance due to 27 was seen growing in the NMR spectrum. Reaction of (28)-d_6 with 30 under conditions where PMe_3 dissociation is known to be negligible was again found to exchange Me groups.

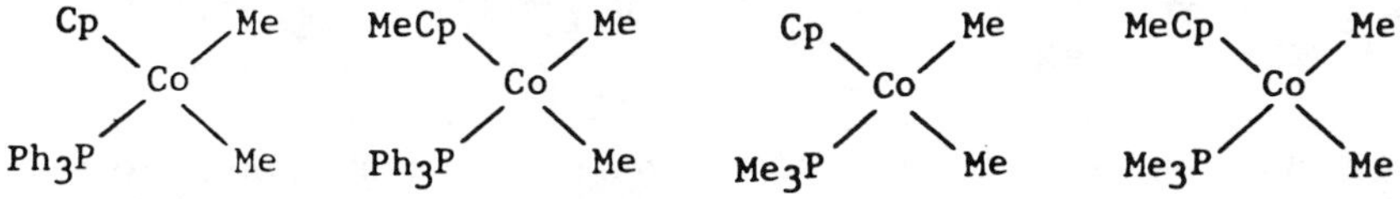

| 27 | 28 | 29 | 30 |

Heating 29-d_6 with 30 revealed no scrambling indicating that dissociation of PR_3 from one metal center is necessary and sufficient for exchange to take place. The rate constants of these processes indicated that phosphine is lost in a rapid preequilibrium step followed by rate determining combination of the unsaturated intermediate so generated with a second saturated complex as shown in Scheme 7.

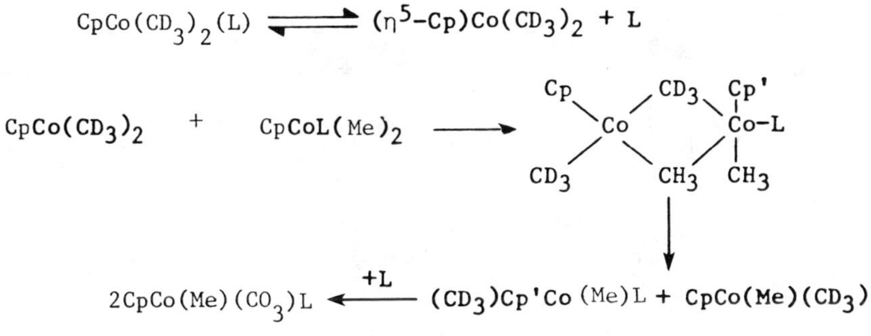

SCHEME 7

Although **27**-d_6 and $CpMoMe_2$ do not exchange methyls, exchange is observed between **27**-d_6 and Cp_2ZrMe_2. A similar mechanism was proposed for the hydrogenolysis of $CpCoMe_2(PPh_3)$ as shown in Scheme 8 (78). The reaction revealed an induction period and was shown *not* to be a free radical process.

$$CpCo(PPh_3)Me \rightleftharpoons CpCo(Me)_2 + PPh_3$$

$$(\textbf{31}) \qquad\qquad (\textbf{32})$$

$$CpCo(PPh_3)_2 \rightleftharpoons CpCo(PPh_3) + PPh_3$$

$$(\textbf{33}) \qquad\qquad (\textbf{34})$$

$$\textbf{33} + H_2 \rightleftharpoons CpCo(PPh_3)H_2$$

$$(\textbf{35})$$

$$\textbf{35} + \textbf{32} \rightleftharpoons 2CH_4 + \textbf{33}$$

SCHEME 8

While examining the trans to cis isomerization process for a variety of dialkylpalladium(II) complexes, intermolecular Me transfer was also observed (79). Examination of the thermolysis product liberated from *cis*-$PdMe_2(PPh_2Me)_2$ derived by isomerizing *trans*-$PdMe_2(PPhMe_2)_2$ in the presence of $Pd(CD_3)_2(PMePh_2)_2$ revealed a $CD_3CD_3 : CH_3CD_3 : CH_3CH_3$ ratio of 0.29 : 0.46 : 0.25, indicating that statistical scrambling had occurred. They proposed Scheme 9 to account for these observations. As the inter-

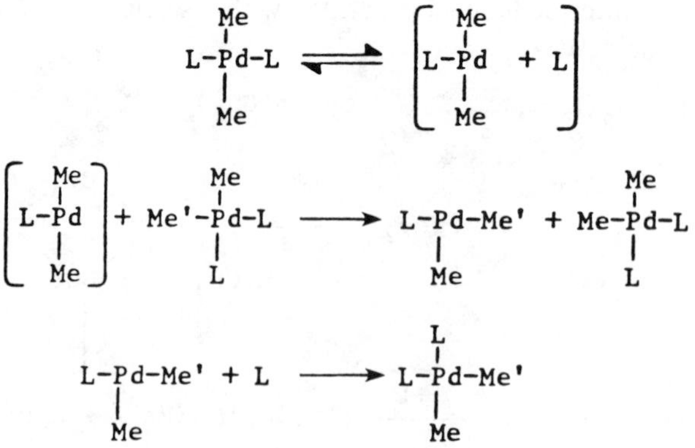

SCHEME 9

mediate in the crucial intermolecular reaction, they postulated an Me bridged species such as **36**. A model for the proposed bridged methyl intermediates **37** was crystallographically characterized by Wilkinson *et al.* as the product from the reaction of $(Me_3P)_3Ru(\mu\text{-}CH_2)_3Ru(PMe_3)_3$ with HBF_4 (*80*).

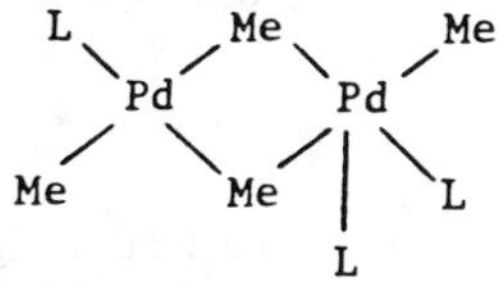

36

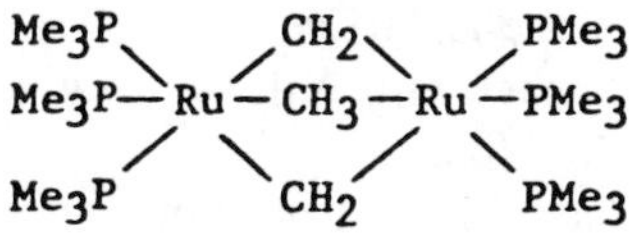

37

Classic model studies for methyl bridged species come from the work of Shapley *et al.* (*81, 82*). They found that treatment of $H_2Os_3(CO)_{10}$ with diazomethane resulted in a material of formulation "$Os_3(CO)_{10}CH_4$." NMR data revealed a mixture of two isomers, $HOs_3(CO)_{10}CH_3$ (**38**) and $H_2Os_3(CO)_{10}CH_2$ (**39**). Spin saturation transfer experiments revealed interconversion of **38** and **39**. After 24 hours at 110°C, **38** and **39** were found to generate **40**. Such equilibria can be envisioned as being involved during methyl transfer between metal centers. Further studies examined the reaction of $D_2Os_3(CO)_{10}$ with CH_2N_2 to prepare $Os_3(CO)_{10}CH_2D_2$. By NMR exchange studies, they concluded that the methyl ligand adopts structure **41** and *not* the previously formulated structure (**38**).

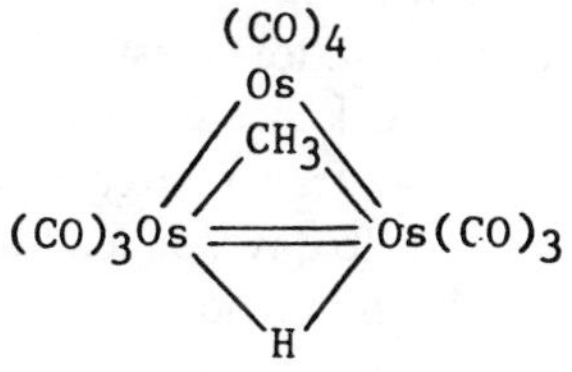

38

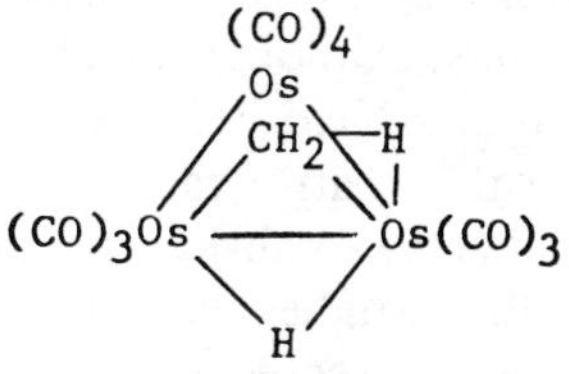

39

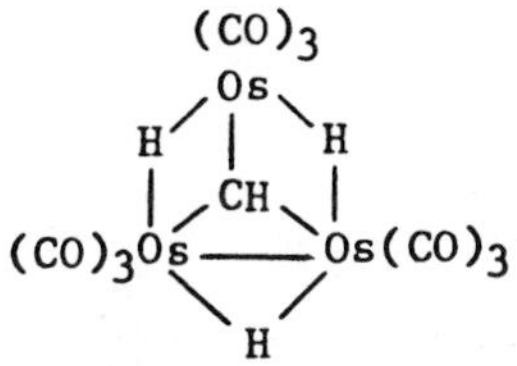

40

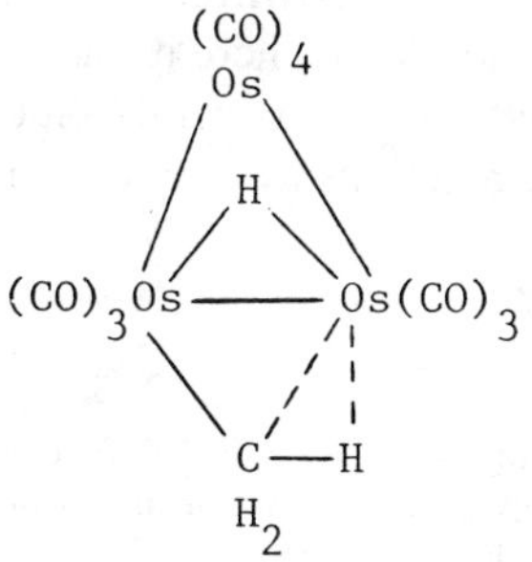

41

I. Miscellaneous

1. η^6-C_6H_6. Transfer of η^6-C_6H_6 during the symmetrization reaction of $(\eta^6$-$C_6H_6)_2Cr$ and $Cr(CO)_6$ results in the formation of $(\eta^6$-$C_6H_6)Cr(CO)_3$ (83).

2. NO. Reaction of $Co(NO)(D)_2$ (D = dimethylglyoxime) with $NiCl_2L_2$ results in the isolation of $[Ni(NO)ClL]_2$ and $CoCl(NO)L$ (84). NO alone does not react with NiL_2X_2 to give the dimer, thus a dissociative transfer process is unlikely.

3. =$CHCMe_3$. Tantalum complexes of the type $Ta(CHCMe_3)X_3L_2$ (L = PR_3, X = Cl or Br) react with $W(O)(OCMe_3)_4$ to give [Ta-$(OCMe_3)_4X]_2$ and $W(O)(CHCMe_3)X_2L_2$ (85). Many intermediates are observed by ^{31}P NMR, but after 3 hours only $W(O)(CHCMe_3)X_2L_2$ crystallizes. No proposed mechanism for this unique transfer of neopentylidene ligand was given.

V

CONCLUSIONS AND FUTURE DIRECTIONS

It is clear that a variety of ligands is capable of being transferred between metal centers. Although much qualitative information exists, few quantitative studies have been carried out. More work needs to be done to determine the relative rates of these bimolecular transfer reactions since current data reveal half-lives ranging from seconds to days, depending on the complex. Basic studies are also needed on type IV "nonproductive" reactions to determine the scope of such reactions and the rate at which they occur.

The future of such transfer reactions lies in catalysis. One can envision a metal complex activating a substrate and transferring the activated species to a second metal where it reacts further. Such reactions could, in principle, be catalytic in both metals. Such reactions could increase the number of catalytic reactions in an exponential fashion.

ACKNOWLEDGMENTS

I wish to acknowledge the helpful comments of A. G. Davies, University College, London; the support of The Dow Chemical Company; and the specific efforts of A. D. Strickler in preparing the manuscript.

REFERENCES

1. J. C. Lockhart, "Redistribution Reactions." Academic Press, New York, 1970.
2. K. Moedritzer, *Adv. Organomet. Chem.* **6**, 171 (1968).
3. P. E. Garrou and G. E. Hartwell, *Inorg. Chem.* **15**, 646 (1976).
4. A. W. Verstuyft, D. A. Redfield, L. W. Cary, and J. H. Nelson, *Inorg. Chem.* **15**, 1128 (1976).
5. J. Evans and J. R. Norton, *J. Am. Chem. Soc.* **96**, 7577 (1974).
6. S. J. Krasmski and J. R. Norton, *J. Am. Chem. Soc.* **99**, 295 (1977).
7. N. H. Alemdaroglu, J. L. M. Penninger, E. Otlay, *Monatsh. Chem.* **107**, 1153 (1976).
8. A. G. Lee, *Organomet. Chem. Rev. A* **6**, 139 (1970).
9. C. N. R. Rao, "Ultraviolet and Visible Spectroscopy." Butterworth, London, 1961.
10. K. Nakamoto, "Infrared Spectra of Inorganic and Coordination Compounds." Wiley (Interscience), New York, 1970.
11. J. A. Pople, W. G. Schneider, H. J. Bernstein, "High Resolution Nuclear Magnetic Resonance." McGraw-Hill, New York, 1959.
12. J. H. Noggle and R. E. Schirmer, "The Nuclear Overhauser Effect." Academic Press, New York, 1971.
13. A. Murray and L. D. Williams, "Organic Synthesis with Isotopes." Wiley (Interscience), New York, 1958.
14. F. W. McLafferty, "Mass Spectroscopy of Organic Ions." Academic Press, New York, 1963.
15. A. N. Nesmeyanov, A. Z. Rubezhov, and S. P. Gubin, *Izv. Akad. Nauk SSSR, Ser. Khim,* p. 149 (1966).
16. A. N. Nesmeyanov, A. Z. Rubezhov, and S. P. Guvin, *J. Organomet. Chem.* **16**, 163 (1969).
17. R. F. Heck, *J. Am. Chem. Soc.* **90**, 317 (1968).
18. D. L. Tibbets and T. L. Brown, *J. Am. Chem. Soc.* **91**, 1108 (1969).
19. H. Werner and A. Kuhn, *Angew. Chem., Int. Ed. Engl.* **16**, 412 (1977).
20. H. Werner and A. Kuhn, *J. Organomet. Chem.* **179**, 421 (1979).
21. J. F. Tinley-Bassett, *J. Chem. Soc.* p. 4784 (1963).
22. S. P. Gubin, A. Z. Rubezhov, B. L. Wiuch, and A. N. Nesemeyanov, *Tetrahedron Lett.,* p. 2881 (1964).
23. H. Yamazaki, T. Nishido, Y. Matsumoto, S. Sumida, and N. Hagihara, *J. Organomet. Chem.* **6**, 86 (1966).
24. G. E. Schroll, United States Patent No. 3,054,815, January 4, 1962.
25. M. D. Rausch, Y. F. Chang, and H. B. Gordon, *Inorg. Chem.* **8**, 1355 (1969).
26. H. E. Bryndza and R. G. Bergman, *Inorg. Chem.* **20**, 2988 (1981).
27. P. M. Maitlis, A. Efraty, and M. L. Games, *J. Am. Chem. Soc.* **87**, 719 (1965).
28. H. Werner and H. J. Kraus, *J. Chem. Soc., Chem. Commun.,* p. **814** (1979).
29. B. T. Heaton and K. J. Timmins, *J. Chem. Soc., Chem. Commun.,* p. 931 (1973).
30. S. Al-Jibori, C. Crocker, and B. L. Shaw, *J. Chem. Soc., Dalton Trans.,* 319 (1981).
31. C. T. Hunt and A. L. Balch, *Inorg. Chem.* **21**, 1641 (1982).
32. P. E. Garrou, unpublished results.
33. B. L. Booth, M. J. Else, R. Fields, H. Goldwhite, and R. N. Haszeldine, *J. Organomet. Chem.* **14**, 417 (1968).
34. J. Powell and B. L. Shaw, *J. Chem. Soc. A,* p. 211 (1968).
35. M. O. Albers, N. J. Coville, T. V. Ashworth, E. Singleton, and H. E. Swanepool, *J. Chem. Soc., Chem. Commun.,* p. 489 (1980).
36. P. E. Garrou, unpublished results.

37. P. M. Maitlis, A. Efraty, and M. L. Games, *J. Organomet. Chem.* **2**, 284 (1964).
38. P. M. Maitlis and A. Efraty, *J. Organomet. Chem.* **4**, 175 (1965).
39. P. M. Maitlis and M. L. Games, *J. Am. Chem. Soc.* **85**, 1887 (1963).
40. P. M. Maitlis and A. Efraty, *J. Organomet. Chem.* **4**, 172 (1964).
41. D. R. Coulson, *J. Chem. Soc., Chem. Commun.*, p. 1530 (1968).
42. R. G. Goodfellow, *J. Chem. Soc., Chem. Commun.*, p. 114 (1968).
43. P. Szabo, L. Fekete, G. Bor, Z. Nagy-Mangos, L. Marko, *J. Organomet. Chem.* **12**, 245 (1968).
44. R. B. King and A. Efraty, *J. Am. Chem. Soc.* **93**, 5260 (1971).
45. W. Strohmeier, W. Rehder-Stirnweiss, and G. Reischig, *J. Organomet. Chem.* **27**, 393 (1971).
46. D. F. Steele and T. A. Stephenson, *J. Chem. Soc., Dalton Trans.*, p. 2161 (1972).
47. J. F. Nixon and J. R. Swain, *J. Chem. Soc., Dalton Trans.*, p. 1044 (1972).
48. R. Mathieu and J. F. Nixon, *J. Chem. Soc., Chem. Commun.*, p. 147 (1974).
49. R. A. Head and J. F. Nixon, *J. Chem. Soc., Chem. Commun.*, p. 135 (1975).
50. J. Chatt and L. M. Venanzi, *J. Chem. Soc.*, p. 4461 (1955).
51. A. Pidcock, R. E. Richards, and L. M. Venanzi, *J. Chem. Soc.*, p. 1707 (1966).
52. H. C. Clark, A. B. Goel, and C. S. Wong, *J. Organomet. Chem.* **190**, C101 (1980).
53. M. Wada, R. Nishiwaki, and Y. Kawasaki, *J. Chem. Soc., Dalton, Trans.*, p. 1443 (1982).
54. H. Munakata and M. L. H. Green, *J. Chem. Soc., Chem. Commun.*, p. 881 (1970).
55. J. VanDongen, C. Masters, and J. P. Visser, *J. Organomet. Chem.* **94**, C29 (1975).
56. A. Musco, R. Naegli, L. M. Venanzi, and A. Albenati, *J. Organomet. Chem.* **228**, C15 (1982).
57. K. Jonas and G. Wilke, *Angew. Chem. Int. Ed. Engl.* **9**, 312 (1970).
58. K. Jonas and G. Wilke, *Angew. Chem. Int. Ed. Engl.* **12**, 943 (1973).
59. J. R. Norton, *Acc. Chem. Res.* **12**, 139 (1979).
60. G. E. Coates, M. L. H. Green, and K. Wade, "Organometallic Compounds of the Transition Elements." Methuen, London, 1968.
61. R. F. Heck, "Organotransitionometal Chemistry," Chapt. 2. Academic Press, New York, 9174.
62. M. D. Johnson, *Acc. Chem. Res.* **11**, 57 (1978).
63. G. W. Parshall, *J. Am. Chem. Soc.* **96**, 2360 (1974).
64. S. Otsuka, A. Nakamura, T. Yoshida, M. Naruto, and K. Ataka, *J. Am. Chem. Soc.* **95**, 3180 (1973).
65. J. P. Visser, W. W. Jager, and C. Masters, *Recl. Trav. Chim. Pays-Bas* **94**, 70 (1975).
66. P. E. Garrou, unpublished results.
67. P. J. Thompson and R. J. Puddephatt, *J. Chem. Soc., Chem. Commun.*, p. 841 (1975).
68. C. P. Casey, C. R. Cyr, R. L. Anderson, and D. F. Martin, *J. Am. Chem. Soc.*, **97**, 3053 (1975).
69. R. E. Dessey, R. J. Pohl, and R. B. King, *J. Am. Chem. Soc.* **88**, 5121 (1966).
70. R. J. Puddephatt and P. J. Thompson, *J. Chem. Soc., Dalton, Trans.*, p. 1811 (1975).
71. R. J. Puddephatt and P. J. Thompson, *J. Chem. Soc., Dalton Trans.*, p. 2091 (1976).
72. R. J. Puddephatt and P. J. Thompson, *J. Chem. Soc., Dalton, Trans.*, p. 1219 (1977).
73. R. J. Puddephatt and P. J. Thompson, *J. Organomet. Chem.* **120**, C51 (1976).
74. C. P. Boekel, A. Jelsma, J. H. Teuben, and H. J. DeLiefde-Meijer, *J. Organomet. Chem.* **136**, 211 (1977).
75. J. A. Marsella, J. C. Huffman, K. G. Caulton, B. Longato, and J. R. Norton, *J. Am. Chem. Soc.* **104**, 6360 (1982).
76. C. M. Mitchell and F. G. A. Stone, *J. Chem. Soc., Dalton Trans.*, p. 102 (1972).
77. H. E. Bryndza, E. R. Evitt, and R. G. Bergman, *J. Am. Chem. Soc.* **102**, 4948 (1980).
78. A. H. Janowicz and R. G. Bergman, *J. Am. Chem. Soc.* **103**, 2488 (1981).
79. F. Ozakwa, T. Ito, Y. Nakamma, and A. Yamamoto, *Bull. Chem. Soc. Jpn.* **54**, 1868 (1981).

80. M. B. Hursthoase, R. A. Jones, K. M. Abdul-Malik, and G. Wilkinson, *J. Am. Chem. Soc.* **101**, 4128 (1979).
81. R. B. Calvert and J. R. Shapley, *J. Am. Chem. Soc.* **99**, 5225 (1977).
82. R. B. Calvert and J. R. Shapley, *J. Am. Chem. Soc.* **100**, 7727 (1978).
83. E. O. Fisher, *Angew. Chem.* **69**, 715 (1957).
84. R. G. Caulton, *J. Am. Chem. Soc.* **95**, 4076 (1973).
85. J. H. Wengrovius and R. R. Schrock, *Organometallics* **1**, 148 (1982).

Silyl, Germyl, and Stannyl Derivatives of Azenes, N_nH_n:
Part I. Derivatives of Diazene, N_2H_2

NILS WIBERG

Institut für Anorganische Chemie
University of Munich
Munich, Federal Republic of Germany

I

INTRODUCTION

So far nine acyclic nitrogen hydrides have been reported (*1*): *ammonia* (azane, NH_3), *hydrazine* (diazane, N_2H_4), *triazane* (N_3H_5) (*2*), and *tetrazane* (N_4H_6) (*3*)—representing saturated "azanes" N_nH_{n+2}—*nitrene* (azene, NH) (*4*), *diazene* (diimine, N_2H_2) (*5–8*), *triazene* (N_3H_3) (*3*), and

131

TABLE I

ACYCLIC NITROGEN HYDRIDES ISOLATED OR DETECTED[a]

	Mono- (n = 1)	Di- (n = 2)	Tri- (n = 3)	Tetra- (n = 4)	General formula
-azane	NH_3	**H_2N-NH_2**	$H_2N-NH-NH_2^b$	$H_2N-NH-NH-NH_2$	N_nH_{n+2}
-azene	NH	**$HN=NH$**	$HN=N-NH_2$	**$H_2N-N=N-NH_2$**	N_nH_n
-azadiene	—	—	**$HN=N=N$**	c	N_nH_{n-2}

[a] Isolated nitrogen hydrides are shown in boldface.

[b] Can be isolated as $N_3H_5 \cdot H_2SO_4$.

[c] N_4H_2 produced from N_2H_4 by radiolysis (1) probably has a cyclic structure.

tetrazene (N_4H_4) (9, 10)—representing simple saturated "azenes" N_nH_n—and *hydrazoic acid* (triazadiene, N_3H)—representing doubly unsaturated "azadienes" N_nH_{n-2} (Table I). Of these, only three, compounds NH_3, N_2H_4, and N_3H, are stable at room temperature. The remaining six may be isolated at low temperatures or in matrix, or have been proposed as reaction intermediates (1–10). In general, the hydrides of nitrogen prove to be more thermolabile than the hydrogen derivatives of its lefthand periodic neighbor, carbon.

The silyl derivatives of nitrogen are usually more stable thermally than the corresponding hydrogen derivatives. Because the silyl groups may be replaced by hydrogen with the help of low-temperature protolysis, the silyl nitrogens can be exploited as possible sources of new thermolabile nitrogen hydrides [cf. the isolation of tetrazene, N_4H_4, from $(Me_3Si)_4N_4$ (10)]. The properties of the silyl and hydrogen derivatives are generally similar, so that the silyl compounds serve as "thermostable" model compounds for the study of thermolabile nitrogen hydrides [cf. thermolysis of $(Me_3Si)_2N_2$, Section IV, A].

The purpose of this article is to summarize preparation and properties of silyl, germyl, and stannyl derivatives of some azenes, N_nH_n ($n = 2$–5). For Group IV derivatives of azanes, NH_3 and N_2H_4, and azadiene, N_3H, consult refs. 11–18. So far, there are no published examples of Group IV derivatives of triazane, N_3H_5, or tetrazane, N_4H_6. Silyl nitrenes, R_3SiN, are probably generated as very short-lived intermediates by photolysis or thermolysis of silylazides (19). Evidently they transform into silylimines, $R_2Si{=}NR$, which themselves are highly reactive (20–22).

Part I of this review deals with preparation and properties of Group IV derivatives of diazene. Part II, which appears in a forthcoming volume of "Advances in Organometallic Chemistry," will cover Group IV derivatives of triazene, tetrazene, and pentazene.

II

PREPARATION

Silyl and germyl derivatives of diazene or organyldiazenes $\geqq E{-}N{=}N{-}E\leqq$ or $\geqq E{-}N{=}N{-}R$ (E = Si, Ge; R = organic group) may be prepared by the following methods: oxidation of appropriate lithium hydrazides (Method A), thermolysis of lithiumarenesulfonic acid hydrazides (Method B), and dehydrogenation of hydrazines (Method C) (Scheme 1). Method C is suited only for the synthesis of silyl- or germylorganyldiazenes.

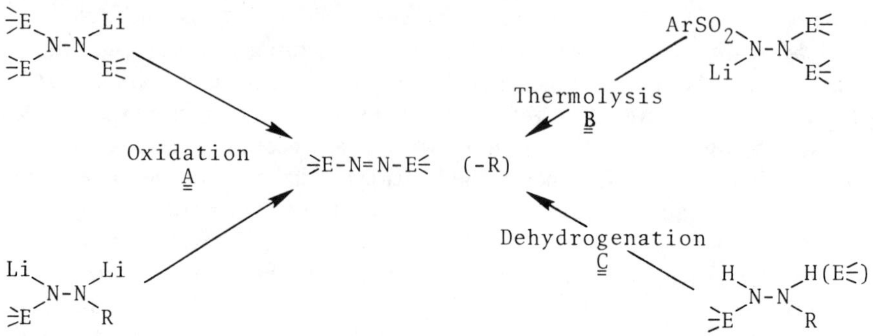

SCHEME 1

In spite of many attempts, stannyl derivatives of diazene or organyldiazenes have not been isolated due to their great instability.

A. Silyl- and Germyldiazenes

Bis(trimethylsilyl)diazene (BSD), $Me_3Si-N=N-SiMe_3$, may be prepared easily by oxidation of lithium tris(trimethylsilyl)hydrazide (**1**) with appropriate oxidizing agents. The BSD so formed does not oxidize further to molecular nitrogen, for example with Br_2, HgO, or $EtOOC-N=N-COOEt$. Nearly quantitative yields of BSD have been obtained by reaction of **1** with sulfonic acid azides, $R'SO_2N_3$ (R' = phenyl or p-tolyl) in ether at $-78°C$ (23) [Eq. (1)]. By analogy with reaction of arenesulfonic acid azides

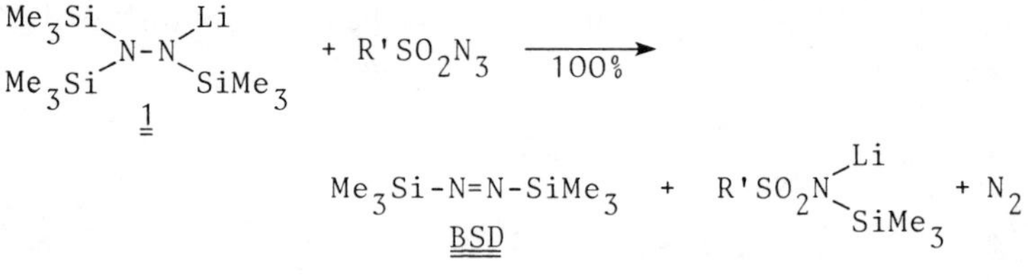

with CH acid methylene derivatives, it appears that the former add to **1** to form a nitrogen chain compound, $(Me_3Si)_2N-N(SiMe_3)-N=N-N(SO_2R')-$(Li), which decomposes into BSD and the triazene $Me_3Si-N=N-N(SO_2R')(Li)$: the latter dissociates and evolves N_2. Investigations with ^{15}N-labeled BSD show that N_2 is eliminated exclusively from the azenesulfonic acid azide (24). Silylated diazenes $(MeO)Me_2Si-N=N-SiMe_2$-(OMe) and $Me_3Si-N=N-SiMe_2-N=N-SiMe_3$ have also been prepared according to Eq. (1) (25).

Oxidation of **1** with arenesulfonic acid chlorides ($R'SO_2Cl$) does not lead exclusively to BSD (*26*) [Eq. (2)]. Evidently the reaction is initiated by

$$
\begin{array}{c}
\underset{Me_3Si}{\overset{Me_3Si}{\diagdown}}N-N\underset{SiMe_3}{\overset{SO_2R'}{\diagup}} \\
\underset{\textbf{2}}{}
\end{array}
$$

ca. 50%

1 + $R'SO_2Cl$ (2)

ca. 50%

$$BSD + R'SO_2Li + Me_3SiCl$$

attack of **1** at the electrophilic centers of the arenesulfonic acid chloride, i.e., the sulfur and chlorine atoms. In the first case, Li/SO_2R' exchange occurs from the hydrazine **2**, whereas in the other case $R'SO_2Li$ and *N*-chloro-tris(trimethylsilyl)hydrazine, $(Me_3Si)_2N-N(Cl)(SiMe_3)$, are formed with the latter immediately decomposing to BSD and Me_3SiCl.

Hexachloroethane will also oxidize **1** to BSD (*27*) [Eq. (3)]. However, the reaction competes with a radical reaction leading to an unwanted side

$$\textbf{1} + Cl_3C-CCl_3 \xrightarrow[\text{ca. 60\%}]{}$$

$$BSD + Me_3SiCl + LiCl + Cl_2C=CCl_2 \quad (3)$$

product, tris(trimethylsilyl)hydrazine. This method is not recommended for the isolation of BSD. Nevertheless, this reaction provides the only known route to bis(trimethylgermyl)diazene (*28*):[1]

$$(Me_3Ge)_3N_2Li + C_2Cl_6 \longrightarrow Me_3Ge-N=N-GeMe_3 + Me_3GeCl + LiCl + C_2Cl_4$$

Finally BSD may also be isolated by thermolysis of lithium bis(trimethylsilyl)arenesulfonic acid hydrazide in a high vacuum at 120°C (*18*) [Eq. (4)]. In an analogous manner, $Me_3Si-N=N-GeMe_3$ is formed by

$$
\begin{array}{c}
\underset{Li}{\overset{R'SO_2}{\diagdown}}N-N\underset{SiMe_3}{\overset{SiMe_3}{\diagup}} \xrightarrow[\text{ca. 90\%}]{} BSD + R'SO_2Li \quad (4)
\end{array}
$$

thermolysis of $R'SO_2(Li)N-N(SiMe_3)(GeMe_3)$ and $Me_3E-N=N-H$ by thermolysis of $R'SO_2(K)N-NH(EMe_3)$ (E = Si, Ge). The thermostable trimethylsilyl- or trimethylgermyldiazenes have been isolated with other products as a mixture upon condensation of the gaseous mixture formed from hydrazide thermolysis on a cold finger at −196°C (*18*). These compounds have been observed only by mass spectrometry.

---

[1] Oxidation of $(Me_3Ge)_3N_2Li$ with $R'SO_2N_3$ or $R'SO_2Cl$ leads to nitrogen via $(Me_3Ge)_2N_2$.

B. Silyl and Germylorganyldiazenes

A good method to prepare silyl or germyl derivatives of organyldiazenes involves dehydrogenation of appropriate silyl- or germylhydrazines (cf. C, Scheme 1). In this way, hydrazines of the type $R_3E—NH—NH—R'$ undergo easy dehydrogenation with diethylazodicarboxylate in organic media (Et_2O, Bu_2O, CH_2Cl_2) to give good yields (29, 30). However, the rate of reaction depends on the nature of the substituents. Reactions of diethyl-azodicarboxylate with hydrazines of the type $R_3E—NH—N(ER_3)(R')$ proceed in an analogous manner but those of $(R_3E)_2N—NHR'$ do not (29) [Eqs. (5) and (6)] (cf. Table II).

$$R_3E\text{-}NH\text{-}NH\text{-}R' \;+\; EtO_2C\text{-}N{=}N\text{-}CO_2Et \;\rightarrow$$

$$R_3E\text{-}N{=}N\text{-}R' \;+\; EtO_2C\text{-}NH\text{-}NH\text{-}CO_2Et \quad (5)$$

$$(R_3E)_2N\text{-}NHR' \;+\; EtO_2C\text{-}N{=}N\text{-}CO_2Et \;\rightarrow$$

$$R_3E\text{-}N{=}N\text{-}R' \;+\; (EtO)(R_3EO)C{=}N\text{-}NH\text{-}CO_2Et \quad (6)$$

The dehydrogenation proceeds by formation of an unstable adduct of hydrazine and diazene having a tetrazane framework [e.g., $EtO_2C—NH—N(CO_2Et)—N(ER_3)—NHR'$] (30a), which decomposes instantaneously to give the end products.

Finally, diazenes of the formula $R_3E—N{=}N—R'$ may also be prepared by dehydrogenation of hydrazines ($R_3E—NH—NH—R'$) with t-butyl-peroxide (31, 32) or chloranil (33) or by the reaction of stannyl derivatives of the above hydrazines with p-benzophenone (16) [Eqs. (7)–(9)] (cf. Table II). Silyl- and germyldiazenes may also be obtained in moderate yields by

$$R_3E\text{-}NH\text{-}NH\text{-}R' \;+\; {}^tBuOO{}^tBu \;\rightarrow\; R_3E\text{-}N{=}N\text{-}R' \;+\; 2{}^tBuOH \quad (7)$$

$$R_3E\text{-}NH\text{-}NH\text{-}R' \;+\; O{=}C_6Cl_4{=}O \;\rightarrow\; R_3E\text{-}N{=}N\text{-}R' \;+\; HO\text{-}C_6Cl_4\text{-}OH \quad (8)$$

$$R_3E(X)N\text{-}N(SnMe_3)R' \;+\; O{=}C_6H_4{=}O \;\rightarrow\; R_3E\text{-}N{=}N\text{-}R' \;+\; HO\text{-}C_6H_4\text{-}OH$$

$$X = H, \; Me_3Sn, \; Me_3Ge \quad (9)$$

oxidation of silylated or germylated dilithium organylhydrazides with a deficiency of bromine in ether at low temperatures [Eq. (10); A, Scheme 1] (32, 34) (cf. Table II). However, bromination is associated with radical side

$$R_3E\text{-}NLi\text{-}NLi\text{-}R' \;+\; Br_2 \;\rightarrow\; R_3E\text{-}N{=}N\text{-}R' \;+\; 2LiBr \quad (10)$$

reactions that lead to significant amounts of amines, $(R_3E)R'NH$, which are inseparable by distillation. As a result the azenes may be isolated at only 30–85% purity.[2]

Finally, method B (Scheme 1) may also be applied for the synthesis of silyl- and germyldiazenes (29). Thus, hydrazine $(Me_3Si)NH{-}NMe(SO_2R')$ in ether at $-78°C$ decomposes on addition of butyllithium [Eq. (11)].

$$(Me_3Si)NH{-}NMe(SO_2R') \xrightarrow[-HBu]{+LiBu}$$

$$(Me_3Si)NLi{-}NMe(SO_2R') \xrightarrow[-R'SO_2Li]{} Me_3Si{-}N{=}N{-}Me \quad (11)$$

C. Stannyldiazenes and -organyldiazenes

All attempts to synthesize stannyl derivatives of diazenes have failed hitherto (17, 29). The oxidation of stannylhydrazines $(Me_3Sn)_2N{-}N(SnMe_3)_2$ and $(Me_3Sn)_2N{-}NPh(SnMe_3)$ with p-benzoquinone and of $(Me_3Sn)_2N{-}NMe(SiMe_3)$ with arenesulfonic acid chloride does not occur as in Eqs. (9) and (2) to form stannyldiazenes $Me_3Sn{-}N{=}N{-}SnMe_3$, $Me_3Sn{-}N{=}N{-}Ph$, and $Me_3Sn{-}N{=}N{-}NMe$, respectively, but instead follows Eqs. (12)–(14). Most probably the diazenes are formed as reaction intermediates that either react further with p-benzoquinone or decompose via $Me_3Sn{-}N{=}N{-}Me \rightarrow Me_3Sn{-}Me + N{\equiv}N$.

$$(Me_3Sn)_2N{-}N(SnMe_3)_2 + 2O{=}C_6H_4{=}O \rightarrow 2Me_3SnO{-}C_6H_4{-}OSnMe_3 + N_2 \quad (12)$$

$$2(Me_3Sn)_2N{-}NPh(SnMe_3) + 3O{=}C_6H_4{=}O \xrightarrow{+2H}$$

$$3Me_3SnO{-}C_6H_4{-}OSnMe_3 + N_2 + 2Ph{-}H \quad (13)$$

$$(Me_3Sn)_2N{-}NMe(SnMe_3) + R'SO_2Cl \rightarrow$$

$$R'SO_2SnMe_3 + Me_3SnCl + SnMe_4 + N_2 \quad (14)$$

Moreover, even the reaction of $Me_3Si{-}N{=}N{-}Ph$ and Me_3SnCl does not lead through

$$Me_3Si{-}N{=}N{-}Ph + Me_3SnCl \rightarrow Me_3SiCl + Me_3Sn{-}N{=}N{-}Ph$$

to trimethylstannylphenyldiazene. Rather, these compounds react as:

$$2Me_3Si{-}N{=}N{-}Ph + Me_3SnCl \rightarrow (Me_3Sn)PhN{-}NPh(SiMe_3) + Me_3SiCl.$$

---

[2] The synthesis of diazenes, $R_3E{-}N{=}N{-}ER_3$, from hydrazides, $R_3E{-}NLi{-}NLi{-}ER_3$, with oxidants (e.g., $O{=}C_6H_4{=}O$, HgO, O_2, or Me_3NO) is not recommended in view of rapid further oxidation to molecular nitrogen.

Therefore, trimethylstannylphenyldiazene may be visualized as a reaction intermediate that would immediately combine further:

$$Me_3Sn—N{=}N—Ph + Me_3Si—N{=}N—Ph \rightarrow (Me_3Sn)PhN—NPh(SnMe_3) + N_2$$

(cf. disproportionation of BSD, Section IV,A).

III

PROPERTIES

A. General Features

Some physical properties of the silyl and germyl derivatives of diazene and organyldiazenes along with their modes of preparation are listed in Table II. Bis(silyl)- and bis(germyl)-substituted diazenes, $R_3E—N{=}N—ER_3$, are thermolabile (slowly decomposing above $-35°C$), hydrolyzable, and extremely air sensitive (inflammable in oxygen) (35). Accordingly, they can be stored only at low temperatures in the absence of air and moisture. Monosubstituted diazenes, $R_3E—N{=}N—H$, are even more reactive, but silyl- and germyl-substituted aryldiazenes, although equally moisture sensitive, are comparatively thermo- and air stable (silylated and germylated alkyldiazenes react with O_2 at room temperature). Silylated and germylated diazenes decompose in UV light in a fashion similar to that of organic azo compounds (for greater details on thermolysis, hydrolysis, etc., cf. Sections IV–VII).

B. Structures

1. Geometric Structure

The parent azo compound, dinitrogen dihydride, N_2H_2, exists in two constitutional isomers, diazene ($H—N{=}N—H$) and isodiazene ($H_2N{=}N$) (36). The former exists in two stereoisomers, cis- and trans-diazene (37).

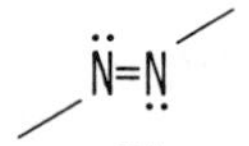

trans-Diazene 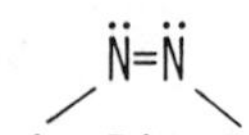   cis-Diazene 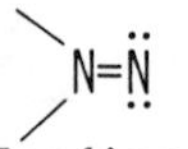   Isodiazene

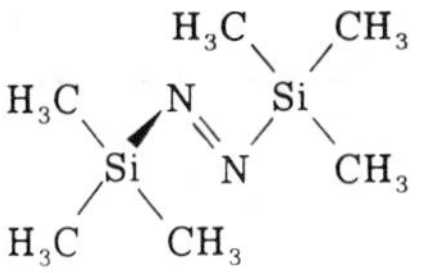

FIG. 1.   Structure of bis(trimethylsilyl)diazene (BSD).

trans-Diazene represents the thermodynamically stable form. Just like bis(amino)- and bis(hydroxo)diazenes, $R_2N-N=N-NR_2$ and $RO-N=N-OR$, bis(organyl)diazenes (*38*) also exist in the trans configuration in ground state. However, the cis form is the preferred configuration in the case of difluorodiazene, $F-N=N-F$ (*39*).

X-Ray analysis of bis(silyl)diazene by Veith and Bärnighausen (*40*) shows that BSD is trans with a planar SiNNSi framework (Fig. 1). The SiNN angle of 120° is comparable to the SiNN angle in Ph_3SiN_3 (*40*). This value suggests sp^2 hybridization of the azo nitrogen atoms, but it is considerably broadened in comparison with the CNN angle in organic azo compounds [$\sim$ 106–113° (*40*)]. The most striking difference between the organic azo compounds and BSD is the latter's extremely short N–N bond length of only 1.17 Å, which lies almost halfway between the values for bis(organyl)diazenes [1.23–1.27 Å (*40*)], and molecular nitrogen (1.10 Å), and corresponds to N–N distances in ionic azides. On the other hand, the Si–N bond length of 1.81 Å in BSD is the longest of all known Si–N bond lengths [the shortest Si–N bond length of 1.64 Å is displayed in $(Me_3Si)_2N^-$ (*40*)]. This is contrary to any real $d\pi p\pi$ bonding relationship between silyl and azo groups.

X-Ray structural determinations of other silyl- and germyldiazenes and organyldiazenes have not yet been reported. It is, however, quite likely that they are also trans and that the bond relationships of $R_3E-N=N$ species correspond to those of BSD.

2. Electronic Structure

A typical energy level diagram of inner π and n molecular orbitals of the bent azo group, N—N, in the ground state is shown in Fig. 2,a (*41*). Here, a π molecular orbital containing two electrons, as well as a vacant π^* orbital, result from the interaction of p_z nitrogen atomic orbitals not involved in the sp^2 hybridization of the N atoms. On the other hand, the n_- and n_+ molecular orbitals, each filled with two electrons, arise from interaction of sp^2 nitrogen

TABLE II. Silyl and Germyl

R^1—N=N—R^2		Best preparation		Mp (°C)	Bp (°C/Torr)	Color	Dec temp. (°C/time[a])
R^1	R^2	Method	cf. Eq.				
Me_3Si	Me_3Si	A	(1)[d]	−3 dec	—	Light blue	−35/d
Me_3Si	Me_3Ge	B	(4)	—	—	Blue	−35/d
Me_3Ge	Me_3Ge	A	(3)	—	−45/HV[e]	Blue	−35/d
Me_2Si	Me_3Si[f]	A	(1)[g]	—	—	Blue	−35/d
Me_3Si	H	B	(4)[g]	—	—	Red	−155/h
Me_3Ge	H	B	(4)[g]	—	—	Red	−155/h
Me_3Si	Me	C	(5), (6)[h]	−49	71/720	Red	160/h
Me_3Ge	Me	C	(6)	−38	25/20	Orange-red	160/h
Me_3Si	ʹBu	C	(6)[h]	−30	25/5	Red	190/h
Me_3Ge	ʹBu	C	(6)[h]	−28	25/1	Orange-red	90/h
Me_3Si	CH_2Ph[i]	C	(6)	−14	30/0.01	Red	80/m
Me_3Si	Ph	C	(6)[h, k]	−2	25/0.01	Deep blue	200/d
Me_3Ge	Ph	C	(9)[k]	—	71/3	Deep blue	—
Et_3Si	Ph	C	(7)	—	162–165/11	Deep blue	—
$PhMe_2Si$	Ph	C	(7)	—	125–130/10	Deep blue	—
Ph_3Si	Ph[m]	C	(8)[m]	112–113.5	—	Deep blue	110/h
$(MeO)_3Si$	ʹBu	C	(6)	−24	25/0.1	Cherry	190/h

[a] Decomposition in days (d), hours (h), or minutes (m).

[b] Absorptions characteristic of color of the compounds only. Shorter wavelength transitions have been shown to be due to Rydberg excitations [for example, BSD: $\tilde{v}/\varepsilon = 40000/260$; $52100/1660$; $57000/2680$; Ref. (41)].

[c] Adiabatic ionization energies (IE), unless otherwise quoted.

[d] Other methods of preparation: Eq. (2), (26), Eq. (3), (27), Eq. (4), (28). In an impure state, the azo compound $(MeO)Me_2Si$—N=N—$SiMe_2(OMe)$ is produced in a way analogous to Eq. (1) (25), and $PhMe_2Si$—N=N—$SiMe_2Ph$ by oxidation of $(PhMe_2Si)_2N_2Li_2$ with $R'SO_2Cl$ (26).

[e] Sublimation point.

[f] Me_3Si—N=N—$SiMe_2$—N=N—$SiMe_2$.

[g] Isolated in impure state.

hybrid orbitals occupied by free electrons. The energy difference between the two π levels is generally much more than 6 eV. For BSD, this difference has been calculated to be ~ 10 eV (41). However, no realistic estimates can be made on the n_-/n_+ energy difference, because the n_- molecular state is strongly mixed with other molecular orbitals (41).

The energetic position of the n_+ molecular orbitals, as the highest electron-filled molecular states, follows from the first ionization potentials of the azo compounds concerned, which for a series of silyl- and germyldiazenes and -organyldiazenes are reproduced in Table II. Starting from the

DERIVATIVES OF DIAZENE (N_2H_2)

| ¹H Chemical shifts (iTMS, ppm) | | | UV Absorption[b] | | | T IE[c] | |
δ (R^1)	δ (R^2)	Solvent	$\tilde{\nu}_{max}$(cm^{-1})	ε	Solvent	(eV)	Reference
0.258	0.258	C_6H_6	12750	5	C_5H_{12}	6.48	(23)
0.187	0.337	C_5H_{12}	13620	—	C_5H_{12}	—	(18)
0.363	0.363	C_6H_6	14310	—	C_5H_{12}	5.95	(28)
0.212	0.198	C_5H_{12}	13000	40	C_5H_{12}	—	(25a)
—	—	—	—	—	—	—	(18)
—	—	—	—	—	—	—	(18)
0.235	4.10	CH_2Cl_2	20350	6	CH_2Cl_2	7.49	(29)
0.375	3.97	CH_2Cl_2	21010	8	CH_2Cl_2	7.26	(29)
0.237	1.13	CH_2Cl_2	20000	9	CH_2Cl_2	6.91	(29)
0.380	1.11	CH_2Cl_2	20850	12	CH_2Cl_2	~6.7	(29)
0.250	5.07[j]	CH_2Cl_2	20280	8	CH_2Cl_2	7.44	(29)
0.350	Multiplet	CH_2Cl_2	17400	26	CH_2Cl_2	7.05	(29)
0.542	Multiplet	CH_2Cl_2	17800	26	C_6H_{14}	6.73	(16)
Multiplet	Multiplet	CCl_4	16890	25	C_6H_{12}	—	(31)
0.59[l]	Multiplet	CCl_4	17210	39	C_6H_{12}	—	(31)
Multiplet	Multiplet	CCl_4	17010	70	C_6H_{12}	—	(33)
3.63	1.22	CH_2Cl_2	19720	7	CH_2Cl_2	7.88[n]	(29)

[h] Also see ref. (30). Other methods of preparation: Eq. (9), (16), Eq. (11), (29).

[i] Owing to its high instability, the azo compound $Me_3Si—N{=}N—CPh_3$ [Eq. (6)] could not be isolated (cf. Section III,C).

[j] δ (CH_2).

[k] The azo compounds $R_3Si—N{=}N—CPh$ (R = Me, Et, Pr), $Me_3Si—N{=}N—p\text{-Tol}$, and $R_3Ge—N{=}N—Ph$ (R = Me, Et, Ph) are produced in an impure state by Eq. (10) (32, 34).

[l] δ (CH_3).

[m] $Ph_3Si—N{=}N—C_6H_4Cl$ (mp 95.5–57°C) and $Ph_3Si—N{=}N—C_6H_4CH_3$ (mp 87.5–89°C) are prepared in the same way.

[n] Vertical IE.

bis(organyl)diazenes, ionization energy decreases stepwise by almost equal amounts on replacing organyl by silyl or germyl groups (cf., e.g., Figs. 2c–e). Apparently the inductive effect of the substituents essentially determines the position of n_+ energy levels.

Thus, the silyl and germyl groups, by their electron-donating properties, destabilize the nitrogen lone pair electrons of the azo group. Accordingly, the ionization energy also increases on moving from t-butyl-substituted to methyl-substituted azo derivatives (Table II) because the methyl group is more electronegative than the t-butyl group. However, inductive effects do

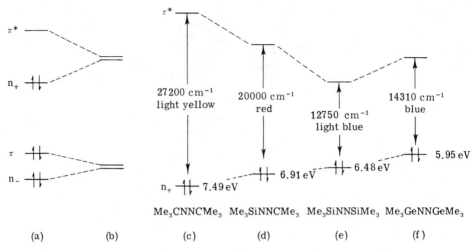

FIG. 2. Energy level scheme of inner molecular orbitals of angular (a) and linear (b) azo groups. The n_+ ionization energies and wavelengths of UV bands responsible for color are also shown (c–f).

not exclusively determine the position of the n_+ energy level because, as the facts show, the Me_3Ge-substituted azo compounds, in spite of the higher electronegativity of the Me_3Ge group, display lower ionization energies than Me_3Si-substituted azo compounds (Fig. 2e and f; $Me_3Ge—N=N—GeMe_3$ exhibits the lowest ionization energy so far discovered for simple azo compounds). It is probable that among other factors, the E–N–N angle plays a role because its widening would cause n_+ and π energy levels to come closer (Fig. 2b) thereby shifting the n_+ level to lower energies.

C. Color

The synthesis of BSD in 1968 as the first fully silyl-substituted diazene (23), with unexpectedly long wavelength absorption and a brilliant light blue color, stimulated a sudden interest in this azo compound. The color-yielding weak intensity absorption occurs at $12750 \, cm^{-1}$, which is already in the invisible infrared spectral region. One "sees" only the short wavelength tail of the absorption band, which by itself represents the longest wavelength band so far observed in any simple azo compound. Correspondingly conspicuous colors are displayed by silyl- and germyldiazenes and -organyldiazenes (E = Si, Ge; Table II):

$$R_3E\text{—}N{=}N\text{—}ER_3 \qquad R_3E\text{—}N{=}N\text{—}Aryl \qquad R_3E\text{—}N{=}N\text{—}Alkyl$$

light blue to blue   dark blue   red

The absorption responsible for the color of BSD and other azo compounds is, in general, attributed to a forbidden $n_+ \to \pi^*$ transition (41, 42). It has been shown experimentally that the bathochromic shift of about 2×7200 cm^{-1}, which occurs on replacement of t-butyl groups in ^{t}Bu—N=N—Bu^t by trimethylsilyl groups, is brought about, as already stated, by the strong raising of n_+ and the weak lowering of π^* energy levels (Fig. 2c–e). Similar changes are observed by substitution of other azo-bound organyl groups with silyl or germyl groups, whereby, in the latter case, the lowering of the π^* energy level is considerably smaller (Fig. 2c and f). The experimental results are explained in a simple manner by inductive and mesomeric effects. The former influence the n_+ as well as π^* orbitals (raising energy levels) and the latter essentially affects only the π^* orbitals (lowering energy levels). The ESR spectrum of the BSD radical anion indicates a real, though small, delocalization of electrons into the π^* molecular orbitals (43).

IV

THERMOLYSIS

A. Silyl- and Germyldiazenes

1. Route of Thermolysis

Azoalkanes (characteristic atomic grouping $\geq$C—N=N—C$\leq$) decompose mainly into hydrocarbons and nitrogen (38):

$$R\text{—}N{=}N\text{—}R \to R\text{—}R + N{\equiv}N$$

However, a corresponding *dissociation* [Eq. (15)] of azosilane BSD

$$R_3E\text{-}N{=}N\text{-}ER_3 \xrightarrow{\text{Dissociation}} R_3E\text{-}ER_3 + N{\equiv}N \qquad (15)$$

(characteristic atomic grouping $\geqslant$Si—N=N—Si$\leqslant$) into hexamethyldisilane and nitrogen (E = Si) does not take place. Instead BSD undergoes thermolysis with *disproportionation* [Eq. (16)] into tetrakis(trimethylsilyl)hydrazine (**3**) and nitrogen (*44*). The BSD decomposition thus resembles

$$2Me_3E\text{-}N{=}N\text{-}EMe_3 \xrightarrow{\text{Disproportionation}} \underset{\underset{=}{3}}{\begin{matrix} Me_3E & & EMe_3 \\ & \diagdown N\text{-}N \diagup & \\ Me_3E & & EMe_3 \end{matrix}} + \ N{\equiv}N \quad (16)$$

not that of azoalkanes but that of the parent azo compound, H—N=N—H, which disproportionates above −180°C into hydrazine and nitrogen: $2N_2H_2 \rightarrow N_2H_4 + N_2$ (6).

The similarity in chemical properties of diazene and BSD is not limited to thermolysis. For example, both react with oxygen at low temperatures to form peroxides (BSD or $N_2H_2 + O_2 \rightarrow Me_3Si$—O—O—$SiMe_3$ or H—O—O—H; cf. Section VI,B), whereas azoalkanes do not react with oxygen apart from poorly characterized combustion reactions at higher temperatures. However, the uncatalyzed hydrogenation of the multiple bonds, generally realizable in case of diazene by the transfer reactions from azo-bound ligands to the double bond system, occurs in case of BSD only in very reactive systems (e.g., ROOC—N=N—COOR), but fails with "normal" multiple bond hydrocarbons such as ethylene:

$$BSD + CH_2{=}CH_2 \ \xrightarrow{\quad\times\quad} \ N_2 + Me_3Si\text{—}CH_2\text{—}CH_2\text{—}SiMe_3$$

One such transfer reaction on phenylacetylene has been observed in the case of monotrimethylsilyldiazene (*45*):

$$Me_3Si\text{—}N{=}N\text{—}H + PhC{\equiv}CPh \rightarrow N_2 + (Me_3Si)PhC{=}CPh(H)$$

In addition to disproportionation, BSD thermolysis at low temperatures proceeds with BSD *dimerization* [Eq. (17)] to form tetrakis(trimethylsilyl)tetrazene (**4**) in low yields (E = Si), (*44*). The proportion of BSD

$$2Me_3E\text{-}N{=}N\text{-}EMe_3 \xrightarrow{\text{Dimerization}} \underset{\underset{=}{4}}{\begin{matrix} Me_3E & & EMe_3 \\ & \diagdown N\text{-}N{=}N\text{-}N \diagup & \\ Me_3E & & EMe_3 \end{matrix}} \quad (17)$$

dimerization increases with temperature at the expense of BSD disproportionation and amounts to 25% at 150°C. Moreover, with increasing temperature BSD decomposes to an increasing extent with *cleavage* [Eq.

(18)] into tris(trimethylsilyl)amine (**5**) and N_2 as well as with *hydrogen abstraction* [Eqs. (19) and (20)] from Me_3Si groups or solvent molecules to give tris(trimethylsilyl)hydrazine (**6**) and bis(trimethylsilyl)amine (**7**) (E = Si). The molecules, damaged through H-abstraction, react further to

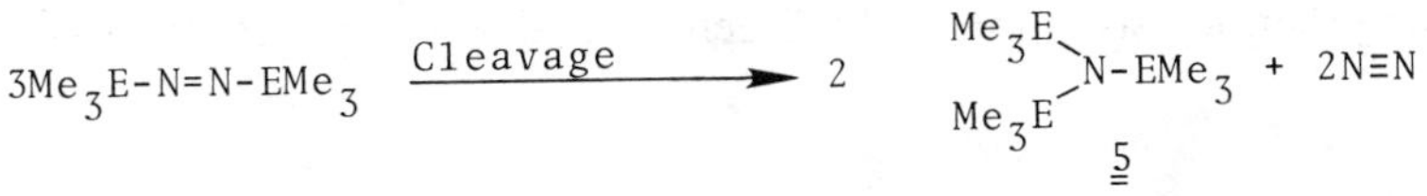

$$3Me_3E-N=N-EMe_3 \xrightarrow{\text{Cleavage}} 2 \quad \begin{matrix} Me_3E \\ \diagdown \\ Me_3E \end{matrix} N-EMe_3 + 2N\equiv N \quad (18)$$

$$\underline{5}$$

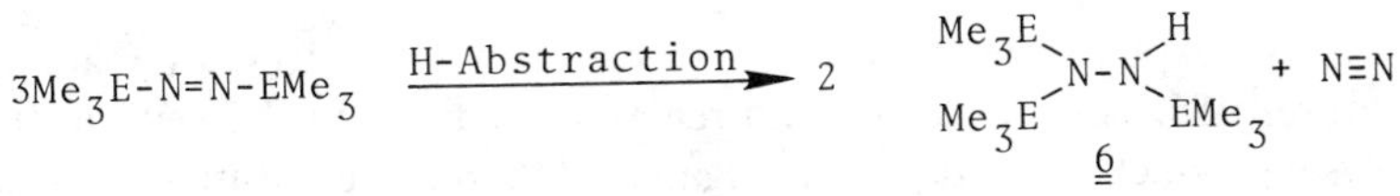

$$3Me_3E-N=N-EMe_3 \xrightarrow{\text{H-Abstraction}} 2 \quad \begin{matrix} Me_3E \\ \diagdown \\ Me_3E \end{matrix} N-N \begin{matrix} H \\ \diagup \\ EMe_3 \end{matrix} + N\equiv N \quad (19)$$

$$\underline{6}$$

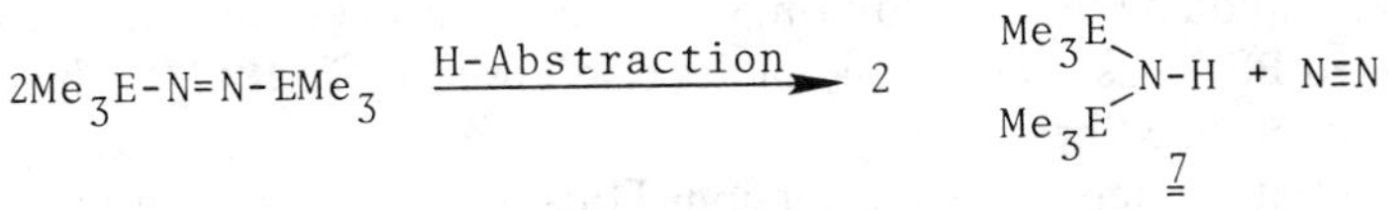

$$2Me_3E-N=N-EMe_3 \xrightarrow{\text{H-Abstraction}} 2 \quad \begin{matrix} Me_3E \\ \diagdown \\ Me_3E \end{matrix} N-H + N\equiv N \quad (20)$$

$$\underline{7}$$

radical secondary products (see below). In BSD thermolysis at 150°C in benzene, the total yield of **3–7** amounts to ~75% (*44, 46*).

Like BSD, bis(trimethylgermyl)diazene, $Me_3Ge-N=N-GeMe_3$, decomposes at low temperatures almost completely by disproportionation [Eq. (16)], and at higher temperatures gives dimerization, cleavage, and H-abstraction [Eqs. (17)–(20), E = Ge] (*28*). In addition, unlike BSD, it dissociates according to Eq. (15) (E = Ge). The percentage yields of the products of thermolysis of bis(trimethylgermyl)diazene in pentane at 80°C are (to compare, the values for BSD are given in parentheses) **2**, 10 (0); **3**, 50 (20); **4**, 20 (20); **5**, 10 (25); **6**, 10 (25); and **7**, < 1 (10). The mixed substituted diazene $Me_3Si-N=N-GeMe_3$ decomposes in a similar fashion, but the product of disproportionation consists mainly of the asymmetric hydrazine derivative $(Me_3Si)_2N-N(GeMe_3)_3$ (*28*).

Like BSD and bis(trimethylgermyl)diazene, the silylated diazenes (MeO)-$Me_2Si-N=N-SiMe_2(OMe)$ and $Me_2Si-N=N-SiMe_2-N=N-SiMe_3$ decompose. The latter rearranges to give a silylated *cis*-tetrazene:

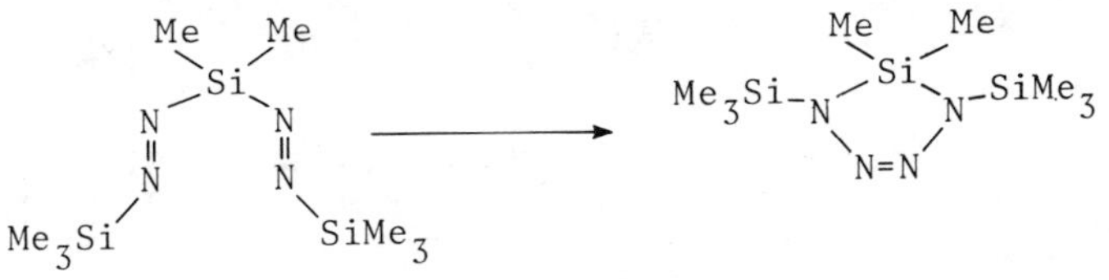

The higher thermolability of $(MeO)Me_2Si—N=N—SiMe_2(OMe)$ (thermolysis near $-50°C$) is interesting as it indicates that electronegative groups on silicon decrease the stability of azo compounds $R_3Si—N=N—SiR_3$. Compounds such as $Cl_3Si—N=N—SiCl_3$ and $(MeO)_3Si—N=N—Si(OMe)_3$ are expected to be very thermolabile. For more information regarding thermolysis of monosilyldiazene $Me_3Si—N=N—H$, see Section V,A.

2. Mechanism of Thermolysis

In order to explain the thermolysis mechanism, the effects of reaction temperature, BSD concentration, and reaction medium on the yield of the thermolysis products have been investigated (44), and the nature of radical side products and the dependence of their yield on the reaction conditions have been studied (46). In addition, the decomposition of undeuterated and deuterated BSD has been carried out and followed by ESR (47). The results are briefly outlined here.

The initiating step in the conversions [Eqs. (16)–(20)], which in contrast to azoalkanes follows not a first order but a higher order, consists of a radical-forming reaction [Eq. (21)]. Thereafter, two molecules of BSD first form

$$2\,BSD \rightarrow [(Me_3Si)_2N-\overset{.}{N}(SiMe_3) + N\equiv N + \cdot SiMe_3]_{cage} \rightarrow \underline{3} + N\equiv N$$

$$\underset{\underline{6a}}{} \overset{-N_2}{} \tag{21}$$

$$\underline{6a} \longleftarrow \longrightarrow \cdot SiMe_3$$

nitrogen and the "radical pair" $\mathbf{6a}/Me_3Si\cdot$ in a solvent cage. The cage radicals either undergo cage recombination to form tetrasilylhydrazine (3) or transform by diffusion from the cage into "kinetically free" radicals. The latter reaction gains increasing importance with increasing temperature.

The free *silyl radicals* formed according to Eq. (21) immediately take up the unreacted BSD molecules to form **6a** [Eq. (22)] so that their stationary concentration during thermolysis remains very low. These have been chemically trapped and identified with the help of cyclohexene [Eq. (23)] and with benzal chloride, in which they abstract a Cl radical [Eq. (24)].

$$Me_3Si\cdot + BSD \rightarrow \underline{6a} \; (\rightarrow \text{succeeding reactions}) \tag{22}$$

$$Me_3Si\cdot + \bigcirc \longrightarrow Me_3Si\diagdown\overset{.}{\bigcirc} \overset{+H}{(\longrightarrow} Me_3Si\diagdown\bigcirc \;) \tag{23}$$

$$Me_3Si\cdot + Cl_2CHPh \rightarrow Me_3SiCl + Cl\dot{C}HPh \; (\xrightarrow[-Me_3SiCl]{+BSD} \; (Me_3Si)_2N\text{-}N\text{=}CHPh) \quad (24)$$

The trimethylgermyl free radicals formed by thermolysis of $Me_3Ge-N=$ $N-GeMe_3$, analogously to Eq. (21), not only add to the diazene but also abstract germyl groups:

$$Me_3Ge-N=N-GeMe_3 + GeMe_3 \rightarrow Me_3Ge + N\equiv N + Me_3Ge-GeMe_3$$

A bit longer lived and thus ESR-spectroscopically identifiable tris(trimethylsilyl)hydrazyl radicals (6a) are able to combine with BSD under Me_3Si abstraction to form tetrasilylhydrazine, 3 [Eq. (25)]. Over and above

$$\underline{6a} + BSD \rightarrow \underline{3} + N\equiv N + \cdot SiMe_3 \quad (25)$$

that, these radicals react (mainly) with the chemical surroundings by H-abstraction to form trisilylhydrazine (6). Accordingly, in thermolysis media benzene, toluene, or acetonitrile, the free radicals phenyl [Eq. (26)], benzyl [Eq. (27)], or CH_2CN [Eq. (28)] are formed. [The hydrazyl radicals formed

$$\underline{6a} + C_6H_6 \rightarrow \underline{6} + \cdot C_6H_5 \quad (\rightarrow u.a. \; C_6H_5\text{-}C_6H_5) \; . \quad (26)$$

$$\underline{6a} + CH_3C_6H_5 \rightarrow \underline{6} + \cdot CH_2C_6H_5 \; (\xrightarrow{+BSD} \; (Me_3Si)_2\dot{N}_2CH_2C_6H_5) \quad (27)$$

$$\underline{6a} + CH_3CN \rightarrow \underline{6} + \cdot CH_2CN \; (\xrightarrow{+BSD} \; (Me_3Si)_2\dot{N}_2CH_2CN) \quad (28)$$

according to Eqs. (27) and (28) become stabilized by Me_3Si- or H-abstraction. In addition, these radicals function as H donors, thereby transforming themselves to hydrazones $(Me_3Si)_2N-N=CHC_6H_5$ and $(Me_3Si)_2-N-N=CHCN$.] Hydrogen bound to silicon in Me_3SiH is also easily abstracted by 6a. The silyl radical formed thereby:

$$6a + Me_3SiH \rightarrow 6 + Me_3Si$$

combines with BSD to give the radical 6a [Eq. (22)] that in turn attacks Me_3SiH and so on. Thus, the thermolysis of BSD in the presence of trimethylsilane runs like an addition reaction [Eq. (29)].

$$BSD + Me_3SiH \rightarrow \underline{6} \quad (29)$$

Because of very fast termination of the radical chain reaction, 6 is formed almost exclusively. The simultaneous addition of toluene to BSD to form

$(Me_3Si)_2HN_2CH_2C_6H_5$ is substantially slower, so that disilylbenzyl-hydrazine is formed in small amounts only.

In addition the hydrazyl radical **6a** reacts with BSD to produce trisilyl-amine (**5**), nitrogen, and the bis(trimethylsilyl)amine radical **7a** [Eq. (30)].

$$\underline{6a} \; + \; BSD \; \rightarrow \; \underline{5} \; + \; N_2 \; + \; (Me_3Si)_2N\cdot \hspace{2cm} (30)$$

$$\underline{7a}$$

Compound **7a** in turn satisfies itself with Me_3Si abstraction from BSD (formation of **5**) or H-abstraction (formation of **7**). In addition, it can add to BSD to give tetrasilyltriazenyl radicals, $(Me_3Si)_2\dot{N}—\dot{N}—\dot{N}(SiMe_3)_2$, which are formed in low concentration and are identifiable by ESR spectroscopy.

Independent of the radical thermolysis reactions of BSD, which are summarized once again in Scheme 2, BSD dimerization to tetrasilyltetrazene

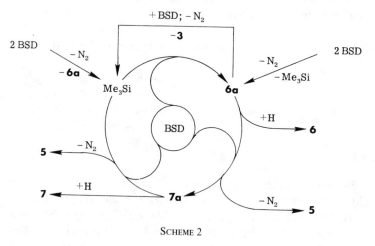

SCHEME 2

(**4**) [Eq. (17)] takes place through a nonradical process involving a double insertion reaction [Eq. (31)]. This process is favored because (i) the formation

$$
\begin{array}{c}
Me_3Si-N=N-SiMe_3 \\
\left(\begin{array}{c} (+) \end{array}\right) \\
Me_3Si-N=N-SiMe_3
\end{array}
\longrightarrow
\begin{array}{c}
Me_3Si-N-N \quad SiMe_3 \\
\;\;|\;\;\;||\;\;| \\
Me_3Si \quad N-N-SiMe_3
\end{array}
\hspace{1cm} (31)
$$

$$\underline{\underline{4}}$$

of **4**, unlike that of **3**, is not retained or suppressed by H-atom donors like toluene; (ii) the yield of **4**, unlike that of **3**, increases with increase in temperature; and (iii) the formation of **4** is catalyzed by Lewis acids like SiF_4 (cf. Section VI,C). In addition, the finding that BSD on irradiation with 250-nm UV light, gives tetra- and trisilylhydrazine (**3** and **6**) and tri- and disilylamine (**5** and **7**) but no tetrasilyltetrazene (**4**), suggests the nonradical formation of **4** (48).[3]

B. *Silyl- and Germylorganyldiazenes*

Silyl- and germylorganyldiazenes are intermediate between azoalkanes and azosilanes and show thermolytic behavior resembling both classes of compounds. Thus, they decompose partly by dissociation and partly by disproportionation. They also show new types of decomposition pathways.

1. *Methyldiazenes*

Trimethylgermylmethyldiazene, $Me_3Ge-N{=}N-Me$ (**8**), decomposes in benzene with evolution of N_2 by a first-order reaction ($t_{1/2}$ at 159°C = 48 minutes) to form tetramethylgermane ($\sim 22\%$), a mixture of isomers of bis(trimethylgermyl)methylhydrazines (**9a** and **9b**, $\sim 32\%$), and other products (Scheme 3) (29). Obviously, the thermolysis of **8** begins, like that

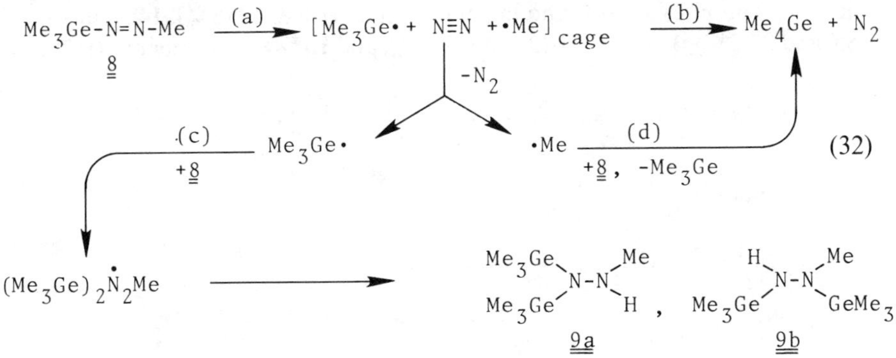

$$Me_3Ge{-}N{=}N{-}Me \xrightarrow{(a)} [Me_3Ge\bullet + N{\equiv}N + \bullet Me]_{cage} \xrightarrow{(b)} Me_4Ge + N_2 \tag{32}$$

SCHEME 3

of azoalkanes, i.e., through a radical reaction [Eq. (32a)]. The radicals Me_3Ge and Me recombine in the solvent cage to form $GeMe_4$ [Eq. (32b)][4] or react after diffusion out of the solvent cage with unreacted **8** [Eq. (32c,d).

Trimethylsilylmethyldiazene, $Me_3Si-N{=}N-Me$ (**10**), decomposes in an analogous manner [Eq. (32)] but the reaction is a bit slower. Thus, nitrogen-containing thermolysis products catalyze rearrangement to hydrazine **11**. Accordingly, the reaction rate increases with time (29). At 160°C the thermolysis ends in a few hours. In all, 95% of **10** decomposes according to Eq. (33) and 5% in other ways (N_2 formation). In the presence of strong

---

[3] The initial step in photolysis of BSD is the reaction $BSD + hv \rightarrow Me_3Si\cdot + N{\equiv}N + Me_3Si\cdot$, cf. Table II, note *b*.

[4] Trimethylstannylmethyldiazene, $Me_3Sn-N{=}N-Me$, decomposes even at low temperatures exclusively to Me_4Sn and N_2 (cf. Section II,C).

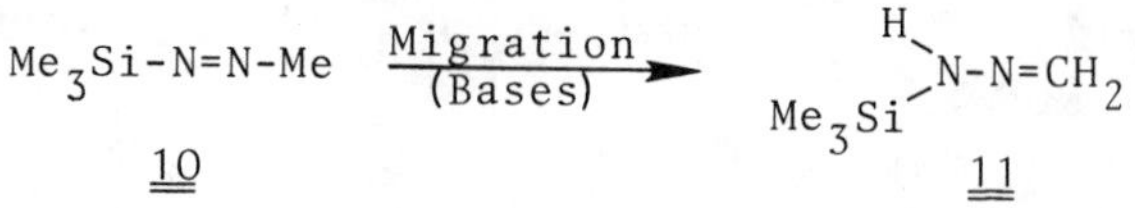

$$Me_3Si-N=N-Me \quad \xrightarrow[\text{(Bases)}]{\text{Migration}} \quad \underset{Me_3Si}{\overset{H}{\diagdown}}N-N=CH_2 \qquad (33)$$

$$\underline{10} \qquad\qquad\qquad\qquad\qquad \underline{11}$$

base such as $(Me_3Si)_2NNa$, the rearrangement occurs promptly and quantitatively at room temperature. The deprotonated product:

$$[Me_3Si-N=N-CH_2 \leftrightarrow Me_3Si-N-N=CH_2]^-$$

formed by the action of a catalytic amount of base on **10**, is the intermediate in the rearrangement, which together with unreacted **10** leads to the formation of **11** and a new deprotonated product, which in turn reacts again with **10**, etc.:

$$Me_3Si-N-N=CH_2 + 10 \rightarrow 11 + Me_3Si-N=N-CH_2$$

The diazene $Me_3Si-N=N-CH_2Ph$ (**12**) undergoes thermolysis at a significantly lower temperature than does **10** (Table II). [The triphenyl substituted diazene $Me_3Si-N=N-CPh_3$ has not yet been isolated because of its thermolability (*29*).] The thermolysis, which is of higher order, is complete in 20 minutes at 90°C and in pentane gives rise to **13** (8 %), **14a** and **14b** (8.6 and 35.2 %), **15a** and **15b** (2.4 and 3.9 %), and **16** (20.0 %) (Scheme 4) (*29*).

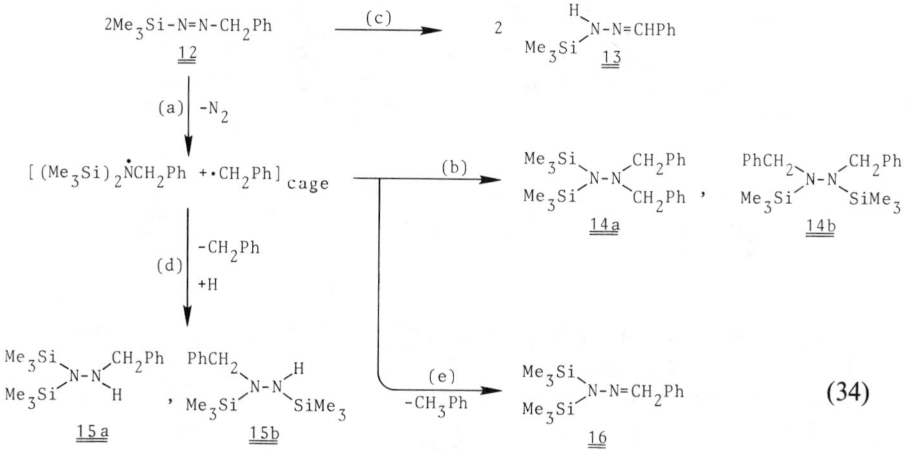

SCHEME 4

Obviously, **12** decomposes substantially analogously to BSD (Section IV,A) with the formation of $(Me_3Si)_2N_2CH_2Ph$ and CH_2Ph radicals [Eq. (34a)], which stabilize through recombination [Eq. (34b)], by H-abstraction [Eq. (34d)], or by H-transfer [Eq. (34e)]. In addition, **12** rearranges in a minor amount to **13** [Eq. (34c)]. Because benzyl radicals are more stable than methyl radicals, the rate-determining step of radical thermolysis of **12** is considerably faster than the corresponding step in the thermolysis of diazene **10**, in which the radical thermolysis, beside rearrangement, does not completely take place.

2. t-Butyldiazenes

Trimethylgermyl-*t*-butyldiazene, $Me_3Ge-N=N-{}^tBu$ (**17**), decomposes in *o*-dichlorobenzene with the evolution of N_2 according to a first-order reaction ($t_{1/2}$ at 96.5°C = 12.2 minutes), giving quantitative yields of iso-butene and trimethylgermane [Eq. (35)] (*29*). A slow hydrogermylation of isobutene [Eq. (36)] competes with the thermolysis. Warming for longer periods leads to almost quantitative formation of $Me_3Ge-CH_2-CHMe_2$.

$$Me_3Ge-N=N-{}^tBu \xrightarrow{\text{Elimination}} Me_3GeH + N\equiv N + CH_2=CMe_2 \quad (35)$$
$$\underline{17}$$

$$Me_3GeH + CH_2CMe_2 \longrightarrow Me_3Ge-CH_2-CHMe_2 \quad (36)$$

The elimination [Eq. (35)] probably occurs as a synchronous reaction involving transfer of an H atom from the ${}^tBu$ group to the germanium atom by simultaneous cleavage of Ge–N and N–C bonds. A radical mechanism is ruled out because of the uniform thermolysis course.

The formation of isobutene in 16% yield as a product of thermolysis of trimethylsilyl-*t*-butyldiazene, $Me_3Si-N=N-{}^tBu$ (**18**), indicates a decomposition mechanism analogous to the elimination [Eq. (35)]. Instead of Me_3SiH, however, $(Me_3Si)_2N-NH^tBu$ (**19**) is formed. Most probably Me_3Si is formed first and adds completely to **18** to form **19**. Since beside **19** the hydrazines **20** and **21** as well as many other products (e.g., Me_3Si^tBu) are formed (*29*), the thermolysis course in Scheme 5 is more likely (cf. Scheme 4).

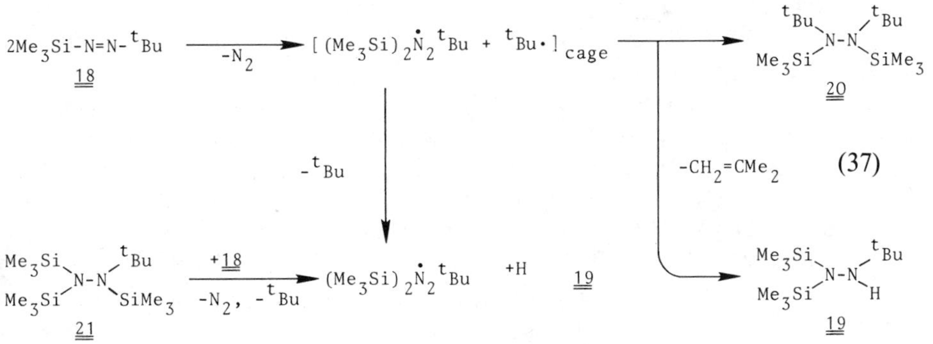

$$2Me_3Si-N=N-{}^tBu \xrightarrow{-N_2} [(Me_3Si)_2\overset{\cdot}{N}_2{}^tBu + {}^tBu\cdot]_{cage} \longrightarrow$$
$$\underline{18}$$

$$-{}^tBu \downarrow$$

$$Me_3Si\diagdown\!\!\diagup{}^tBu \xrightarrow[-N_2,\ -{}^tBu]{+\underline{18}} (Me_3Si)_2\overset{\cdot}{N}_2{}^tBu \xrightarrow{+H} \underline{19}$$
$$Me_3Si^{N-N}\diagdown SiMe_3$$
$$\underline{21}$$

$$-CH_2=CMe_2 \quad (37)$$

$$\overset{{}^tBu\diagdown\!\!\diagup{}^tBu}{Me_3Si^{N-N}\diagdown SiMe_3}$$
$$\underline{20}$$

$$\overset{Me_3Si\diagdown\!\!\diagup{}^tBu}{Me_3Si^{N-N}\diagdown H}$$
$$\underline{19}$$

SCHEME 5

3. *Phenyldiazenes*

Trimethylsilylphenyldiazene, $Me_3Si-N=N-Ph$, is by far the most stable of the silylated and germylated organyldiazenes hitherto investigated. Even at 200°C, only about 60% of the compound decomposes in 2 days to form a large number of products, many of which are not yet well characterized (*29*). In contrast, trimethoxysilylphenyldiazene, $(MeO)_3Si-N=N-Ph$, unexpectedly decomposes rapidly near 0°C, also producing a large number of poorly characterized products (*29*). Triphenylsilylphenyldiazene, $Ph_3Si-N=N-Ph$, is converted slowly to Ph_3SiCl and $PhCl$ at 110°C in carbon tetrachloride (*33*).

V

HYDROLYSIS

A. *Silyl- and Germyldiazenes*

Of the hydrolyses of silylated and germylated diazenes, only the hydrolysis of BSD has been investigated in detail (*48–50*). The hydrolysis of other azosilanes and azogermanes is expected to be analogous to that of BSD.

The protolysis of BSD with active proton compounds HX (X = OH, OR, Cl, etc.) takes place according to Eq. (38), through mono(trimethylsilyl)diazene (MSD), and ultimately to diazene, which disproportionates instantaneously into hydrazine and nitrogen (*6, 8*). Since both BSD and, to an even

$$Me_3Si-N=N-SiMe_3 \xrightarrow[-Me_3SiX]{+HX} Me_3Si-N=N-H \xrightarrow[-Me_3SiX]{+HX} H-N=N-H \quad (38)$$
$$\text{BSD} \qquad\qquad\qquad\qquad \text{MSD} \qquad\qquad\qquad \text{Diazene}$$

greater extent, MSD are thermolabile azo compounds, thermolysis of BSD and especially of MSD takes place along with the protolysis reaction because the latter does not progress very fast. BSD in pentane or benzene decomposes comparatively slowly (in hours) with *water* at room temperature. In competition with hydrolysis of BSD, 20% of the initial amount of BSD thermolyzes to **3, 4, 5, 6,** and **7** [cf. Section IV,A, Eqs. (16–(20)]. The hydrolytically formed MSD from 80% of the starting BSD reacts further with the unhydrolyzed BSD in a fashion similar to the reactions of BSD with BSD with disproportionation, dimerization, cleavage, and H-abstraction [Eqs. (39)–(43)] (*48–51*).

Because these reactions are rapid, the steady-state concentration of MSD is very small (the red color of MSD fails to appear). Therefore, thermolysis

$$\text{MSD} + \text{BSD} \xrightarrow{\text{Disproportionation}} \underset{\underline{6}}{\overset{Me_3Si}{\underset{Me_3Si}{>}}N-N\overset{H}{\underset{SiMe_3}{<}}} + N_2 \quad (39)$$

$$\text{MSD} + \text{BSD} \xrightarrow{\text{Dimerization}} \underset{\underline{22}}{\overset{Me_3Si}{\underset{Me_3Si}{>}}N-N=N-N\overset{H}{\underset{SiMe_3}{<}}} \quad (40)$$

$$\text{MSD} + 2\text{BSD} \xrightarrow{\text{Cleavage}} 2 \underset{\underline{5}}{\overset{Me_3Si}{\underset{Me_3Si}{>}}N-SiMe_3} + \underset{\underline{7}}{\overset{Me_3Si}{\underset{Me_3Si}{>}}N-H} + 2N_2 \quad (41)$$

$$\text{MSD} + 2\text{BSD} \xrightarrow{\text{H-Abstraction}}$$

$$2 \underset{\underline{6}}{\overset{Me_3Si}{\underset{Me_3Si}{>}}N-N\overset{H}{\underset{SiMe_3}{<}}} + \underset{\underline{23}}{\overset{H}{\underset{Me_3Si}{>}}N-N\overset{H}{\underset{SiMe_3}{<}}} + N_2 \quad (42)$$

$$\text{MSD} + \text{BSD} \xrightarrow{\text{H-Abstraction}} \underset{\underline{7}}{\overset{Me_3Si}{\underset{Me_3Si}{>}}N-H} + \underset{\underline{24}}{Me_3Si-N\overset{H}{\underset{H}{<}}} + N_2 \quad (43)$$

of MSD, which obviously proceeds like that of BSD, occurs only to a minor extent. Also, subsequent hydrolysis of MSD, other than the reactions shown in Eqs. (39)–(43), does not take place. On the other hand, the primary products 3–7 and 22–24 hydrolyze further, 23 and 24 fast, 5–7 and 22 slowly, and 3 and 4 very slowly. Beside Me_3SiOH, $(Me_3Si)_2O$, and N_2, the hydrolysis products finally formed (after weeks) are the nitrogen hydrides ammonia, hydrazine, and hydrazoic acid.[5] The formation of the last product emanates from tetrazene (N_4H_4) (as well as its silyl derivative $Me_3SiN_4H_3$), which under the reaction conditions decomposes into hydrazine and nitrogen as well as into ammonia and hydrazoic acid (25, 52, 53) [Eq. (44)].

$$H_2N-NH_2 + N\equiv N \longrightarrow H_2N-N=N-NH_2 \longrightarrow NH_4^+ + N=N=N^- \quad (44)$$

The rate of reaction of BSD with water increases with addition of *base* or *acid*. Protolysis accelerates in the same direction with replacement of water by *alcohol*, in which—unlike water—BSD is soluble, with the result that the

---

[5] After 2 weeks of hydrolysis of 10 mmol BSD, the products are 11.1 mmol Me_3SiOH, 2.9 mmol $(Me_3Si)_2O$, 2.8 mmol N_2, 2.7 mmol NH_3, 2.7 mmol N_2H_4, 1.7 mmol HN_3, 0.7 mmol 3, and 0.08 mmol 4 (48).

protolysis may evolve in a homogeneous phase. The acidic as well as alkaline hydrolysis and alcoholysis of BSD terminate in a few minutes at room temperature, with the consequence that these reactions are not accompanied by BSD thermolysis. Naturally, the acidic as well as the alkaline alcoholysis occur much faster at higher temperatures. Under these conditions, the decomposition of MSD, formed from BSD by protolysis, takes place not only according to Eqs. (39)–(43), but also occurs by protolysis, to a large extent to diazene, N_2H_2 [Eq. (38)], which in itself reacts with BSD or MSD in a way analogous to Eqs. (39)–(43) or undergoes thermolysis (6, 8). The intermediate formation of diazene is supported by the finding that, just as in characterization of diazene (8), the addition of BSD to fumaric acid dissolved in alcohol–ether hydrogenates it to succinic acid (49) [Eq. (45)].

$$HOOC-CH=CH-COOH \xrightarrow[\substack{-N\equiv N}]{\substack{+(BSD \xrightarrow{Alcoholysis}) \ HN=NH}} HOOC-CH_2-CH_2-COOH \quad (45)$$

The major product formed from "brisk" protolysis, occurring with the evolution of nitrogen, is hydrazine in 80–90% yield, which among other products arises from disproportionation of diazene: $2N_2H_2 \rightarrow N_2H_4 + N_2$ (6, 8). Moreover, ammonia and hydrazoic acid are formed, with additional formation of hydrogen in a strong alkaline milieu. The formation of the latter is characteristic of base-catalyzed dissociation of diazene (8):

$$N_2H_2 \xrightarrow{OH^-} N_2 + H_2$$

which again supports the formation of N_2H_2 as an intermediate product of brisk BSD protolysis. The same is valid for hydrazoic acid, whose formation from diazene is acid catalyzed (54):

$$2N_2H_2 \xrightarrow{H^+} NH_3 + HN_3$$

Accordingly, the share of HN_3 along with that of NH_3 increases with increasing acidity of reaction medium and amounts to 24% if BSD is protolyzed with 0.46 M ethereal HCl at $-90°C$.[6]

Acid-catalyzed N_2H_2 decomposition develops in the process of N_2H_2 dimerization, giving rise to tetrazenium ion $N_4H_5{}^+$

$$2N_2H_2 \xrightarrow{+H^+} N_4H_5{}^+$$

[6] The formation of NH_3 and HN_3 does not arise exclusively from acid-catalyzed decomposition of N_2H_2, but also to some extent by the protolysis of silylated tetrazenes formed through BSD–MSD, MSD–MSD, BSD–N_2H_2, as well as MSD–N_2H_2 dimerization. Accordingly, HN_3 is also formed by the alkaline BSD hydrolysis.

which decomposes above $-20°C$ into ammonium ion and hydrazoic acid $(N_4H_5^+ \rightarrow NH_4^+ + HN_3)$. Since the latter reacts further with diazene, $N_2H_2 + HN_3 \rightarrow NH_3 + 2N_2$ (49), the yield of HN_3 at room temperature $(<10\%)$ is less than that at low temperatures.

B. Silyl- and Germylorganyldiazenes

The protolysis of silylated and germylated organyldiazenes Me_3E—$N{=}N$—R with proton active substances, HX (X = OH, OR, Cl, etc.), leads to monoorganyldiazenes [Eq. (46)] which in turn undergo rapid thermal decomposition into hydrocarbons and nitrogen (57, 58) [Eq. (47)] or disproportionation with excess Me_3E—$N{=}N$—R to give a protolysis-sensitive hydrazine, $(Me_3E)RN$—NHR (29, 32, 55, 56) [Eq. (48)].

$$Me_3E\text{-}N{=}N\text{-}R \xrightarrow[-Me_3EX]{+HX} H\text{-}N{=}N\text{-}R \tag{46}$$

$$H\text{-}N{=}N\text{-}R \xrightarrow{\text{Decomposition}} H\text{-}R + N{\equiv}N \tag{47}$$

$$Me_3\text{-}N{=}N\text{-}R + R\text{-}N{=}N\text{-}H \longrightarrow \underset{R}{\overset{Me_3E}{\diagdown}}N\text{-}N\underset{R}{\overset{H}{\diagup}} \xrightarrow[-Me_3EX]{+HX} \underset{R}{\overset{H}{\diagdown}}N\text{-}N\underset{R}{\overset{H}{\diagup}} \tag{48}$$

Since Me_3E—$N{=}N$—R reacts comparatively slowly with water according to Eq. (46), the steady-state concentration of H—$N{=}N$—R remains small. With methanol, protolysis is more rapid so that Me_3Si—$N{=}N$—Me and Me_3Si—$N{=}N$—Ph react near $-100°C$ (Me_3Si—$N{=}N$—^tBu, however, reacts at room temperature) (29). The organyldiazenes formed at low temperatures are recognizable because of their yellow color and are UV-spectroscopically detectable (58). In the methanolysis of Me_3Si—$N{=}N$—Me, the ultimate products are those from the thermolysis of H—$N{=}N$—Me: CH_4 and N_2 [Eq. (47)]. Phenyldiazene, H—$N{=}N$—Ph, formed by the methanolysis of Me_3Si—$N{=}N$—Ph, provides benzene and nitrogen as well as trimethylsilyldiphenylhydrazine, $(Me_3Si)PhN$—NH(Ph) [Eq. (48)] (29).[7]

---

[7] In toluene (17%) and in methylene chloride (9%), hydrazine is formed (29).

VI

REACTIVITY

A. General Features

BSD is exceedingly reactive (50), and the same is generally true for silyl- and germyldiazenes. Silylated or germylated organyldiazenes, with regard to their reactivity, occupy a position between azosilanes (azogermanes) and azoalkanes.

The remarkable reactivity of BSD (and other silylated and germylated diazenes) may be understood from the energy level diagram in Fig. 2. The inner molecular orbitals, significant for chemical reactions of BSD, are the highest occupied n_+ and the lowest vacant π^* molecular orbitals, which unlike the corresponding orbitals of azoalkanes lie energetically very high and low, respectively. Hence, they may easily gain or lose electrons, either fully in the case of an *electrochemical redox reaction* or partially in case of *Lewis acid–base reaction*. The tendency of BSD to accept or donate electrons is impressively shown by the easy course of reduction to molecular anions BSD^- and BSD^{2-} as well as by the appearance of an unusually intense peak for the molecular cation BSD^+ in the mass spectrum.

In reactions involving partial electron donation, BSD čan act as a base toward a Lewis acid A to form an adduct $BSD \rightarrow A$, which may be isolated in some cases and appears in the partially rearranged BSD framework in others (cf. Section VII). If the BSD-linked Lewis acid also contains a basic center B near the acidic center A, then the complex is, in general, not isolable because the acidic center of BSD, activated due to adduct formation, reacts with the basic center of AB (Scheme 6). The total reaction goes beyond an *addition* of AB at the azo double bond or a *substitution* of an azo-bound silyl

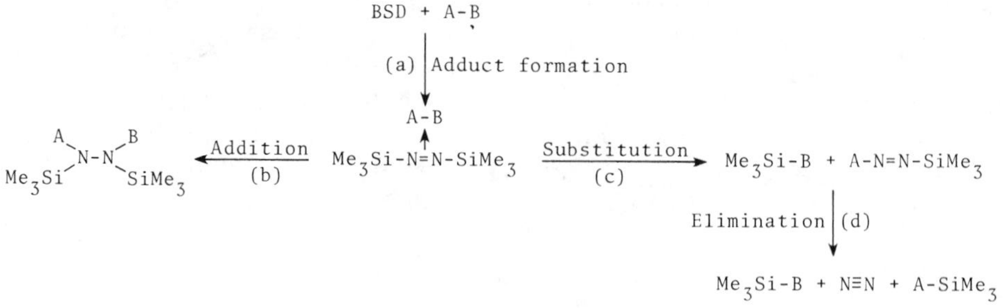

$$BSD + A\text{-}B$$

(a) Adduct formation

$$A\text{-}B$$

$$Me_3Si\text{-}N\overset{\cdot}{=}N\text{-}SiMe_3$$

$$\underset{Me_3Si}{\overset{A}{\diagdown}}N\text{-}N\underset{SiMe_3}{\overset{B}{\diagup}} \quad \overset{Addition}{\underset{(b)}{\longleftarrow}} \quad Me_3Si\text{-}N\overset{\cdot}{=}N\text{-}SiMe_3 \quad \overset{Substitution}{\underset{(c)}{\longrightarrow}} \quad Me_3Si\text{-}B + A\text{-}N\text{=}N\text{-}SiMe_3$$

Elimination (d)

$$Me_3Si\text{-}B + N\equiv N + A\text{-}SiMe_3$$

SCHEME 6

group by A. Often the substitution product itself is unstable and decomposes with the *elimination* of R_3Si—A (Scheme 6).

Beside the above-mentioned nonradically occurring reactions in condensed phase, one observes many *radical reactions* of silyl- and germyldiazenes and -organyldiazenes. Examples have already been discussed in Section IV. More reactions of this type are exemplified by oxidation with oxygen as well as by reactions with nitric oxide and halohydrocarbons (Section VI,B,C,D). The reactivity of BSD has been investigated quite exhaustively. Reactions of this diazene therefore become the focal point for subsequent sections of this chapter.

B. *Reactions with Elements*

As indicated below, BSD and other Group IV azo compounds react with alkali and alkaline earth metals and halogens and chalcogens but not with the elements of main Groups III–V and hydrogen (50).[8] Hydrogen may, of course, react in the presence of hydrogen carriers, as shown in Eq. (49), by the easy hydrogenation of trialkylsilylphenyldiazene in the presence of palladium catalyst (55).

$$R_3Si-N=N-Ph + H_2 \xrightarrow{[Pd]} R_3Si-NH-NH-Ph \tag{49}$$

1. *Electropositive Metals*

Alkali metals M (Li, Na, K) reduce BSD in the presence of diethyl ether at low temperatures to form the dianion BSD^{2-} [Eq. (50)]. In this way ether

$$BSD + 2M \xrightarrow{Et_2O} M_2BSD \tag{50}$$

insoluble, inflammable, thermally stable, colorless di(alkalimetal)bis(trimethylsilyl)hydrazides, also obtainable from $(Me_3Si)_2N_2H_2 + MR$, are formed (59). The reduction of BSD occurs through the BSD monoanion to the BSD dianion. Accordingly, on combining BSD with alkali metals in diethyl ether, a quintet signal typical of BSD^- is observed in the ESR spectrum of the reaction mixture (43, 60). The mono(alkalimetal)bis(trimethylsilyl)diazenide (MBSD) so formed exists in a disproportionational

---

[8] From the existing complexes of the type $M(BSD)_n$, where M is a transition element, it may be visualized that the active forms of the metal may be able to combine directly with BSD.

equilibrium with BSD and M_2BSD [Eq. (51)], which because of the insolubility of M_2BSD in diethyl ether lies considerably on the righthand side.

$$
\begin{array}{ccc}
\underset{\substack{\text{Me}_3\text{Si} \diagdown \diagup \text{SiMe}_3 \\ N \vdots N \\ M \\ + \quad M \\ N \vdots N \\ \text{Me}_3\text{Si} \diagup \diagdown \text{SiMe}_3 \\ \text{MBSD}}}{} &
\rightleftharpoons &
\underset{\substack{\text{Me}_3\text{Si} \diagdown \diagup \text{SiMe}_3 \\ N \vdots N \\ M \qquad M \\ N \vdots N \\ \text{Me}_3\text{Si} \diagup \diagdown \text{SiMe}_3 \\ [\text{MBSD}]_2}}{} \quad \rightleftharpoons \quad
\underset{\substack{\text{Me}_3\text{Si} \diagdown \diagup \text{SiMe}_3 \\ N = N \\ M \ + \ M \\ N - N \\ \text{Me}_3\text{Si} \diagup \diagdown \text{SiMe}_3 \\ M_2\text{BSD}}}{} &
(51)
\end{array}
$$

Thus, the compounds M_2BSD are not end products of direct BSD–alkali metal reduction but are formed indirectly by redox–disproportionation of MBSD. The reaction intermediate MBSD is in fact formed through a route involving a dark brown ethereal solution (stable at $-78°C$) of $Li(BSD)_3$, $Na(BSD)_2$, and $K(BSD)_2$, which probably contain the BSD radical anion and BSD coordinated alkali metal cations $Li(BSD)_2{}^+$, $NaBSD^+$, and $KBSD^+$ (61).[9] Accordingly, if BSD is reacted with alkali metals in diethyl ether at $-78°C$, the light blue color of BSD changes quite rapidly (in hours), with partial consumption of alkali metal, first into a dark brown colored solution (primary reduction) from which, with further consumption of alkali metal, M_2BSD starts precipitating out (secondary reduction) very slowly (Li in months and Na or K in days). MBSD is formed in small concentrations in the intervening period during secondary reduction.

For *tetrahydrofuran* (THF), the equilibrium [Eq. (51)] lies completely (M = Li) or partially (M = Na, K) to the left because M_2BSD is highly (M = Li) or partially (M = Na, K) soluble in the solvent. Consequently, MBSD may be obtained through reduction of BSD with alkali metals in THF [Eq. (52)] and also by comproportionation of M_2BSD and BSD [Eq.

$$\text{BSD} + \text{M} \xrightarrow{\text{THF}} \text{MBSD} \tag{52}$$

(51)] (61). Of course, the compounds MBSD are so thermolabile even at $-78°C$ that their preparation from M and BSD at this temperature is accompanied by partial (M = Li) or considerable (M = Na, K) decomposition. In the case of LiBSD, decomposition products are nitrogen and lithium bis(trimethylsilyl)amide [Eq. (53)].

$$2\text{MBSD} \longrightarrow \text{N} \equiv \text{N} + 2\text{MN}(\text{SiMe}_3)_2 \tag{53}$$

Of the total free nitrogen expected according to Eq. (53), only one-half is liberated at first in a fast reaction (in a few hours at $-40°C$), while the second

[9] The dark brown compound $LiBSD_3$ is a bit more stable than others, and after removal of ether in a vacuum at $-60°C$, it can be recrystallized from pentane at low temperature. It starts decomposing with the elimination of BSD at $-15°C$ in a high vacuum.

half is formed very slowly (in a few weeks at $-40°C$). This happens because $LiN(SiMe_3)_2$ formed in this way combines with LiBSD to give a $1:1$ complex $LiBSD \cdot LiN(SiMe_3)_2$ (**25**), which is thermostable and dissociates on a negligible scale: $LiBSD + LiN(SiMe_3)_2 \rightleftarrows$ **25**, into $LiN(SiMe_3)_2$ and the thermolabile LiBSD.

The solution of LiBSD in THF contains essentially the dimer $[LiBSD]_2$ because the formation of **25**, which requires at least the participation of two molecules of the compound, follows exactly the first-order rate law $(t_{1/2} = 80.4$ minutes at $-48°C).$[10] The total course of BSD thermolysis can be formulated briefly $(M = Li)$ as shown in Eq. (54). Probably $(LiBSD)_2$

$$\tag{54}$$

undergoes a linkage isomerization to form a tetrazane derivative that decomposes into bis(trimethylsilyl)amide and nitrogen:

$$(LiBSD)_2 \rightleftharpoons (Me_3Si)_2N-NLi-NLi-N(SiMe_3)_2 \longrightarrow$$

$$(Me_3Si)_2NLi + N\equiv N + LiN(SiMe_3)_2$$

The bis(trimethylsilyl)amide would then transfer to reaction intermediate **25** in a fast secondary reaction: $(LiBSD)_2 + 2 LiN(SiMe_3)_2 \rightarrow 2$ (**25**).

Unlike that of LiBSD, the THF solution of NaBSD does not decompose exclusively according to Eq. (53) but additionally as shown in Eq. (55) $(M = Na)$.[11] KBSD thermolyzes in THF only according to Eq. (55).

$$4 MBSD \rightarrow MN(SiMe_3)_2 + 2 MN_2(SiMe_3)_3 + MN_3 \tag{55}$$

Obviously, a silyl group transfer takes place between MBSD molecules: $2 MBSD \rightarrow M-N=N-SiMe_3 + (Me_3Si)MN-N(SiMe_3)_2$. As inferred from independent attempts (cf. Section VI,E), the trimethylsilyldiazenide $M-N=N-SiMe_3$ so formed will decompose to bis(trimethylsilyl)amide and azide: $2 M-N=N-SiMe_3 \rightarrow MN(SiMe_3)_2 + MN_3$.

---

[10] The rate (which does not follow first order) of the slow decomposition reaction decreases considerably because of equilibrium $LiBSD + LiN(SiMe_3)_2 \rightleftarrows$ **25**, which naturally will shift increasingly to the right due to the formation of $LiN(SiMe_3)_2$.

[11] In addition, MBSD $(M = Na, K)$ decomposes in traces according to: $4 MBSD \rightarrow 2 M_2BSD + (Me_3Si)_4N_2 + N_2$ (**60**).

2. Electronegative Nonmetals

In contrast to the BSD anion-forming reductions of BSD with the more electropositive metals, the oxidation of BSD with the most electronegative nonmetals does not lead to the BSD cation. The products of reaction of BSD with *halogens* (even at very low temperatures) are nitrogen and trimethylhalosilane (*50*) [Eq. (56)].

$$\text{BSD} + \text{Hal}_2 \rightarrow \text{Me}_3\text{Si-Hal} + \text{N}\equiv\text{N} + \text{Hal-SiMe}_3 \qquad (56)$$

Their formation follows the path (a), (c), and (d) of Scheme 6.

Oxygen reacts with BSD as well as with bis(trimethylgermyl)diazene even at $-78°C$ to form nitrogen and the bis(trimethylmetal)peroxide (*28, 50*) [Eq. (57)]. Silylated and germylated alkyldiazenes react with oxygen at $0°C$ to form peroxides (*29*). Trimethylsilylphenyldiazene, $\text{Me}_3\text{Si}-\text{N}{=}\text{N}-\text{Ph}$,

$$\text{Me}_3\text{E-N=N-X} + \text{O}_2 \xrightarrow[\text{X = EMe}_3\text{, R}]{\text{E = Si, Ge}} \text{Me}_3\text{E-O-O-X} + \text{N}_2 \qquad (57)$$

on the other hand, does not react with oxygen at room temperature or even above. Because of the radical nature of (triplet) oxygen, the redox reaction [Eq. (57)] must initiate through a radical forming step, e.g.,

$$\text{Me}_3\text{E}-\text{N}{=}\text{N}-\text{EMe}_3 + \text{O}_2 \longrightarrow [\text{Me}_3\text{E}-\text{O}-\text{O} + \text{N}\equiv\text{N} + \text{EMe}_3]_{\text{cage}}$$

The radicals recombine in the solvent cage to form peroxide (cf. reaction with radical NO, Section VI,D,2).

Like oxygen, *sulfur* and red *selenium* oxidize BSD to nitrogen (*50*) [Eq. (58)]. The oxidation reactions [Eqs. (57) and (58)] become slower with

$$\text{BSD} + \tfrac{1}{8}\text{Y}_8 \xrightarrow{\text{Y = S, Se}} \text{Me}_3\text{Si-Y-SiMe}_3 + \text{N}_2 \qquad (58)$$

an increase in atomic mass of the chalcogen and hence are driven increasingly backward by the BSD thermolysis as shown in the following tabulation:

	O_2	S_8	Se_8	Te_n
Oxidation (%)	100	47	2	0
Thermolysis (%)	0	53	98	100

C. Reactions with Halides

1. Boron Halides

Boron halides such as BCl_3, $RBCl_2$, R_2BCl, $(R_2N)_2BCl$, and $(RO)_2BCl$ react with azosilanes and -germanes, but mostly yield mixtures of products.

However, reactions with *diorganylboron halides* are simple and are discussed in this section.

Diphenylboron chloride combines with BSD in methylene chloride to give shiny yellow bis(diphenylboryl)diazene (**26**) in about 30% yield (*62, 63*) [Eq. (59)]. Thermostable and acid-resistant compound **26** forms the first

$$\text{BSD} + 2Ph_2BCl \rightarrow \underset{\underline{\underline{26}}}{Ph_2B\text{-}N{=}N\text{-}BPh_2} + 2Me_3SiCl \qquad (59)$$

known example of an azoborane. It is polymeric in nature and hence nonvolatile and insoluble in common solvents.

Just like BSD, silylated organyldiazenes react with diorganylboron halides in methylene chloride to give boryl organyldiazenes (*29*) [Eq. (60)].

$$Me_3Si\text{-}N{=}N\text{-}R + R'_2BBr \rightarrow \underset{\underline{\underline{27}}}{R'_2B\text{-}N{=}N\text{-}R} + Me_3SiBr \qquad (60)$$

According to Eq. (60), $Me_2B—N{=}N—{}^tBu$ (mp 24–25°C, sublimes in high vacuum), $Ph_2B—N{=}N—{}^tBu$ (mp 142–144°C, decomposes), $Me_2B—N{=}N—Ph$ (mp 74–76°C, sublimation point 70°C at 0.01 Torr), and $Ph_2B—N{=}N—Ph$ (mp 159–160°C, decomposes) have been isolated as crystalline, shiny yellow compounds ($\tilde{\nu}_{max}$ = 25500–28000 cm^{-1}), soluble in CH_2Cl_2. They exist as dimers in vapor as well as in solution, and on the basis of NMR studies, structure **27a** or **27b** may be assigned. [$Ph_2B—N{=}N—{}^tBu$

27a 27b

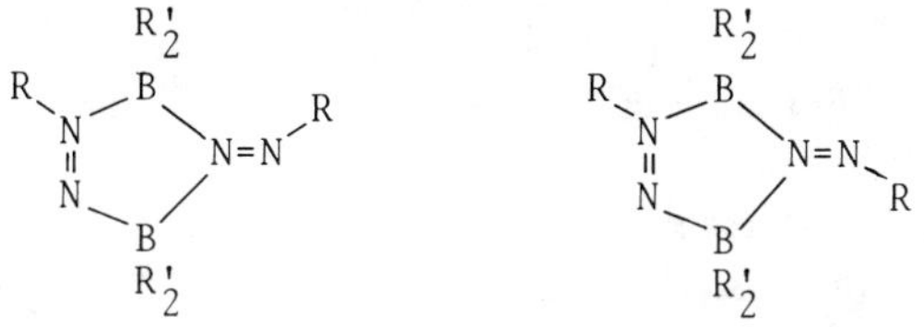

is obtained as a mixture of configurational isomers **27a** and **27b**, which probably contains additional constitutional isomers (*29*)].

The substitution of silyl groups in BSD as well as $Me_3Si—N{=}N—R$ takes place according to Scheme 6, modes a and c. The primary formation of an adduct of the azo compounds with borane is indicated in the reaction of $Me_3Si—N{=}N—Me$ with trimethylborane in methylene chloride, which at −78°C leads to a yellow adduct [Eq. (61)], insoluble in CH_2Cl_2. This adduct

$$\underset{\text{red}}{Me_3Si\text{-}\ddot{N}{=}\ddot{N}\text{-}Me} + \underset{\text{colorless}}{BMe_3} \underset{-60^\circ C}{\overset{-78^\circ C}{\rightleftharpoons}} \underset{\text{yellow}}{Me_3Si\text{-}\ddot{N}{=}N(BMe_3)\text{-}Me} \qquad (61)$$

decomposes above $-60°C$ to starting materials with the appearance of the red color of $Me_3Si-N=N-Me$ (29) [Eq. (61)]. In reaction of BSD with Ph_2BCl, the formation of an adduct is indicated by the appearance of an intermediate deep blue color typical of a BSD adduct (cf. Section VII).

2. Carbon Halides

Monohalomethanes, CH_3X (X = Cl, Br, I), do not react with BSD. If X is replaced by a more nucleophilic leaving group such as sulfate, that is, if BSD is treated with dimethylsulfate, then beside BSD thermolysis, a nucleophilic substitution of a silyl group of BSD by a methyl group takes place (29) [Eq. (62)].

$$BSD + CH_3X \xrightarrow{X = SO_4Me} Me_3Si-N=N-Me + Me_3SiX \qquad (62)$$

Dihalomethanes CH_2X_2 (X = Cl, Br, I) react with BSD, as shown in Eq. (63), with the formation of bis(silyl)hydrazone **28** (27). Similarly, BSD

$$2BSD + CH_2X_2 \rightarrow (Me_3Si)_2N-N=CH_2 + 2Me_3SiX + N_2 \qquad (63)$$
$$\underline{\underline{28}}$$

converts $PhCHCl_2$ to $(Me_3Si)_2N-N=CHPh$ [cf. Eq. (24)[12]]. Thermolysis of BSD competes with Eq. (63). The yield of **28** increases with increase in temperature, and obeys the order $CH_2Cl_2 < CH_2Br_2 < CH_2I_2$, being 10% at 70°C for CH_2Cl_2, 90% at 70°C and 45% at 20°C for CH_2Br_2, and 90% at 20°C for CH_2I_2. The rate of reaction of Eq. (63), X = Cl, is negligibly small at 0°C, so that methylene chloride may be used as a medium for reactions with BSD.

The formation of **28** follows a radical pathway (27). The initiating step is probably liberation of silyl radicals from BSD [$2BSD \rightarrow (Me_3Si)_3N_2 + N_2 + Me_3Si$, Eq. (21)]. These radicals abstract halogen from methylene halide to convert it to CH_2X radicals [Eq. (64)] which add to BSD to form hydrazyl radicals, $(Me_3Si)_2N-NCH_2X$, which in turn react with additional BSD to form **28** [Eq. (65)]. The mechanism shown in Eqs. (64) and (65)

$$Me_3Si\cdot + CH_2X_2 \rightarrow Me_3SiX + \cdot CH_2X \qquad (64)$$

$$\cdot CH_2X \xrightarrow{+BSD} (Me_3Si)_2N-\overset{\cdot}{N}-CH_2X \xrightarrow[-N_2, -Me_3SiX]{+BSD} \underline{\underline{28}} + Me_3Si\cdot \qquad (65)$$

---

[12] Hydrazones are also formed by the action of BSD on $PhCH_3$ as well as CH_3CN (Section IV,A).

easily explains why bis(trimethylsilyl)dibromomethane, $(Me_3Si)_2CBr_2$, in its reaction with BSD does not form the hydrazone $(Me_3Si)_2N$—N=C-$(SiMe_3)_2$ but leads instead to bis(trimethylsilyl)bromomethane, $(Me_3Si)_2$-$CHBr$ (27):

$$2(Me_3Si)_2CBr_2 + BSD + 2H \rightarrow 2(Me_3Si)_2CHBr + 2Me_3SiBr + N_2$$

In this case, the addition of radicals [formed as in Eq. (64)] to BSD is slower than H-abstraction for steric reasons.

Trihalomethanes, CHX_3, react with BSD in an unpredictable manner (64). However, silyl derivatives of bromoform, Me_3SiCBr_3, react smoothly at 20°C with the formation of *N*-isocyanide, **29** [Eq. (66)]. Apparently a

$$2BSD + Me_3SiCBr_3 \rightarrow (Me_3Si)_2N\text{-}N\text{≡}C + 3Me_3SiBr + N_2 \qquad (66)$$
$$\underline{29}$$

hydrazone, $(Me_3Si)_2N$—N=$CBr(SiMe_3)$, is first formed analogously to Eq. (63) and then decomposes with elimination of Me_3SiBr to form **29**.

Tetrahalomethanes, CX_4 (X = Cl, Br), react with BSD in analogy to dihalomethanes, CH_2X_2 [Eq. (63)], at temperatures below 0°C to form hydrazones quantitatively (27, 65) [Eq. (67)]. Mixed perhalogenated

$$2BSD + CX_4 \rightarrow (Me_3Si)_2N\text{-}N\text{=}CX_2 + 2Me_3SiX + N_2 \qquad (67)$$
$$\underline{30}$$

methanes such as CBr_2F_2 or CCl_3F are also transformed by BSD to hydrazones, $(Me_3Si)_2N$—N=CF_2 or $(Me_3Si)_2N$—N=$CClF$. The hydrazones **30**, generated likewise by radical reactions,[13] are able to react further with BSD (in comparison with BSD thermolysis) to form *N*-isocyanide **29**: $BSD + 30 \rightarrow 29 + 2Me_3SiX + N_2$ [yield of **29** from $(Me_3Si)_2N$—N=CCl_2 at 70°C is negligible but is 90 % from $(Me_3Si)_2N$—N=CBr_2 at 70°C]. Overall, carbon tetrahalides are converted through BSD to **29** according to: $3BSD + CX_4 \rightarrow 29 + 4Me_3SiX + 2N_2$.

Analogously to $(Me_3Si)_2N$—N=CCl_2, *phosgene,* O=CCl_2, is reduced to carbon monoxide by BSD at temperatures below -100°C (27) [Eq. (68)].

$$BSD + Cl_2C\text{=}O \rightarrow C\text{≡}O + 2Me_3SiCl + N_2 \qquad (68)$$

---

[13] Conceivable radical initiation reaction: $BSD + CX_4 \rightarrow Me_3Si + N_2 + Me_3SiCl + CX_3$ (followed because of comparatively higher stability of CX_3 at lower temperatures). Radical chain, cf. Eqs. (64) and (65).

In contrast to the radical reactions described above in Eqs. (63), (66), and (67), the last reaction [Eq. (68)] is expected to follow a nonradical path as in Scheme 6, modes a, c, and d.

Hexachloroethane, C_2Cl_6, reacts with BSD above $+20°C$ according to Eq. (69) (27). Simultaneous thermolysis of BSD occurs to a minor extent (10%).

$$\text{BSD} + Cl_3C-CCl_3 \rightarrow Cl_2C=CCl_2 + 2Me_3SiCl + N_2 \qquad (69)$$

3. Other Main Group IV Halides

With the exception of carbon tetrafluoride and silicon tetrachloride, all other halides, EX_4, of Group IV metals react with BSD (27, 50, 63). Reactions with SiF_4, $GeCl_4$, and $SnCl_4$ are discussed separately here because they run quite differently.

Silicon tetrafluoride, SiF_4, interacts with BSD in ether at temperatures below $-40°C$ to form an intense blue complex, $BSD \cdot SiF_4$ (31), which decomposes above $-40°C$ into its components (66). In a bomb tube, the existing equilibrium mixture $BSD + SiF_4 \rightleftarrows$ 31 changes, according to Scheme 7, slowly to tetrazenes 4 and 34, hydrazine 33, amine 35, as well as

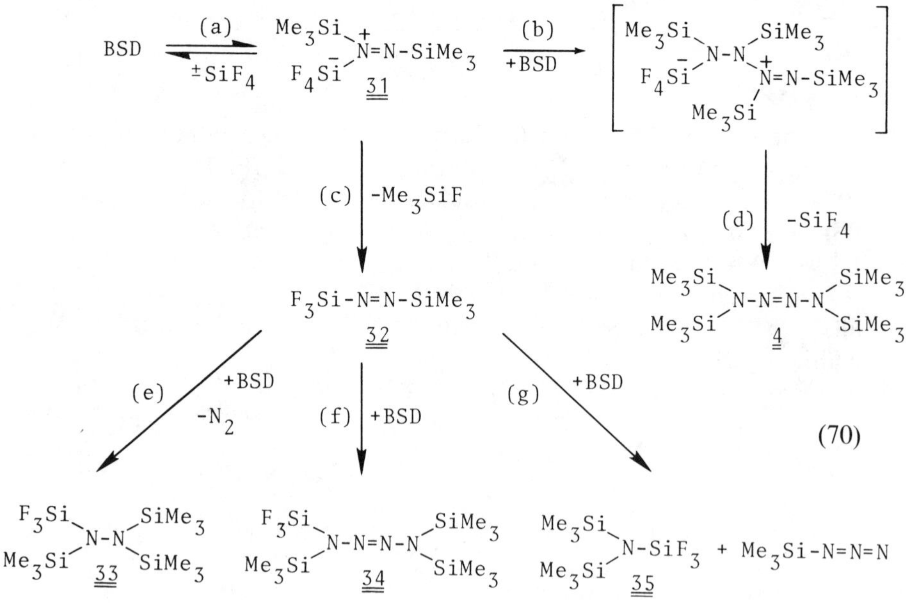

$$(70)$$

SCHEME 7

trimethylsilyl azide and fluoride (66). On using SiF_4 in lesser amounts (molar ratio SiF_4:BSD $= \frac{1}{10}$), the side products **33–35**, Me_3SiN_3, and Me_3SiF are formed in traces, whereas with an excess of SiF_4 (molar ratio SiF_4:BSD >4), the yields are modest.

The conspicuous catalysis of BSD dimerization to tetrazene **4** by the Lewis acid SiF_4 depends on the fact that primary adduct **31** combines further with the uncomplexed BSD to form $(BSD)_2 \cdot SiF_4$, which undergoes SiF_4 cleavage to form **4** [Eq. (70a,b, and d)]. The formation of SiF_3-containing side products may be attributed to irreversible decomposition of **31** to Me_3SiF and $F_3Si—N{=}N—SiMe_3$ (**32**) [Eq. (70c)]. The azosilane **32** itself reacts with BSD under disproportionation [Eq. (70e)] and dimerization [Eq. (70f)] as well as with the formation of **35** and Me_3SiN_3 [Eq. (70g)]. In this way, the general importance of dimerization of inorganic diazenes to tetrazenes $—N{=}N— + —N{=}N— \rightarrow {>}N—N{=}N—N{<}$, is indicated.

Like BSD + BSD and BSD + $F_3Si—N{=}N—SiMe_3$, the azo compounds BSD + $Me_3Si—N{=}N—H$, $Me_3Ge—N{=}N—GeMe_3 + Me_3Ge—N{=}N—GeMe_3$ and both azo groups of $Me_3Si—N{=}N—SiMe_3—N{=}N—SiMe_3$ combine to form tetrazenes (cf. Sections IV and V).

Germanium tetrachloride, $GeCl_4$, reacts with BSD in methylene chloride at $-78°C$ to form a series of nitrogen-containing products. The hydrazine **36** is the major product in 50% yield if BSD and $GeCl_4$ are reacted in a 1:2 molar ratio (63) [Eq. (71)]. If, on the other hand, BSD and $GeCl_4$ are reacted

$$2\,BSD \;+\; 2\,GeCl_4 \;\rightarrow\; \begin{array}{c} Cl_3Ge \diagdown \qquad \diagup GeCl_3 \\ N{-}N \\ Me_3Si \diagup \qquad \diagdown SiMe_3 \end{array} \;+\; 2\,Me_3SiCl \;+\; N_2 \qquad (71)$$

$$\underline{36}$$

in a molar ratio of 2:1, then tetrakis(trimethylsilyl)tetrazene, $(Me_3Si)_2$-$N—N{=}N—N(SiMe_3)_2$, is formed as the main product in $\sim 75\%$ yield.

The dimerization of BSD is triggered by the catalytic action of Lewis acid $GeCl_4$ (or $GeCl_2$, see below) and runs analogously to Eq. (70a,b, and d). The mode of formation of **36**, on the other hand, is not clear. It is probably formed by means of a silyl-group substitution leading to diazene $Cl_3Ge—N{=}N—SiMe_3$, which can disproportionate into **36** and N_2 [cf. Eq. (70a,c, and e)]. It is also conceivable that BSD brings about reduction of $GeCl_4$ to $GeCl_2$ [cf. Eq. (72)], which can form an adduct, $BSD \cdot GeCl_2$, that on addition of $GeCl_4$ can in turn lead to **36**. The last route is supported by formation of additional, structurally unexplained Si-, Ge-, and N-containing, but Cl-free products (63).

Tin tetrachloride, $SnCl_4$, reacts vigorously with BSD in methylene chloride at $-78°C$ to give a quantitative reaction involving reduction to tin dichloride (*50, 63*) [Eq. (72)].

$$BSD + SnCl_4 \rightarrow SnCl_2 + 2Me_3SiCl + N_2 \qquad (72)$$

4. *Main Group V Halides*

a. Monohaloamines. Dimethylchloramine, Me_2NCl, reacts with BSD in CH_2Cl_2 at $-78°C$ almost exclusively according to Eq. (73) (R′ = Me).

$$BSD + R'_2NCl \rightarrow Me_3Si-NR'_2 + Me_3SiCl + N_2 \qquad (73)$$

Reaction of $(Me_3Si)_2NCl$ with BSD in CH_2Cl_2 at about 0°C is in harmony with Eq. (73) (R′ = Me_3Si). Another major product of the latter reaction is $(Me_3Si)_2NH$ (*67*). Typical products of radical BSD thermolysis are also formed (Section IV). Evidently, the reaction depicted in Eq. (73) follows, at least in the latter case, a radical path. The initiation reaction is represented as: $BSD + R'_2NCl \rightarrow [Me_3Si \cdot + N_2 + Me_3SiCl + R'_2N \cdot]_{cage}$. The radicals so formed recombine in the solvent cage to form $Me_3SiNR'_2$, or after leaving the cage they abstract chlorine ($Me_3Si + R'_2NCl \rightarrow Me_3SiCl + R'_2N$ and $Me_3Si + CH_2Cl_2 \rightarrow Me_3SiCl + CH_2Cl$) or hydrogen or silyl groups ($R'_2N + H \rightarrow R'_2NH; R'_2N + BSD \rightarrow R'_2NSiMe_3 + N_2 + Me_3Si$).

b. Monohalophosphanes. Reactions of phosphanes, R'_2PCl, with BSD differ from those of chloramines because instead of a partial positive charge, chlorine has a partial negative charge in these compounds. This cannot lead to reduction of substituted chlorine but only of the central phosphorus atom. In fact, phosphanes of the form R'_2PCl react with BSD in methylene chloride to form tetrakis(organyl)diphosphanes (*63*) [Eq. (74)]. The reaction shown

$$BSD + 2R'_2PCl \xrightarrow{CH_2Cl_2} R'_2P-PR'_2 + 2Me_3SiCl + N_2 \qquad (74)$$

in Eq. (74), according to the general Scheme 6 (see Section VI,A), occurs via the phosphorus azo compound **37**, formation of which may be observed by the appearance of a deep greenish-blue color during the course of the reaction [Eq. (75a)]. Compound **37** decomposes with cleavage of nitrogen [Eq. 75b)],

$$BSD + R'_2P-Cl \xrightarrow[-Me_3SiCl]{(a)} Me_3Si-N=N-PR'_2 \xrightarrow[-N_2]{(b)} Me_3Si-PR'_2 \qquad (75)$$
$$\underline{\underline{37}} \qquad\qquad\qquad \underline{\underline{38}}$$

to bis(organyl)trimethylsilylphosphanes **38** (isolable if $R' =$ mesityl), which ultimately react with starting material R'_2PCl according to Eq. (76).

$$R'_2P\text{-}Cl \ + \ Me_3Si\text{-}PR'_2 \ \rightarrow \ R'_2P\text{-}PR'_2 \ + \ Me_3SiCl \qquad (76)$$

According to Eqs. (75) and (76), tetraphenyldiphosphane is formed in quantitative amounts from Ph_2PCl in CH_2Cl_2 at $-30°C$. The yield is slightly less for tetramesityldiphosphane formed from Mes_2PCl. Bis(*t*-butyl)chlorophosphane, on the other hand, does not react with BSD.

Reaction of R'_2PCl with BSD in diethyl ether is slower than in CH_2Cl_2 (*63*). The product of this reaction with Ph_2PCl is phosphamidine **39** [Eq.

$$2\,BSD \ + \ Ph_2PCl \ \xrightarrow{\ Et_2O\ } \ Ph_2P{\displaystyle \begin{array}{c} \diagup N(SiMe_3) \\ \diagdown N(SiMe_3)_2 \end{array}} \ + \ 2\,Me_3SiCl \ + \ N_2 \qquad (77)$$

$$\underline{39}$$

(77)], which can react further with an excess of Ph_2PCl [e.g., **39** $+ 4\,Ph_2PCl \rightarrow$ $[(Ph_2P)_2N\text{---}PPh_2{=}N\text{---}PPh_2\text{---}PPh_2]Cl + 3\,Me_3SiCl]$. Surprisingly, a compound with isolated nitrogen atoms is formed in this way. The BSD must have undergone cleavage in the course of formation of **39**.

Most probably, the reaction in Eq. (77) proceeds at first through Eq. (75a) to form the phosphorus azo compound **37** ($R' =$ Ph), which undergoes [2 + 3] cycloaddition with BSD to form adduct **40** [Eq. (78); cf. ref. (*68*)]. Intermediate **40** decomposes to **39** and N_2 as a result of migration of two silyl

$$Ph_2P{\displaystyle \begin{array}{c} \diagup N{=}N{\diagdown}^{SiMe_3} \\ + \\ \diagup N{=}N{\diagdown}^{SiMe_3} \end{array}}Me_3Si \qquad \longrightarrow \qquad Ph_2P{\displaystyle \begin{array}{c} \diagup N{\diagdown}_{N{-}SiMe_3} \\ \| \quad \quad | \\ \diagdown N{-}N{\diagdown}_{SiMe_3} \end{array}} \qquad \xrightarrow{-N_2} \quad \underline{\underline{39}} \qquad (78)$$

$$\underline{\underline{40}} \qquad \qquad SiMe_3$$

groups to nitrogen atoms bound to phosphorus. In contrast to Ph_2PCl, Mes_2PCl reacts in ether according to Eq. (74).

The assumption that reaction of BSD with chlorophosphanes is initiated through Eq. (75a) (formation of phosphanylsilyldiazene) is especially supported by reactions of chlorophosphanes of the form R'_2PCl ($R' =$ Me, Ph) with silylated organyldiazenes $Me_3Si\text{---}N{=}N\text{---}R$ ($R =$ Me, ^tBu, Ph). In methylene chloride at $-30°C$ to $+30°C$, they lead to yellow- or red-colored *phosphanylorganyldiazenes*, $R'_2P\text{---}N{=}N\text{---}R$ (*29*). These are more or less thermolabile, and some of the closely characterized compounds are listed in Table III. In an analogous fashion (*29*), the yellow- or red-colored

TABLE III

PHOSPHANYL AND ARSANYL DERIVATIVES OF ORGANYLDIAZENES

$R^1 - N = N - R^2$

R^1	R^2	Preparation [Eq. (79)[a]] (Temperature/time/yield)	Mp (°C)	Bp (°C/Torr)	Decomposition[b] (Temperature/time)[c]	Color	UV absorption[a] $\tilde{v}_{max}$(cm^{-1})	$\approx \varepsilon$	IE[d] (eV)
Me$_2$P	Me	25°C/1 h/20%	—	0/20	—	Yellow	23850	—	7.68
Me$_2$As	Me	−78°C/30 m/90%	—	—	−50°C/2 h	Yellow	22960	40	—
Me$_2$P	^{t}Bu	30°C/5 h/60%	—	0/5	84°C/1 h	Yellow	—	—	—
Me$_2$As	^{t}Bu	−78°C/10 m/0%[e]	—	—	<−78°C/<m	—	—	—	—
Me$_2$P	Ph	−10°C/1 h/0%[f]	—	—	—	—	—	—	—
Me$_2$As	Ph	−78°C/10 m/90%	−20/dec	—	−10°C/5 h[g]	Red	20130	100	7.75
Ph$_2$P	Me	−10°C/1 h/75%	—	80–90/0.01	50°C/3 h[h]	Yellow	23430	120	7.91
Ph$_2$P	^{t}Bu	−25°C/5 h/90%	−10/dec	—	0°C/24 h[i]	Yellow	23030	80	—
Ph$_2$P	Ph	−30°C/5 h/30%	60/dec	—	70°C/3 h	Red	20750	220	—

[a] In CH$_2$Cl$_2$.

[b] Unless otherwise stated, decomposition follows the equation $R'_2E - N = N - R \rightarrow R'_2E - R + N \equiv N$.

[c] Yield of dec >90%.

[d] Adiabatic ionization energies.

[e] The decomposition of Me$_2$As−N=N−^{t}Bu is faster than its formation.

[f] Instead of Me$_2$P−N=N−Ph, unidentified products are found.

[g] Beside Me$_2$AsPh, the arsanes Me$_3$As and MeAsPh$_2$ are formed.

[h] Beside Ph$_2$PMe, the H-migration product Ph$_2$PNH−N=CH$_2$ is found.

[i] Beside the normal decomposition ($\rightarrow$ N$_2$ + Ph$_2$P^{t}Bu), the reaction Ph$_2$P−N=N−^{t}Bu $\rightarrow$ Ph$_2$PH + N$_2$ + CH$_2$=CMe$_2$ (Ph$_2$PH $\rightarrow$ secondary reaction) takes place.

thermolabile *arsanylorganyldiazenes*, $R'_2As-N=N-R$, have been obtained at $-78°C$ [Eq. (79)] (Table III). The yields of **41** are, of course, low

$$Me_3Si-N=N-R + R'_2ECl \xrightarrow[E = P, As]{} R'_2E-N=N-R + Me_3SiCl \quad (79)$$
$$\underline{41}$$

because side reactions such as thermal decompositions (especially those of $Me_3E-N=N-{}^tBu$ and $Me_2As-N=N-Me$) may accompany the substitution reaction [Eq. (79)] (cf. Table III).

c. *Dihalophosphanes.* Reactions of dichlorophosphanes, $R'PCl_2$, with BSD are more complicated and faster than those of chlorophosphanes, R'_2PCl (*63*). For example, $MesPCl_2$ in its reaction with BSD provides phosphane **42** [Eq. (80)]. Evidently, $MesPCl_2$ first reacts according to Eq. (75) to form $(Me_3Si)MesPCl$, which combines further with BSD to give

$$3BSD + MesPCl_2 \xrightarrow{CH_2Cl_2} MesP\begin{smallmatrix}N(SiMe_3)_2\\ \\N(SiMe_3)_2\end{smallmatrix} + Me_3SiCl + 2N_2 \quad (80)$$
$$\underline{42}$$

$(Me_3Si)MesP-N=N-SiMe_3$. The latter compound is transformed with BSD to adduct **43**, which finally forms **42** [Eq. (81)].

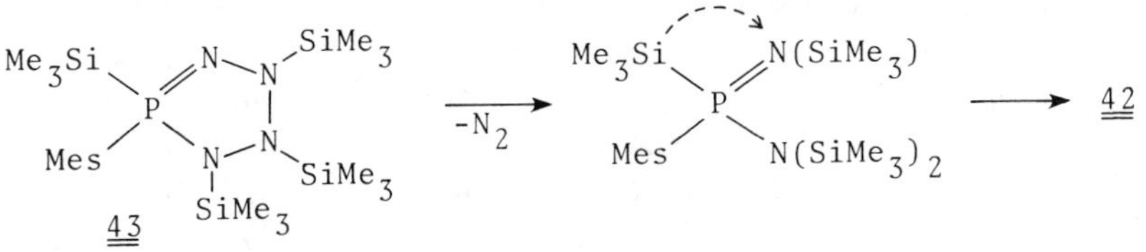

$$\underline{43} \xrightarrow{-N_2} \longrightarrow \underline{42} \quad (81)$$

d. *Trihalophosphanes, -arsanes, and -stibanes.* Phosphorus trichloride, PCl_3, combines with BSD in methylene chloride at $-78°C$ with evolution of N_2 in an extremely vigorous reaction accompanied by yellow coloration of the reaction mixture and formation of insoluble products [the sum total formula approximates to $P_2N_2Cl_2(SiMe_3)$]. On the other hand, reaction of $Me_3Si-N=N-Ph$ with phosphorus trifluoride, PF_3, in the complex $RhCl(PPh_3)_2(PF_3)$ runs smoothly in benzene with loss of fluoride to form $RhCl(PPh_3)_2(F_2P-N=N-Ph)$ (*69*).

Arsenic trichloride, $AsCl_3$, and antimony tribromide, $SbBr_3$, react differently from PCl_3 and are reduced by BSD to elemental arsenic and antimony.

5. Main Group VI Halides

Monochlorosulfanes, $R'SCl$, react with BSD in diethyl ether or methylene chloride at low temperatures to form silylsulfanes Me_3SiSR' [Eq. (82)], which in turn, may react further with an excess of monochlorosulfanes according to Eq. (83) to generate disulfanes (27).[14] Reaction intermediates

$$\text{BSD} + \text{R'S-Cl} \xrightarrow[-\text{Me}_3\text{SiCl}]{\text{(a)}} \text{Me}_3\text{Si-N=N-SR'} \xrightarrow[-\text{N}_2]{\text{(b)}} \text{Me}_3\text{Si-SR'} \quad (82)$$
$$\underline{44}$$

$$\text{R'-SCl} + \text{Me}_3\text{Si-SR'} \rightarrow \text{R'S-SR'} + \text{Me}_3\text{SiCl} \quad (83)$$

in the formation of $\text{Me}_3\text{SiSR'}$ may well be the thermolabile azo compounds **44** [Eq. (82a)], which decompose instantaneously with the evolution of N_2 [Eq. (82b)] (cf. Scheme 6). If R' represents an amino group with bulky substituents, then the compounds $\text{Me}_3\text{SiSR'}$ are isolable [e.g., if $\text{R'} = (\text{Me}_3\text{Si})_2\text{N}$, $(\text{Me}_3\text{Si})_3\text{N}_2$, and $(\text{Me}_3\text{Si})_2\text{NS}$].[15] Otherwise, elemental sulfur is formed as shown in Eq. (84). In this way, the reaction of BSD and Me_2NSCl

$$\text{Me}_3\text{Si-S-NR}_2' \rightarrow \text{Me}_3\text{Si-NR}_2' + \tfrac{1}{8}\text{S}_8 \quad (84)$$

proceeds with the elimination of N_2 and Me_3SiCl to form $\text{Me}_3\text{SiNMe}_2$ and sulfur. The intermediate formation of thiohydroxylamine $\text{Me}_3\text{SiSNMe}_2$ has been identified by NMR spectroscopy at $-60°\text{C}$; it decomposes, however, above $-30°\text{C}$ in a short time. A similar decomposition according to Eq. (84) has been observed for $\text{Me}_3\text{SiSN(SiMe}_3)_2$ above $180°\text{C}$ (27).

Sulfur dichloride, SCl_2, reacts with BSD according to Eq. (82) to displace N_2 and Me_3SiCl, thereby forming trimethylsulfur chloride, Me_3SiSCl, which is not isolable because it decomposes to sulfur and trimethylsilyl chloride or combines further with BSD to form $(\text{Me}_3\text{Si})_2\text{S}$ [cf. Eq. (82)]. Consequently, the reaction following addition of SCl_2 solution dropwise to a BSD solution takes place according to Eq. (85). The yield of $(\text{Me}_3\text{Si})_2\text{S}$

$$(\text{Me}_3\text{Si})_2\text{S} + 2\text{Me}_3\text{SiCl} + 2\text{N}_2 \xleftarrow[+2\text{BSD}]{\text{(a)}} \text{SCl}_2 \xrightarrow[+\text{BSD}]{\text{(b)}} \tfrac{1}{8}\text{S}_8 + 2\text{Me}_3\text{SiCl} + \text{N}_2 \quad (85)$$

and S_8 in Et_2O are 35 and 65 % and in CH_2Cl_2 75 and 25 % (27, 50, 63). On the other hand, if BSD is added dropwise to SCl_2 solution, then, as expected,

---

[14] For preparation of $\text{Me}_3\text{SiSR'}$, R'SCl is purposely added dropwise to a BSD solution and not the reverse.

[15] $\text{Me}_3\text{SiSN(SiMe}_3)_2$ and $\text{Me}_3\text{SiSN}_2(\text{SiMe}_3)_3$ represent the silyl derivatives of the unknown thiohydroxylamine (HSNH_2) and thiohydroxylhydrazine (HSN_2H_3) (27).

sulfur is formed quantitatively [Eq. (85b)]. *Selenium dichloride*, $SeCl_2$, and *tellurium dichloride*, $TeCl_2$, are also reduced by BSD to selenium and tellurium as observed in the reactions of $SeCl_4$ and $TeCl_4$ with BSD, which must proceed through the dichlorides (63).

In a way analogous to Eq. (85), *disulfur dichloride*, S_2Cl_2, reacts with BSD to form sulfur and bis(trimethylsilyl)disulfane, with the latter decomposing slowly: $(Me_3Si)_2S_2 \rightarrow (Me_3Si)_2S + \frac{1}{8}S_8$ (27, 50, 63, 69a) [Eq. (86)]. Of

$$(Me_3Si)_2S_2 + 2Me_3SiCl + 2N_2 \xleftarrow[+2BSD]{(a)} S_2Cl_2 \xrightarrow[+BSD]{(b)} \tfrac{1}{4}S_8 + 2Me_3SiCl + N_2 \quad (86)$$

course, the reaction follows the path of Eq. (86a) because of rapid silylation, and the yield is quantitative if S_2Cl_2 is added dropwise to a BSD solution in methylene chloride. Quantitative formation of sulfur is observed if BSD is added dropwise to the S_2Cl_2 solution. *Diselenium dichloride*, Se_2Cl_2, reacts with BSD in a way corresponding to Eq. (86). Instead of the expected bis(trimethylsilyl)diselenane, $(Me_3Si)_2Se_2$, the decomposition products formed are $(Me_3Si)_2Se$ and Se_x (63).

Like the above-mentioned sulfenyl chlorides, —S—Cl, sulfinyl and sulfonyl chlorides, —SO—Cl and SO_2—Cl, react with BSD and silylated organyldiazenes to form an azo compound first. Because subsequent reactions occur rapidly, the azo compounds cannot be isolated. The action of sulfonic acid chloride, $R'SO_2Cl$, (R' = Tol) on BSD below $-10°C$ in diethyl ether, e.g., leads quantitatively to **45** (29) [Eq. (87a)]. Apparently

$$2BSD + TolSO_2Cl \longrightarrow \underset{Me_3Si}{\overset{TolSO_2}{>}} N-N \underset{SiMe_3}{\overset{SiMe_3}{<}} + Me_3SiCl + N_2 \quad (87a)$$
$$\underset{\mathbf{45}}{}$$

an azo compound, $TolSO_2$—N=N—$SiMe_3$, is formed first which, in combination with BSD, changes to **45**. Likewise, the azo compounds $R'SO_2$—N=N—R, generated by reaction of Me_3Si—N=N—R (R = Me, Ph) with $R'SO_2Cl$ (R' = Me, Tol), could not be isolated because of their decomposition to a variety of products under reaction conditions (70). A uniform course is followed only by the reaction of Me_3Si—N=N—'Bu with $R'SO_2Cl$ (29) [Eq. (87b)].

$$Me_3Si-N=N-{}^tBu + R'SO_2Cl \rightarrow Me_3SiSO_2R' + {}^tBuCl + N_2 \quad (87b)$$

Thionyl chloride, $SOCl_2$, and *sulfuryl chloride*, SO_2Cl_2, react with BSD in methylene chloride at $-78°C$ according to Eq. (88) and (89) (*63*).

$$2BSD + SOCl_2 \rightarrow 2Me_3SiCl + (Me_3Si)_2O + \tfrac{1}{8} S_8 + 2N_2 \tag{88}$$

$$2BSD + SO_2Cl_2 \rightarrow 2Me_3SiCl + (Me_3SiO)_2S + 2N_2 \tag{89}$$

Their reactions appear to take the route

$$BSD + ClSO_nCl \rightarrow ClSO_n-N{=}N-SiMe_3 + Me_3SiCl \rightarrow SO_n + 2Me_3SiCl + N_2$$

The sulfur oxides SO and SO_2 react further with BSD to form very unstable products, $Me_3S-OSiMe_3$ [$\rightarrow(Me_3Si)_2O + \tfrac{1}{8}S_8$] or $Me_3SiO-S-OSiMe_3$ (metastable below $-40°C$) (Section VI,D).

6. *Transition Metal Halides*

Transition metal halides L_mMX_n (L = ligand, X = halogen, m = 0, 1, 2 . . .) undergo multiple reduction with BSD. The reaction products are transition metal halides L_mMX_{n-p} in lower oxidation states, complexes such as L_mM, or the metal itself (*49*). The products, L_mMX_{n-p}, L_mM, or M can, in some cases, react further with BSD to form complexes (cf. Section VII). In this way, BSD in methylene chloride transforms *titanium tetrachloride*, $TiCl_4$, to titanium dichloride which, being a very mild oxidizing agent, is incapable of further reduction (with BSD) to the metallic state (*50*) [Eq. (90)]. Complete reduction to the metallic state, on the other hand, has been observed

$$TiCl_4 + BSD \rightarrow 2Me_3SiCl + N_2 + TiCl_2 \tag{90}$$

by the action of BSD on *iron trichloride*, $FeCl_3$, *cuprous halides*, CuX (X = Cl, Br, I), and *mercuric dichloride*, $HgCl_2$ (*49, 71*). Iron, formed in this way, takes up BSD to form $FeBSD_x$ (*49*) [Eq. (91) (M = Fe, Cu, or Hg; n = 1, 2, or 3)].

$$2MX_n + nBSD \rightarrow 2nMe_3SiX + nN_2 + M \xrightarrow[M\,=\,Fe]{+xBSD} FeBSD_x) \tag{91}$$

Bis(cyclopentadienyl)titanium dichloride, Cp_2TiCl_2, reacts with BSD in THF to generate Cp_2Ti, which, in turn, is stabilized by complex formation with BSD (*72*). The same is true for reactions of BSD with other cyclopentadienylmetal halides, such as $CpTiCl_3$, $CpVCl_2$, $CpVCl_3$, Cp_2CrI, and $CpCrCl_2$ (*49, 73*) [Eq. (92)].

$$2Cp_mMX_n + nBSD \rightarrow 2nMe_3SiX + nN_2 + Cp_mM \xrightarrow{+xBSD} Cp_mMBSD_x) \tag{92}$$

The reactions shown in Eqs. (90)–(92) evidently take the course mentioned in Scheme 6, wherein the primary adduct $L_mMX_n \cdot BSD$ transforms into a diazenido complex, $L_mMX_{n-1}-N{=}N-SiMe_3$, which with the cleavage of trimethylhalosilane and nitrogen leads to the reduction product L_mMX_{n-2}. Therefore, cyclopentadienylmetal halides such as Cp_2CoCl or Cp_2VCl_2, which are not Lewis acids, fail to react with BSD. In contrast to Cp_2TiCl_2, the complex bis(trimethylsilylcyclopentadienyl)titanium dichloride, $(C_5H_4SiMe_3)_2TiCl_2$, in which titanium is sterically hindered, does not react with BSD (49).

The postulated intermediate diazenido complexes are indirectly suggested by the reaction of BSD with *zinc dichloride*, $ZnCl_2$, in diethyl ether. The halide functions as a very weak oxidizing agent (normal potential for $Zn/Zn^{2+} = -0.763$ V) and hence is not reduced by BSD (49), but behaves instead as a dimerization catalyst. In addition, it transforms BSD to Me_3SiCl, Me_3SiN_3, and zinc chloride bis(trimethylsilyl)amide (cf. SiF_4, Section VI,C,3) [Eq. (93)]. The formation of the latter compounds occurs through

$$2\,BSD \quad \xrightarrow[-Me_3SiCl,\ N_2]{+ZnCl_2} \quad Me_3Si-N{=}N{=}N + (Me_3Si)_2NZnCl$$

$$\Big\downarrow (ZnCl_2) \tag{93}$$

$$(Me_3Si)_2N-N{=}N-N(SiMe_3)_2$$

a path involving the BSD substitution product $Me_3Si-N{=}N-ZnCl$ ($BSD + ZnCl_2 \rightarrow Me_3Si-N{=}N-ZnCl + Me_3SiCl$), which reacts further with excess BSD:

$$Me_3Si-N{=}N-ZnCl + BSD \rightarrow (Me_3Si)_2N-N{=}N-N(SiMe_3)ZnCl \rightarrow$$
$$Me_3SiN_3 + (Me_3Si)_2NZnCl$$

Diazenido complexes were isolated as products from the reaction of trimethylsilylphenyldiazene, $Me_3Si-N{=}N-Ph$ (instead of BSD), with *cyclopentadienyltitanium trichloride* ($CpTiCl_3$) (74), or *pentacarbonylmanganese bromide* [$(CO)_5MnBr$] (75) [Eqs. (94) and (95)].

$$Me_3Si-N{=}N-Ph + CpTiCl_3 \rightarrow CpTiCl_2-N{=}N-Ph + Me_3SiCl \tag{94}$$

$$2\,Me_3Si-N{=}N-Ph + 2(CO)_5MnBr \rightarrow [(CO)_4Mn-N{=}N-Ph]_2 + 2\,Me_3SiBr \tag{95}$$

D. *Reactions with Oxides, Sulfides, and Imides*

1. *Oxides, Sulfides, and Imides of Carbon*

Carbon dioxide, CO_2, and *carbon disulfide*, CS_2, insert into the Si–N bonds of silylamines (76). In this way, BSD should react with CY_2 (Y = O, S) to give the unknown azocarbonic acid esters **46** and **47** [Eq. (96)]. Indeed,

$$
BSD \xrightarrow{+CY_2} \underset{\underset{\underline{46}}{Me_3SiY}}{\overset{Y}{\diagdown}}C-N=N-SiMe_3 \xrightarrow{+CY_2} \underset{\underset{\underline{47}}{Me_3SiY}}{\overset{Y}{\diagdown}}C-N=N-C{\overset{Y}{\diagup}}_{YSiMe_3} \tag{96}
$$

BSD solution in diethyl ether reacts with carbon dioxide and disulfide below 0°C (Scheme 8). However, instead of following Eq. (96), reactions occur with

$$
2BSD + CY_2 \xrightarrow[\;-N_2\;]{(CO_2,\ COS,\ CS_2)} \underset{\underline{48a}}{\underset{YSiMe_3}{Me_3Si\diagdown}}Y=C{\overset{N-N\diagup SiMe_3}{\diagup}}_{SiMe_3} \rightleftarrows Me_3SiY-C{\overset{N-N\diagdown SiMe_3}{\diagup}}_{\underset{\underline{48b}}{\underset{YSiMe_3}{SiMe_3}}} \tag{97}
$$

$$
\xrightarrow[-(Me_3Si)_2Y]{(CO_2,\ CS_2)} \underset{\underline{49a}}{\overset{YSiMe_3}{N{\diagdown}}}\!\!\!\!N{\overset{C}{=}}N\text{-}SiMe_3 \rightleftarrows \underset{\underline{49b}}{Me_3Si} \rightleftarrows Me_3Si-N{\overset{Y}{=}}\!\!N\text{-}SiMe_3 \tag{98}
$$

$$
\xrightarrow[-N_2]{(CS_2)} \underset{\underline{50}}{\overset{Me_3Si\diagdown}{Me_3Si}}N\text{-}S\text{-}SiMe_3 + Me_3Si\text{-}N=C=S \tag{99}
$$

$$
\xrightarrow[-N_2]{(CS_2)} \underset{\underline{51}}{\overset{Me_3Si\diagdown}{Me_3Si}}N\text{-}S\text{-}C{\equiv}N + (Me_3Si)_2S \tag{100}
$$

SCHEME 8

the evolution of N_2 and formation of $(Me_3Si)_2Y$ to give hydrazine **48** [Eq. (97)] as well as five-membered ring compounds **49** [Eq. (98)] in 84 and 15% (Y = O) and 26 and 27% yield (Y = S). In addition, CS_2 reacts with the evolution of nitrogen to form $Me_3Si-N=C=S$ (23%) as well as $(Me_3Si)_2S$ (14%) and the amines **50** and **51** [Eqs. (99) and (100)] beside other compounds in smaller amounts (27, 77). *Carbonyl sulfide*, COS, reacts with BSD exclusively to form hydrazine **48** (27).

As illustrated in Scheme 8, a number of constitutional isomers are possible for compounds **48** and **49**. NMR and IR spectroscopy (27, 29, 77) show the existence of equilibrium mixtures **48a** $\rightleftarrows$ **48b** and **49a** $\rightleftarrows$ **49b** in the case of CO_2 reaction products, the presence of **48a** for COS reaction products, and the existence of **48a** and **49c** for CS_2 reaction products.

Formation of products **48–51** (Scheme 9) progresses through intermediate **46**, which with the addition of BSD may transform to adducts **52** and **53** which, in turn, would decompose to the observed end products. Reaction of **46** with more BSD evidently takes place so fast that the further storage of the CY_2 in **46** in the form of **47** does not happen even with a large excess of CY_2.

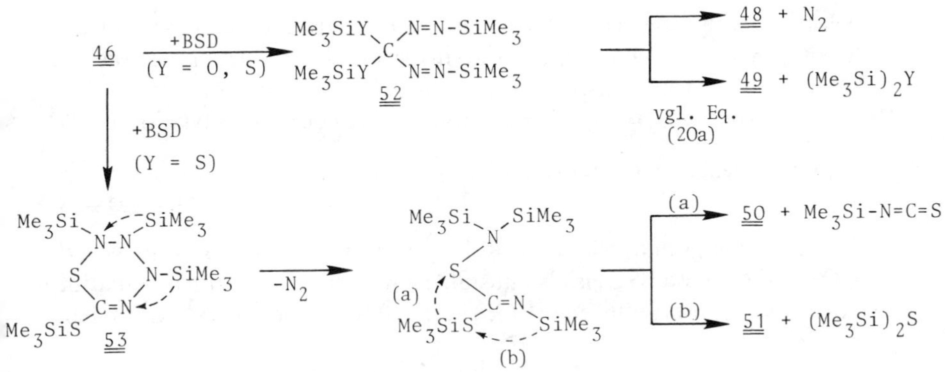

SCHEME 9

Formation of **46** as a result of reaction of BSD with CO_2, COS, and CS_2 is supported by the reaction of BSD with *phenylisocyanate*, $PhN{=}C{=}O$, which in the sense of Eq. (96) leads to **54** and **55** (*78*) [Eq. (101)]. Inter-mediate formation of **46** is further supported by the reaction of BSD with

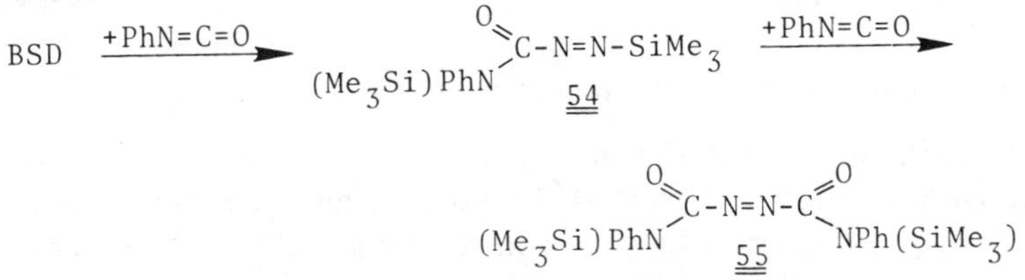

acetone, $Me_2C{=}O$, which inserts itself in the Si–N bond to form the red or yellow compounds **56** or **57** (*50*) [Eq. (102)].

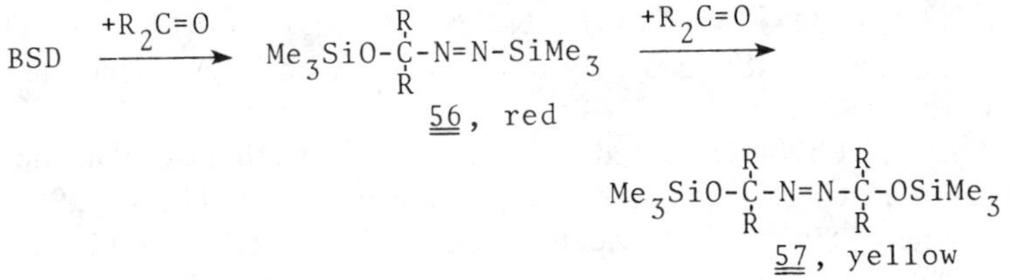

The same is true for other carbonyl compounds, $R_2C{=}O$, so long as R is an alkyl group. If R is a phenyl group, e.g., the reaction of BSD with

benzophenone, $Ph_2C=O$, then **58** is formed with the evolution of nitrogen
(50) [Eq. (103)]. Probably an azo compound of type **56** or **57** is formed,

$$BSD + 2Ph_2C=O \rightarrow Me_3SiO\text{-}\underset{\underset{Ph}{|}}{\overset{\overset{Ph}{|}}{C}}\text{-}\underset{\underset{Ph}{|}}{\overset{\overset{Ph}{|}}{C}}\text{-}OSiMe_3 + N_2 \qquad (103)$$

$$\underline{58}$$

which then, with cleavage of N_2, decomposes to the recombining radicals,
$(Me_3SiO)Ph_2C$. Formation of **56** as an intermediate product of the reaction
of BSD with Ph_2CO is also supported by the appearance of side product
$Ph_2C=N—N(SiMe_3)_2$, which may arise in the course of silylation of **56**:

56 + BSD → $(Me_3SiO)Ph_2C—N_2(SiMe_3)_3 + N_2 \rightarrow$

$$Ph_2C=N—N(SiMe_3)_2 + (Me_3Si)_2O + N_2$$

The azo compound **59**, formed by reaction of the diketone *benzil*,
PhCO—COPh, with BSD, is also unstable and decomposes with elimination
of N_2 to give the *trans*-ethylene **60** (27) [Eq. (104)]. A corresponding ethylene

$$BSD + \underset{Ph}{\overset{O}{\diagdown}}C\text{-}C\underset{Ph}{\overset{O}{\diagup}} \longrightarrow Me_3SiO\text{-}\underset{\underset{Ph}{|}}{\overset{\overset{N\diagdown^{N\text{-}SiMe_3}}{||}}{C}}\text{-}C\underset{Ph}{\overset{O}{\diagup}} \xrightarrow{-N_2} Me_3SiO\text{-}\underset{\underset{Ph}{|}}{C}=\underset{\underset{Ph}{|}}{C}\text{-}OSiMe_3 \qquad (104)$$

$$\underline{59} \qquad\qquad\qquad \underline{60}$$

is also formed by the action of BSD on *oxalylchloride*, ClCO—COCl
[i.e., Cl in place of Ph in Eq. (104)].

2. *Oxides and Imides of Nitrogen and Phosphorus*

Nitric oxide, NO, and BSD react vigorously in pentane at $-78°C$ with
quantitative formation of hexamethyldisiloxane, nitrogen, and dinitrogen
oxide (79) [Eq. (105)]. In analogy with the reaction of BSD with diradical O_2

$$BSD + 2NO \rightarrow (Me_3Si)_2O + N_2 + N_2O \qquad (105)$$

(Section VI,B,2), the reaction with NO radical [Eq. (105)] perhaps takes
place as $BSD + NO \rightarrow [Me_3SiNO + N\equiv N + SiMe_3]_{cage} \rightarrow (Me_3Si)_2NO + N_2$. The radicals $(Me_3Si)_2NO$ may react further with NO to generate the
reaction products.

Nitrobenzene, $PhN=O$, is reduced by BSD in methylene chloride or
toluene at $-78°C$ to bis(trimethylsilyl)-*N*-phenylhydroxylamine (**59**) and
azoxybenzene in 80 and 20% yields, respectively (50, 63) [Eq. (106a)].[16]

---

[16] Nitrobenzene ($PhNO_2$) is reduced by BSD almost exclusively to azoxybenzene (63).

$$\underset{\substack{Me_3Si \quad \underline{\underline{59}} \quad SiMe_3}}{\overset{Ph}{\underset{}{\diagdown}}N-O} \quad \xleftarrow[-N_2]{+PhNO} \quad BSD \quad \xrightarrow[-(Me_3Si)_2O, \ -N_2]{+2PhNO} \quad \underset{O}{\overset{Ph}{\diagdown}}N=N\diagdown_{Ph}$$

$$(106a)$$

The reaction is apparently initiated through insertion of the nitroso group at the Si–N bond [Eq. (106b)]. In this respect the reaction resembles that

$$BSD \quad \xrightarrow{+PhN=O} \quad \underset{Me_3SiO}{\overset{Ph}{\diagdown}}N-N=N-SiMe_3 \quad \xrightarrow{+PhN=O} \quad \underset{Me_3SiO}{\overset{Ph}{\diagdown}}N-N=N-N\overset{Ph}{\underset{OSiMe_3}{\diagup}}$$
$$\underline{\underline{61}} \qquad\qquad\qquad \underline{\underline{62}}$$

$$(106b)$$

with ketones [Eq. (102)]. The intermediate product **61** decomposes with N_2 elimination to **59**, whereas **62** undergoes cleavage of N_2, dimerization of $(Me_3SiO)PhN$ radicals so formed, and elimination of $(Me_3Si)_2O$ to form azoxybenzene [cf. the formation of **58** from BSD + Ph_2CO, Eq. (103)].

As compared to nitrosobenzene, BSD reacts more slowly with *methyl nitrite*, MeO—NO, but faster with *nitrosyl chloride*, Cl—NO. The compounds $(Me_3Si)_2O$, Me_3SiX (X = OMe or Cl), Me_3SiN_3, and N_2 are formed as reaction products (63). *Dimethylnitrosamine*, Me_2N—NO, on the other hand, is inert toward BSD. Probably, reactions with X—NO (X = MeO, Cl) occur analogously to reactions with Ph—NO, to form first a hydroxylamine derivative $(Me_3Si)XN$—$OSiMe_3$, which must dissociate to give $(Me_3Si)_2O$ and Me_3Si—NO. The nitrososilane Me_3Si—NO so formed can insert into the Si–N bond of BSD to form $(Me_3SiO)(SiMe_3)N$—N=N—$SiMe_3$, which decomposes with the elimination of $(Me_3Si)_2O$. However, other mechanisms are also possible.

Addition of BSD occurs not only at the C=O and N=O double bonds of ketones and nitroso derivatives but also at the "electrophilic" N=N double bond of azodicarbonic acid esters (solvent CH_2Cl_2, $-78°C$):

$$BSD + ROOC—N=N—COOR \rightarrow Me_3Si(ROOC)N—N(COOR)—N=N—SiMe_3$$

The adduct decomposes almost at once with N_2 cleavage and silyl-group transfer **63** (29, 63) [Eq. (107)].

$$BSD + \underset{RO}{\overset{O}{\diagdown}}C-N=N-C\overset{O}{\underset{OR}{\diagup}} \quad \xrightarrow{-N_2} \quad \underset{Me_3SiO}{\overset{RO}{\diagdown}}C=N-N=C\overset{OR}{\underset{OSiMe_3}{\diagup}} \qquad (107)$$
$$\underline{\underline{63}}$$

In contrast, the compound R—P=N—R' [R = $(Me_3Si)_2N$; R' = Me_3Si, formally an "imide of phosphorus"] cannot be silylated. Instead, BSD adds to [bis(trimethylsilyl)aminotrimethylsilyl]phosphazene in ether at $-30°C$ with the oxidation of phosphorus as shown in Eq. (108). Reaction of BSD with phosphazenes, leading to compounds with $P=N-N(SiMe_3)_2$ groups, is capable of generalization (80) [Eq. (109)] (cf. Section VII).

$$BSD + (Me_3Si)_2N-P=N-SiMe_3 \rightarrow (Me_3Si)_2N-P\begin{smallmatrix} \diagup N-SiMe_3 \\ \underset{\underline{\underline{64}}}{\diagdown} N-N(SiMe_3)_2 \end{smallmatrix} \qquad (108)$$

$$BSD + Me_2P-N=CPh_2 \rightarrow Me_2P\begin{smallmatrix} \diagup N=CPh_2 \\ \diagdown N-N(SiMe_3)_2 \\ \underline{\underline{64}} \end{smallmatrix} \qquad (109)$$

3. Oxides and Imides of Sulfur

BSD reacts (in ether or methylene chloride) with *sulfur dioxide*, (SO_2), *sulfur imide oxides* (sulfinylimines), $RN=S=O$ ($R = SiMe_3$, Ph, $PhSO_2$), and *sulfur diimides*, $RN=S=NR$ ($R = $ Ph, Me_3Si), in the same way [Eq. (110)] (X, Y = O, NR) with reduction of sulfur from oxidation state

$$BSD + X=S=Y \rightarrow Me_3Si-X-S-Y-SiMe_3 + N_2 \qquad (110)$$
$$\underline{\underline{65}}$$

four to two (27, 50, 63). Reduction occurs near $-78°C$ for SO_2 and $RN=S=O$ and at $-30°$ for $RN=S=NR$. The formation of **65** is expected to occur through insertion of SXY in a Si–N bond of BSD with subsequent N_2 cleavage from the insertion product.

Bis(trimethylsilyl)sulfoxylate **65** (X, Y = O), made accessible for the first time according to Eq. (110), decomposes above $-40°C$ to yield SO_2, S_8, and $(Me_3Si)_2O$ [Eq. (111)]. In the presence of SO_3, this decomposition

$$2Me_3Si-O-S-O-SiMe_3 \rightarrow SO_2 + \tfrac{1}{8}S_8 + 2(Me_3Si)_2O \qquad (111)$$

occurs near $-78°C$, whereby an SO_3 adduct of hexamethyldisiloxane, bis(trimethylsilyl)sulfate, $(Me_3Si)_2SO_4$, is produced (63).

Sulfur trioxide, SO_3, is likewise reduced by BSD. The expected product, bis(trimethylsilyl)sulfite, corresponding to Eq. (112), is not isolable because

$$BSD + SO_3 \rightarrow Me_3SiO-\overset{O}{\underset{\shortmid}{S}}-OSiMe_3 + N_2 \qquad (112)$$

it is thermally unstable and reacts further with SO_3 to form SO_2 and $(Me_3Si)_2SO_4$. Since SO_2, according to Eq. (110), can react first with BSD and then with SO_3, the total reaction of BSD with SO_3 ends up with the formation of $(Me_3Si)_2SO_4$, N_2, and sulfur as shown in Eq. (113).

$$3BSD + 4SO_3 \rightarrow (Me_3Si)_2SO_4 + \tfrac{1}{8}S_8 + 3N_2 \qquad (113)$$

Sulfur monoxide, SO, which is required as an intermediate reaction product of BSD with $SOCl_2$ (Section VI,C,5), is silylated by BSD: BSD + SO → $Me_3Si—S—O—SiMe_3 + N_2$. Trimethylsiloxytrimethylsulfane, $Me_3Si—S—X$ (X = $OSiMe_3$), as expected, is very unstable and decomposes under condensation to Me_3SiX and sulfur so that the result of reaction of BSD with thionyl chloride, even at very low temperatures, leads only to products formed according to the equation $2\,BSD + SOCl_2 → 2\,Me_3SiCl + (Me_3Si)_2O + \frac{1}{8}S_8 + 2\,N_2$ (50, 63).

E. *Other Reagents*

1. *Azides*

Among reactions of BSD, conversion of azides to amines, **66**, deserves special attention (50, 81, 82) [Eq. (114)]. Below 0°C, BSD reacts only with

$$BSD + X-N=N=N → \underset{\underline{66}}{X-N(SiMe_3)_2} + 2N\equiv N \tag{114}$$

electrophilic azides, XN_3 [e.g., X = acyl, tosyl, $(RO)_3Si$, but not Ph or R_3Si], and the rate of the reaction increases with the polarity of the reaction medium ($Et_2O < CH_2Cl_2$). The reaction is expected to begin by electrophilic attack of the azides at the azo group of BSD to form **67**. This intermediate can then undergo silyl group transfer and nitrogen cleavage to form triazene **68**, which changes easily to **66** [Eq. (115)]. The rate of the reaction

$$BSD \xrightarrow{+XN_3} \underset{\underline{67}}{X-\overset{..}{N}=\overset{..}{N}-\overset{..}{\underset{|}{N}}-\overset{+}{\underset{SiMe_3}{N}}=\overset{..}{N}-SiMe_3} \xrightarrow{-N_2} \underset{\underline{68}}{X-\overset{..}{N}=\overset{..}{N}-\overset{..}{N}\overset{SiMe_3}{\underset{SiMe_3}{<}}} \xrightarrow{-N_2} \underline{66} \tag{115}$$

[Eq. (114)] increases in the following order of azides: $(MeO)_3Si—N_3 < p\text{-}TolSO_2—N_3 < {}^tBuOCO—N_3$. The postulated intermediate product **68** has not yet been trapped in any of these cases.

2. *Organolithium Compounds*

Organolithium compounds react with BSD in the sense of Scheme 6 with addition at the double bond to form hydrazides **69** [Eq. (116a)] as well as with substitution of silyl by organyl groups to give tetraorganylsilanes and lithium trimethylsilyldiazenide (**70**) (83). However, instead of compound

70, only the secondary products LiN_3 and $LiN(SiMe_3)_2$ are isolable [Eq. (116b,c)]. Addition increases at the expense of substitution in the order LiMe < LiBu < LiPh.

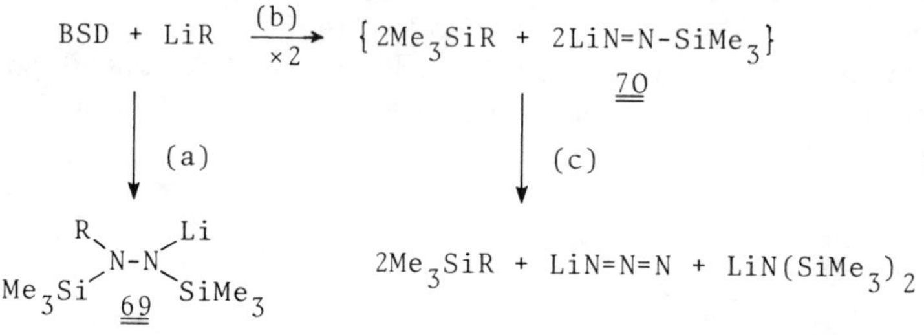

$$BSD + LiR \xrightarrow[\times 2]{(b)} \{2Me_3SiR + 2LiN=N-SiMe_3\}$$

$$\underset{70}{}$$

$$\downarrow (a) \qquad\qquad \downarrow (c) \qquad\qquad (116)$$

$$\underset{\underset{69}{Me_3Si}}{\overset{R}{\diagdown}}\underset{SiMe_3}{N-N}\overset{Li}{\diagup} \qquad 2Me_3SiR + LiN=N=N + LiN(SiMe_3)_2$$

Likewise, trimethylsilylphenyldiazene reacts with organolithium compounds by addition at the azo group and by cleavage of the Si–N bond (29). Once again, the substitution product is not isolable and decomposes with the elimination of nitrogen [Eq. (117)]. With regard to the reaction sequence shown in Eq. (117b,c), phenyllithium consequently acts only as a catalyst for the decomposition: $Me_3Si—N=N—Ph \rightarrow Me_3SiPh + N_2$. On interposing other lithium organyls (e.g., LiMe), phenyllithium is a reaction product and can, in turn add, so that two hydrazines (e.g., with R = Me and Ph) are formed according to Eq. (117a). The reaction of $Me_3Si—N=N—Ph$ with NaOMe (30) proceeds in conformity with Eq. (117b,c). Thus, the system $Me_3SiNNPh/NaOMe$ acts as a generator of phenyl anion.

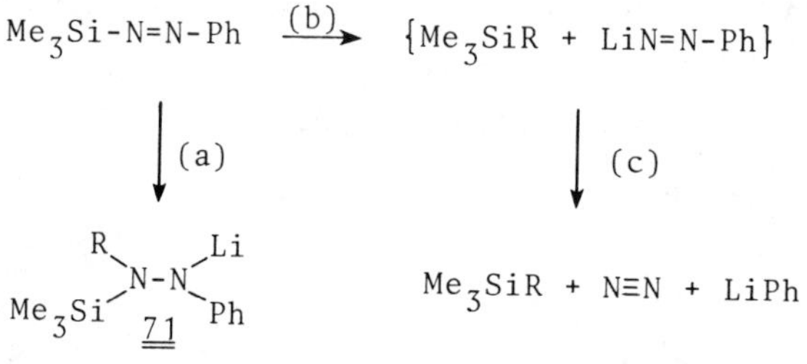

$$Me_3Si-N=N-Ph \xrightarrow{(b)} \{Me_3SiR + LiN=N-Ph\}$$

$$\downarrow (a) \qquad\qquad \downarrow (c) \qquad\qquad (117)$$

$$\underset{\underset{71}{Me_3Si}}{\overset{R}{\diagdown}}\underset{Ph}{N-N}\overset{Li}{\diagup} \qquad Me_3SiR + N\equiv N + LiPh$$

The lithium organyl LiMe does not add to trimethylsilylmethyldiazene, $Me_3Si—N=N—Me$ (29), but acts as a catalyst for the isomerization of this diazene to hydrazone **11** (for the mechanism, cf. Section IV,B) as well as for the decomposition to tetramethylsilane and nitrogen [for mechanism, cf. Eq. (117b,c)]. With an increasing amount of LiMe, the rate of catalytic

$$\underset{11}{Me_3SiNH-N=CH_2} \xrightarrow[(LiMe)]{(a)} Me_3Si-N=N-Me \xrightarrow[(LiMe)]{(b)} Me_3SiMe + N_2 \quad (118)$$

decomposition of the compound increases faster than the rate of isomerization, so that in the presence of 50 mol % LiMe at $-78°C$, the reaction [Eq. (118b)] is complete in a few minutes in 100% yield. On the other hand, the other reaction [Eq. (118a)] runs almost quantitatively if a very limited amount of LiMe is used (e.g., 80% yield in the presence of 8% LiMe) because catalyst is abstracted from reaction medium as

$$11 + MeLi \rightarrow Me_3Si-NLi-N=CH_2.$$

The lithium salt of **11** catalyzes the isomerization of $Me_3Si-N=N-Me$ (Section IV,B) and not decomposition.

Analogously to lithium organyls, other main group metal organyls also react with silylated and germylated diazenes and organyldiazenes, primarily with addition to the azo system. Even trimethylboron may add to $Me_3Si-N=N-Me$ or $Me_3Si-N=N-Ph$ in CH_2Cl_2 at room temperature or $60°C$ quantitatively with the formation of hydrazines **72a** (R = Me) or **72a** and **72b** (R = Ph) (29) [Eq. (119)].

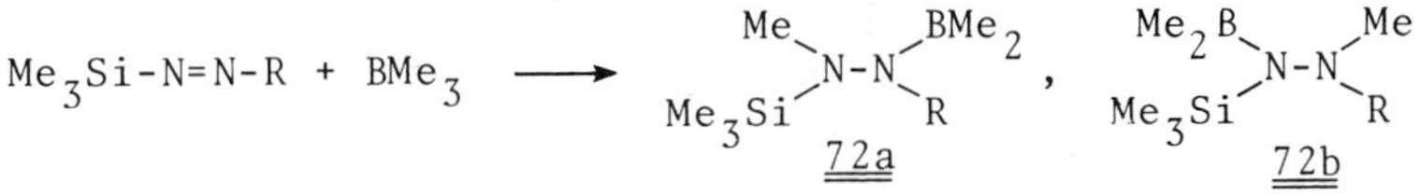

$$Me_3Si-N=N-R + BMe_3 \longrightarrow \underset{\underline{72a}}{\overset{Me}{\underset{Me_3Si}{\diagdown}}N-N\overset{BMe_2}{\underset{R}{\diagup}}} , \underset{\underline{72b}}{\overset{Me_2B}{\underset{Me_3Si}{\diagdown}}N-N\overset{Me}{\underset{R}{\diagup}}} \qquad (119)$$

VII

BSD Complexes

A. General Features

Bis(organyl)diazenes (organic azo compounds), $\geq C-\ddot{N}=\ddot{N}-C\leq$, may use their free n-electron pairs to form σ complexes and their π electron pair to give π complexes with transition metals and other elements (84). In bis(silyl)diazenes the HOMO (n) lies much higher and the LUMO (π^*) somewhat lower than for organic azo compounds (see Section III,B). On this basis silyl diazenes should be good ligands, acting as strong n donors and moderate π acceptors. In fact, BSD does serve as a strong Lewis base. BSD combines with Lewis acid SiF_4 to give a blue σ complex, $SiF_4 \cdot BSD$ (**31**) (Section VI,C,3), and it combines with Ph_2BCl to form a deep blue complex, $Ph_2BCl \cdot BSD$ (Section VI,C,1), and with Li^+ to form a brown complex, $Li(BSD)_2{}^+$ (Section VI,B,1). BSD also forms a CT complex with trinitrobenzene (v_{max} of CT absorption = 21000 cm^{-1}).

Cuprous halides, CuX (*49, 71*), suspended in diethyl ether at $-50°C$ (X = Cl), $-30°C$ (X = Br), or room temperature (X = I), react with BSD to form solid complexes: BSD · 2CuCl (reddish brown, insoluble in Et_2O, decomposes $> -30°C$ in Cu, N_2, Me_3SiCl), BSD · 2CuBr (reddish brown, insoluble in Et_2O, decomposes $> -10°C$ in Cu, N_2, Me_3SiBr), and BSD · CuI (lemon yellow, soluble in Et_2O, decomposes $> -20°C$ in Cu, N_2, Me_3SiI). In high vacuum, BSD · CuI loses half of the BSD at $-10°C$ to form BSD · 2CuI (reddish brown, insoluble in Et_2O, decomposes $>0°C$). In the complexes BSD · 2CuX, the BSD molecules may be, as in the case of the azomethane complex Me_2N_2 · 2CuCl (*85*), coordinated to two Cu^+ ions which are each joined by Cl^- ions.

The σ complexes **73** of BSD with Lewis acids L_nM are mostly unstable (Scheme 10). The complexed silylated diazene $Me_3Si—N{=}N—SiMe_3$ rearranges with silyl group migration to complexed isodiazene $\ddot{N}—N(SiMe_3)_2$ (**75**) [Eq. (120b)], adds BSD to form complexed tetrazene (**74**) [Eq. (120a)], or cleaves into two complexed nitrenes, $\ddot{N}—SiMe_3$ (**76**) [Eq. (120c)]. The

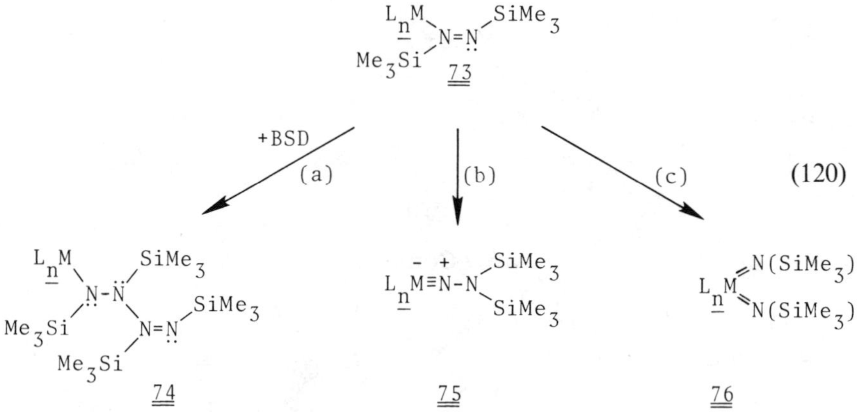

SCHEME 10

complexes **74** could be isolated only in the form of their dissociation products L_nM and $(Me_3Si)_2N—N{=}N—N(SiMe_3)_2$ (cf. Section VI,C,3).

B. *BSD Reactions with Cp_2M*

BSD combines with metallocenes Cp_2Ti, Cp_2V, Cp_2Cr, and Cp_2Mn to form coordination compounds. Cp_2Fe does not react, whereas Cp_2Co and Cp_2Ni react in a complex fashion.

1. *Bis(cyclopentadienyl)titanium and -vanadium*

Vanadocene and BSD react in diethyl ether to form a dark brown, oxygen-sensitive paramagnetic complex, $Cp_2V \cdot BSD$ (77), with a terminal isodiazene ligand (72)[17] [Eq. (121)]. Compound 77 is stable to $100°C$ and can be

$$Cp_2V \xrightarrow{+BSD} \underset{\underline{77}}{Cp_2V \equiv N - \overset{..}{N}} \overset{- \quad +}{\underset{SiMe_3}{\overset{SiMe_3}{<}}} \tag{121}$$

sublimed at $60°C$ in vacuum. Characteristic structural features of 77 are a linear VNN and a planar $NNSi_2$ atomic network, a short V–N internuclear distance (166.6 pm) indicative of multiple bonding, and an almost normal N–N internuclear distance characteristic of a single bond (86). Structurally, 77 is identical with a dark red, air-sensitive, paramagnetic complex $Cp_2V \cdot BSD_{0.5}$ (78), which is formed from Cp_2V, Me_3SiN_3, and BSD in diethyl ether. The reaction occurs through $(Cp_2V)_2NSiMe_3$ (metallic dark red) to form 78 as well as 77 (87), as shown in Eq. (122).

$$2Cp_2V \xrightarrow{Me_3SiN_3} \underset{Cp_2V}{\overset{Cp_2V}{>}} N-SiMe_3 \xrightarrow[-\underline{77}]{+BSD} \underset{\underline{78}}{Cp_2 \overset{- \quad +}{V} \equiv N-SiMe_3} \tag{122}$$

Compound 78 exhibits a linear VNSi framework with a short V–N internuclear distance (166.5 pm). This finding, especially in view of the determined Si–N distance in the range of a single bond, indicates that like 77, 78 also has a triple bond between vanadium and nitrogen. In the mesomeric arrangement [Eq. (123), $R = SiMe_3$ or $N(SiMe_3)_2$], the righthand limiting

$$[Cp_2V = \overset{..}{N} - R \longleftrightarrow Cp_2 \overset{- \quad +}{V} \equiv N - R] \tag{123}$$

structure therefore has more weight; the nitrene ligand functions as a four-electron donor.[18]

The dark violet, oxygen-sensitive, diamagnetic complex 79 is prepared by reaction of BSD with Cp_2Ti in THF, obtained *in situ* by reaction of Cp_2TiCl_2 with BSD (cf. Section VI,C,6) according to Eq. (124). The structure is apparently like that of 77 (72).[19] A titanium complex, $Cp_2Ti = N - SiMe_3$,

---

[17] The formation of reddish-brown complex 77 occurs through a dark green isomer $Cp_2V \cdot BSD$, stable below $-20°C$, which perhaps contains unrearranged σ (or π) bound BSD (72).

[18] If the left-hand limiting structure had a higher weight, the nitrene would have acted only as two-electron donor, and as a result the VNR framework would have been angular.

[19] The reaction of Cp_2TiCl_2 with BSD leads to, beside 79, Me_3SiCP, red $Cp_3Ti_2H_2Cl \cdot BSD$ (structure unknown), and dimeric Cp_2Ti in small amounts. In contrast to Cp_2TiCl_2, Cp_2VCl_2 does not react with BSD.

corresponding to the vanadium complex **78** could not be synthesized in spite of many attempts (*88*).

$$Cp_2TiCl_2 \xrightarrow[-2Me_3SiCl, \ -N_2]{+BSD} Cp_2Ti \xrightarrow{+BSD} Cp_2\overset{-}{Ti}\equiv\overset{+}{N}-N\overset{SiMe_3}{\underset{SiMe_3}{\diagdown}} \quad (124)$$
$$\underline{79}$$

Neither **77** nor **79** could be converted to diazene complexes $Cp_2M=N-NH_2$ (M = Ti, V) by desilylation with ethereal hydrochloric acid because along with the Me_3Si/H exchange, the diazene ligand is dislodged (*49*). The protolysis [Eq. (125)] takes place in 90% (Ti) and 75%

$$Cp_2\overset{-}{M}\equiv\overset{+}{N}-N(SiMe_3)_2 + 6HCl \rightarrow Cp_2MCl_2 + N_2H_6Cl_2 + 2Me_3SiCl \quad (125)$$

(V) yield, and it also leads to the formation of $CpTiCl_3$ (10%) and red $[CpVClBSD]_2$ (25%). The latter products are evidently formed through the protolytic displacement of a Cp ligand and in case of **79**, the isodiazene ligand also. The methanolysis of **77** occurs exclusively with Cp cleavage and formation of brown $[CpVXBSD]_2$ (*71*) [Eq. (126)]. Reaction of **77** with

$$2 \ \underline{77} \ + \ 2HX \xrightarrow{X \ = \ Cl, \ OMe} [CpVXBSD]_2 + 2CpH \quad (126)$$

Me_3SnX likewise leads to cleavage of $CpSnMe_3$ yielding $(CpMXBSD)_2$ [regarding the structure of $(CpVXBSD)_2$, see Section VII,C,2].

2. Bis(cyclopentadienyl)chromium, -manganese, and -iron[20]

Chromocene and BSD react together at low temperatures in diethyl ether to form a metallic black, paramagnetic complex, Cp_2CrBSD (**80**), which decomposes above $-20°C$, and with ethereal HCl transforms to a compound with the formula $[CpCrBSD]_2^{2+} \cdot 2Cl^-$. In an analogous manner, manganocene and BSD combine at $-60°C$ in Et_2O to form an olive green, paramagnetic complex, Cp_2MnBSD (**81**), which thermolyzes at $-50°C$ [Eq. (127)].

$$Cp_2Cr \ (Cp_2Mn) \xrightarrow[-60°]{+BSD} Cp_2CrBSD \ (Cp_2MnBSD) \quad (127)$$
$$\underline{80} \qquad\qquad \underline{81}$$

There are no X-ray reports for complexes **80** and **81** as yet, but they most probably have structures similar to **77**.

---

[20] For a more complete discussion see refs. *49* and *72*.

The instability of **80** and the even greater instability of **81** may be interpreted in the following way: Nitrenes like $\ddot{N}$—$N(SiMe_3)_2$ act as four-electron donors [cf. Eq. (123)]. The metallic centers of Cp_2Ti (14 outer Ti electrons), Cp_2V (15 electrons), Cp_2Cr (16 electrons), and Cp_2Mn (17 electrons), on the addition of isodiazene ligands, therefore, possess 18, 19, 20, and 21 electrons, respectively. Even in the case of **77**, the excess electron (of the 18-electron shell), which may be partly transferred to the Cp ligand, makes the easier replaceability of Cp ligands noticeable. The Cp ligands in the case of isodiazene complexes of Cp_2Cr and Cp_2Mn are expected to be even more reactive. Actually the addition of BSD—according to thermolysis of **80** and **81**—is associated with the removal of a Cp ligand [cf. Section VII,C,1, Eq. (133)]. This idea is supported by the finding that ferrocene, Cp_2Fe, whose central Fe atom has a noble gas configuration, does not react with BSD. Cp_2Co^+, which is isoelectronic with Cp_2Fe, is also inert to BSD. In those cases of metallocenes Cp_2Co and Cp_2Ni that react with BSD, the reaction probably begins with attack of BSD on the Cp ring.

3. Bis(cyclopentadienyl)cobalt and -nickel

Cobaltocene reacts with BSD (excess) in diethyl ether near room temperature to form an orange-red, air-stable, diamagnetic complex, **82**. Nickelocene, under similar conditions, provides a red, highly air-sensitive, diamagnetic complex, **83** (49). According to spectroscopic investigations, the complexes **82** and **83** are believed to have the structures shown in Eqs. (128) and (129).

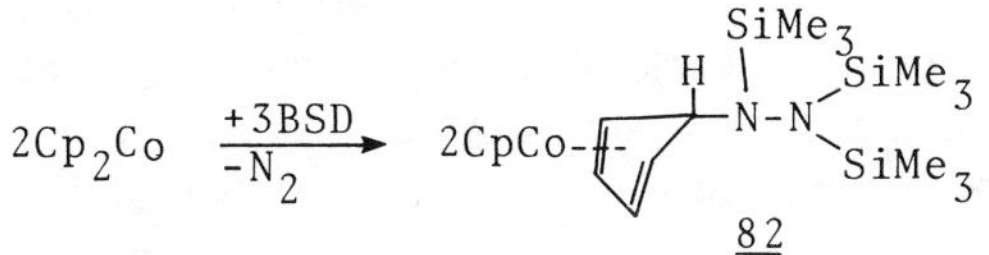

$$2Cp_2Co \xrightarrow[-N_2]{+3BSD} 2CpCo-\underset{\underline{82}}{\overset{H}{\diagdown}}\,N-N\overset{SiMe_3}{\underset{SiMe_3}{\diagup}} \qquad (128)$$

$$Cp_2Ni \xrightarrow[-N_2,\ -CpSiMe_3]{+BSD} CpNi-N-N\underset{\underline{83}}{\overset{SiMe_3}{\diagup}}SiMe_3 \qquad (129)$$

C. BSD Reactions with CpM and CpMCl

1. CpMBSD$_n$ Complexes.

In diethyl ether, BSD reacts with cyclopentadienyltitanium trichloride, $CpTiCl_3$, at $-78°C$ and with cyclopentadienylvanadium trichloride,

$CpVCl_3$, and cyclopentadienylchromium dichloride, $CpCrCl_2$, at room temperature with the evolution of nitrogen to form complexes $[CpTiBSD_2]$ (**84**) [Eq. (130)], $[CpVBSD_2]$ (**85**), $[CpVBSD]_2$ (**86**), $[CpVBSD_{0.75}]_2$ (**87**) [Eq. (131)], and $[CpCrBSD]_2$ (**88**) [Eq. (132)] (*49, 73*). The progress of

$$CpTiCl_3 \xrightarrow[-3Me_3SiCl, \ -N_2]{+1.5BSD} \text{"CpTi"} \xrightarrow{+2BSD} [CpTiBSD_2] \quad (130)$$
$$\underline{84}$$

$$CpVCl_3 \xrightarrow[-3Me_3SiCl, \ -1.5N_2]{+3BSD} \text{"CpV"} \xrightarrow{+nBSD}$$

$$[CpVBSD_2], \ \tfrac{1}{2}[CpVBSD]_2, \ \tfrac{1}{2}[CpVBSD_{0.75}]_2 \quad (131)$$
$$\underline{85} \qquad\qquad \underline{86} \qquad\qquad \underline{87}$$

$$CpCrCl_2 \xrightarrow[-2Me_3SiCl, \ -N_2]{+BSD} \text{"CpCr"} \xrightarrow{+BSD} \tfrac{1}{2}[CpCrBSD]_2 \quad (132)$$
$$\underline{88}$$

reaction indicates that at first BSD reduces the cyclopentadienylmetal chlorides to cyclopentadienyl fragments (cf. Section VI,C,6), which then react with BSD to form the complexes mentioned ($CpTiCl_3$ adds BSD at $-78°C$ in Et_2O without the evolution of N_2 to form at first a dark green solution, which probably contains $CpTiCl_3 \cdot BSD$, but on warming loses N_2 and becomes deep violet).

In case of manganese, there are no appropriate cyclopentadienyl-manganese halides which can supply CpMn. A CpMn-containing BSD complex $[CpMnBSD]_2$, a dark green oxygen-sensitive solid stable to $100°C$ (**89**), is formed along with other side products as a result of reaction of Cp_2Mn with BSD at room temperature in diethyl ether (*72*), that is, conditions under which Cp_2MnBSD decomposes [Eq. (133); cf. Section VII,B,2]. Similarly, the chromium complex **88** is formed from Cp_2Cr and BSD in diethyl ether at room temperature (*73*).

$$2Cp_2Mn \xrightarrow{+3BSD} [CpMnBSD]_2 + 2CpSiMe_3 + N_2 \quad (133)$$
$$\underline{89}$$

Cyclopentadienyliron chloride, $CpFeCl_2$, is unexpectedly not converted by BSD to a complex $(CpFeBSD)_n$. Instead, Cp_2Fe and $Fe(BSD)_n$ are formed according to Eq. (134). $Fe(BSD)_n$ is also generated by the action of BSD on

$$2CpFeCl_3 \xrightarrow[-4Me_3SiCl, \ -2N_2]{+2BSD} 2\text{"CpFe"} \xrightarrow[-Cp_2Fe]{+nBSD} (FeBSD)_n \quad (134)$$

$FeCl_2$ or $FeCl_3$ (49). Until now, X-ray structural analyses existed only for complexes **88** and **89** (72, 73). The diamagnetic manganese compound $[CpMnBSD]_2$ contains bridged isodiazene ligands (cf. **91**). Characteristic structural features are a planar MnNMnN bicyle, with an Mn–Mn distance typical for an Mn–Mn double bond, and Mn–N distances indicative of Mn–N bonds with multiple bonding contributions. The isomerized BSD ligand is similar to that in Cp_2VBSD (see above). Since each manganese atom with seven outer electrons has five electrons from the cyclopentadienyl ligand, two each from $\ddot{N}$—$N(SiMe_3)_2$ ligands (that is, four electrons), and two electrons from the neighboring doubly bonded manganese atom, formally it has $7 + 5 + 4 + 2 = 18$ electrons (noble gas configuration).

The chromium complex $[CpCrBSD]_2$ is a dark violet, oxygen-sensitive, diamagnetic compound stable up to 100°C. In contrast to the manganese complex **88**, **89** contains four trimethylsilylnitrene ligands, of which two are terminal and two bridged, bound to both of the chromium atoms (cf. **90**). The formation of complex $[CpCrBSD]_2 = [CpCr(NSiMe_3)_2]_2$ from BSD and CpCr (formed *in situ* from $CpCrCl_2$ and BSD) follows Scheme 10, Eq. (102c) with BSD cleavage and formation of dimeric $CpCr(NSiMe_3)_2$. Each chromium atom in this complex has a distorted tetrahedral environment of four ligands (Cp as well as three Me_3SiN). Both ligand tetrahedrons possess a common edge along with two common corners, which are occupied by two Me_3SiN ligands. The central ring made up of two Cr and two N atoms is almost planar, and with the help of a Cr–Cr single bond is divided into a bicyclic system. The bond distances between the Cr atoms and bridged N atoms are somewhat shorter and those between the Cr atoms and the terminal N atoms much shorter than a Cr–N single bond. The terminal grouping CrNSi is bent ($\measuredangle$ CrNSi = 160°), which means that the silyl-nitrene ligands probably act as 2-electron donors only [cf. Eq. (123)]. If so, each chromium atom in the complex has formally 6 (Cr) + 5 (Cp) + 3 × 2 (three $NSiMe_3$ ligands) + 1 (Cr–Cr bond) = 18 electrons.

If $[CpCrBSD]_2$ were placed in the same structure as $(CpMnBSD)_2$, then chromium would not have a sufficient number of electrons; the electron deficiency forces the BSD cleavage to give trimethylsilylnitrene. A structure similar to **90** is possibly displayed by the vanadium complex **86** as well as by a complex, $[CpCr(NH)(NSiMe_3)]_2$, obtained by the methanolysis of **90**.

The compounds obtained according to Eqs. (130) and (131), $CpTiBSD_2$ [violet, also synthesizeable from $CpTiCl_3 + Li_2N_2(SiMe_3)_2$] and $CpVBSD_2$, deserve special mention. Appropriate crystals for X-ray analysis have not yet been obtained for these compounds. For the vanadium complex, structure **92** with isodiazene ligands is quite likely. The complexes $CpVBSD_2 = CpV[\equiv N-N(SiMe_3)_2]_2$ and $[CpVBSD]_2 = [CpV(\equiv N-SiMe_3)_2]_2$ could then be structurally related to each other. (In contrast to the second complex, the first is monomeric perhaps because of steric considerations.) In complex $CpVBSD_2$, vanadium acquires a noble gas configuration, if structure **92** is correct. Isoelectronic with this complex are the titanium complex $CpTi(H)BSD_2$ and the chromium complex $(C_5H_6)CrBSD_2$. Compounds of these compositions have actually been found as reaction products of BSD with $CpTiCl_3$ and $CpCrCl_2$ (*49*). The titanium complex nevertheless changes thermally to $CpTiBSD_2$ (*88*).

Complex $[CpVBSD_{0.75}]_2$, which is formed in small amounts by Eq. (131) and in good yield by reaction of Me_3SiN_3 with Cp_2V, has been assigned structure **93** on the basis of spectroscopic investigations (*87*).

$$CpM(\equiv N-N(SiMe_3)_2)$$
$$\underline{\underline{92}}$$

$$\underline{\underline{93}}$$

2. $CpMClBSD_n$ Complexes

Reaction of BSD with $CpVCl_3$ does not lead exclusively to the products **85**–**87** but also yields red $[CpVClBSD]_2$ (**94**) as well as red $[CpVClBSD_{0.5}]_2 = [CpVCl(NSiMe_3)]_2$ (**95**) [Eq. (135)]. The complex **94** is also

$$CpVCl_3 \xrightarrow[-2Me_3SiCl, -N_2]{+BSD} \text{``CpVCl''} \xrightarrow{+nBSD}$$

$$[CpVClBSD]_2, \quad [CpVClBSD_{0.5}]_2 \quad (135)$$
$$\underline{\underline{94}} \qquad\qquad \underline{\underline{95}}$$

obtained by the action of ethereal HCl or of Me_3SnCl on Cp_2VBSD (see Section VII,B,1). With the addition of excess ethereal HCl, it changes to its brown isomer (*49*). On heating, the reactive brown complex transforms back to the less reactive red isomer. In an exactly corresponding way, **95** displays two isomers: a reactive green form, obtainable by the action of Me_3SnCl on $Cp_2VBSD_{0.5} = Cp_2VNSiMe_3$, and a red form, obtainable thermally from the green form and also according to Eq. (135) (*87*).

Structures of isomers **94** and **95** are still unknown, but may be based on the structural rearrangements **94a** and **94b** as well as **95a** and **95b**. Analogous

structures are probably accurate for the brown to reddish-brown complexes $[CpV(OMe)BSD]_2$ [Eq. (126)] and $[CpV(OMe)(NSiMe_3)]_2$ (**96**), obtained by the methanolysis of Cp_2VBSD (**77**) as well as $[CpVBSD_{0.75}]_2$ (**93**). Formation of **96** according to: $93 + 2MeOH \rightarrow 96 + Me_3SiNH_2$ shows a structure of type **95a** with bridged nitrene groups. The more reactive forms of isomers **94** and **95** may be dehalogenated with BSD to give $CpVBSD_2$ (**85**) and $[CpVBSD]_2$ (**86**), respectively.

ACKNOWLEDGMENTS

I wish to express my appreciation to my co-workers, who have enthusiastically carried out all the experiments described here: Drs. W. Baumeister, H. Bayer, G. Fischer, H.-W. Häring, G. Hübler, W.-Ch. Joo, R. Meyers, H. J. Pracht, W. Schneid, G. Schwenk, W. Uhlenbrock, S. K. Vasisht, M. Veith, E. Weinberg, and G. Ziegleder.

REFERENCES

1. N. Wiberg, *Chimia* **30**, 426 (1976).
2. F. Feher and K. H. Linke, *Z. Anorg. Allg. Chem.* **344**, 18 (1966).
3. E. Hayon and M. Simic, *J. Am. Chem. Soc.* **94**, 42 (1972).
4. W. Lwowski, "Nitrenes." Wiley, New York, 1970.
5. N. Wiberg, G. Fischer, and H. Bachhuber, *Z. Naturforsch. B.: Anorg. Chem., Org. Chem.* **34**, 1385 (1979).
6. N. Wiberg, G. Fischer, and H. Bachhuber, *Chem. Ber.* **107**, 1456 (1974).
7. C. Willis, R. A. Back, J. M. Parson, and J. G. Purdon, *J. Am. Chem. Soc.* **99**, 4451 (1977), and references therein.
8. S. Hünig, H. R. Müller, and W. Thier, *Angew. Chem. Int. Ed. Engl.* **4**, 271 (1965).
9. N. Wiberg, H.-W. Häring, and S. K. Vasisht, *Z. Naturforsch. B.: Anorg. Chem., Org. Chem.* **34**, 356 (1979).
10. N. Wiberg, H. Bayer, and H. Bachhuber, *Angew. Chem. Int. Ed. Engl.* **14**, 177 (1975).
11. U. Wannagat, *Adv. Inorg. Radiochem.* **6**, 225 (1964).
12. B. J. Aylett, *Prep. Inorg. Reactions* **2**, 93 (1965).
13. J. S. Thayer and R. West, *Adv. Organomet. Chem.* **5**, 169 (1967); R. M. Pike and N. Sobinsky, *J. Organomet. Chem.* **253**, 183 (1983).
14. M. F. Lappert and H. Pyszora, *Adv. Inorg. Radiochem.* **9**, 133 (1966).

15. R. West, *Adv. Organomet. Chem.* **16**, 1 (1977).
16. N. Wiberg and M. Veith, *Chem. Ber.* **104**, 3176 (1971).
17. N. Wiberg and M. Veith, *Chem. Ber.* **104**, 3191 (1971).
18. N. Wiberg, S. K. Vasisht, G. Fischer, and E. Weinberg, *Chem. Ber.* **109**, 710 (1976).
19. D. W. Klein and J. W. Connolly, *J. Organomet. Chem.* **33**, 311 (1971).
20. D. R. Parker and L. H. Sommer, *J. Am. Chem. Soc.* **98**, 618 (1976).
21. D. R. Parker and L. H. Sommer, *J. Organomet. Chem.* **110**, C1 (1976).
22. M. Elseikh and L. H. Sommer, *J. Organomet. Chem.* **186**, 301 (1980).
23. N. Wiberg, W.-Ch. Joo, and W. Uhlenbrock, *Angew. Chem. Int. Ed. Engl.* **7**, 640 (1968).
24. R. Meyers, Dissertation, University of München, 1978.
25. G. Ziegeleder, Dissertation, University of München, 1977; N. Wiberg and G. Ziegeleder, *Chem. Ber.* **111**, 2123 (1978).
26. E. Weinberg, Dissertation, University of München, 1974.
27. G. Hübler, Dissertation, University of München, 1977.
28. N. Wiberg, S. K. Vasisht, and G. Fischer, *Angew. Chem. Int. Ed. Engl.* **15**, 236 (1976).
29. W. Schneid, Dissertation, University of München, 1977.
30. J. C. Bottaro, *J. Chem. Soc., Chem. Commun.* p. 990 (1978).
30a. E. E. Smissman, and M. Paquin, *J. Org. Chem.* **38**, 1652 (1973).
31. H. Watanabe, K. Inoue, and Y. Nagai, *Bull. Chem. Soc. Jpn.* **43**, 2660 (1970).
32. M. Rivière-Baudet and J. Satgé, *Bull. Soc. Chim. Fr.*, p. 549 (1973).
33. H. Watanabe, M. Matsumoto, and Y. Cho, *Org. Prep. Proced. Int.* **6**, 25 (1974).
34. U. Wannagat and C. Krüger, *Z. Anorg. Allg. Chem.* **326**, 288, 296 (1964).
35. N. Wiberg, *Angew. Chem. Int. Ed. Engl.* **7**, 640 (1968).
36. N. Wiberg, G. Fischer, and H. Bachhuber, *Angew. Chem. Int. Ed. Engl.* **15**, 385 (1976).
37. N. Wiberg, G. Fischer, and H. Bachhuber, *Angew. Chem. Int. Ed. Engl.* **16**, 780 (1977).
38. S. Patai, "*The Chemistry of the Hydrazo, Azo and Azoxy Groups.*" Wiley, New York, 1975.
39. S. H. Bauer, *Inorg. Chem.* **6**, 309 (1967).
40. M. Veith and H. Bärnighausen, *Acta Crystallogr., Sect. B* **30**, 1806 (1974).
41. H. Bock, K. Wittel, M. Veith, and N. Wiberg, *J. Am. Chem. Soc.* **98**, 109 (1976).
42. H. Seidl, H. Bock, N. Wiberg, and M. Veith, *Angew. Chem. Int. Ed. Engl.* **9**, 69 (1970).
43. U. Krynitz, F. Gerson, N. Wiberg, and M. Veith, *Angew. Chem. Int. Ed. Engl.* **8**, 755 (1969).
44. N. Wiberg and W. Uhlenbrock, *J. Organomet. Chem.* **70**, 239 (1974).
45. J. D. Wuest, *J. Org. Chem.* **45**, 3120 (1980).
46. N. Wiberg and W. Uhlenbrock, *J. Organomet. Chem.* **70**, 249 (1974).
47. N. Wiberg, W. Uhlenbrock, and W. Baumeister, *J. Organomet. Chem.* **70**, 259 (1974).
48. W. Uhlenbrock, Dissertation, University of München, 1971.
49. H.-W. Häring, Dissertation, University of München, 1978.
50. N. Wiberg, *Angew. Chem. Int. Ed. Engl.* **10**, 374 (1971).
51. N. Wiberg and W. Uhlenbrock, *Chem. Ber.* **105**, 63 (1972).
52. J. Kroner, N. Wiberg, and H. Bayer, *Angew. Chem. Int. Ed. Engl.* **14**, 178 (1975).
53. H. Bayer, Dissertation, University of München, 1979.
54. N. Wiberg, H.-W. Häring, and S. K. Vasisht, *Z. Naturforsch. B.: Anorg. Chem., Org. Chem.* **34**, 356 (1979).
55. U. Wannagat and C. Krüger, *Z. Anorg. Allg. Chem.* **326**, 304 (1964).
56. H. Watanabe, K.-I. Awano, M. Ohmori, N. Kodama, J.-I. Sakamoto, Y. Onodera, and Y. Nagai, *J. Organomet. Chem.* **186**, 7 (1980).
57. M. N. Ackermann, J. L. Ellenson, and D. H. Robinson, *J. Am. Chem. Soc.* **90**, 7173 (1968).
58. E. M. Kosower, *Acc. Chem. Res.* **4**, 193 (1971).
59. N. Wiberg, E. Weinberg, and W.-Ch. Joo, *Chem. Ber.* **107**, 1764 (1974).
60. N. Wiberg and W.-Ch. Joo, *Z. Naturforsch. B.: Anorg. Chem., Org. Chem.* **26**, 512 (1971).

61. N. Wiberg, W.-Ch. Joo, and E. Weinberg, *J. Organomet. Chem.* **73**, 49 (1974).
62. N. Wiberg and G. Schwenk, *Angew. Chem. Int. Ed. Engl.* **8**, 755 (1969).
63. G. Schwenk, Dissertation, University of München, 1971.
64. N. Wiberg and W. Uhlenbrock, unpublished results (1971).
65. N. Wiberg and W. Uhlenbrock, *Chem. Ber.* **104**, 3989 (1971).
66. N. Wiberg, S. K. Vasisht, H. Bayer, and R. Meyers, *Chem. Ber.* **112**, 2718 (1979).
67. W. Baumeister, Dissertation, University of München, 1971.
68. A. Schmidpeter and W. Zeiss, *Angew. Chem. Int. Ed. Engl.* **10**, 396 (1971).
69. J. F. Nixon and M. Kooti, *J. Organomet. Chem.* **149**, 71 (1978).
69a. N. Wiberg and E. Kühnel, unpublished results (1983).
70. J. L. Kice and R. S. Gabrielsen, *J. Org. Chem.* **35**, 1004 (1970).
71. N. Wiberg, H.-W. Häring, and O. Schieda, *Angew. Chem. Int. Ed. Engl.* **15**, 386 (1976).
72. N. Wiberg, H.-W. Häring, G. Huttner, and P. Friedrich, *Chem. Ber.* **111**, 2708 (1978).
73. N. Wiberg, H.-W. Häring, and U. Schubert, *Z. Naturforsch. B.: Anorg Chem., Org. Chem.* **33**, 1365 (1978).
74. J. R. Dilworth, H. J. de Liefde Meijer, and J. H. Teuben, *J. Organomet. Chem.* **159**, 47 (1978).
75. E. W. Abel and C. A. Burton, *J. Organomet. Chem.* **170**, 229 (1979).
76. H. Breedervelt, *Recl. Trav. Chim. Pays-Bas* **81**, 276 (1962).
77. N. Wiberg and G. Schwenk, *Chem. Ber.* **104**, 3986 (1971).
78. N. Wiberg, W. Uhlenbrock, and G. Hübler, unpublished results (1977).
79. N. Wiberg and W.-Ch. Joo, unpublished results (1967).
80. W. Zeiss, Dissertation, University of München, 1972.
81. H. Pracht, Dissertation, University of München, 1971.
82. N. Wiberg and H. Bayer, unpublished results (1979).
83. N. Wiberg and W.-Ch. Joo, unpublished results (1968).
84. M. Herberhold, W. Golla, and K. Leonhard, *Chem. Ber.* **107**, 3209 (1974), and references therein.
85. I. D. Brown and J. D. Dunitz, *Acta Crystallogr.* **13**, 28 (1960).
86. M. Veith, *Angew. Chem. Int. Ed. Engl.* **15**, 387 (1976).
87. N. Wiberg, H.-W. Häring, and U. Schubert, *Z. Naturforsch. B.: Anorg. Chem., Org. Chem.* **35**, 599 (1980).
88. N. Wiberg, H.-W. Häring, and D. Boeckh, unpublished results (1979).

ADVANCES IN ORGANOMETALLIC CHEMISTRY, VOL. 23

Polarization Transfer NMR Spectroscopy for Silicon-29: The INEPT and DEPT Techniques

THOMAS A. BLINKA,
BRADLEY J. HELMER,
and
ROBERT WEST

Department of Chemistry
University of Wisconsin
Madison, Wisconsin

I
INTRODUCTION

Although silicon-29 NMR is a highly valuable form of spectroscopy for organosilicon compounds, it is only now becoming a routine technique.[1] Several characteristics of ^{29}Si make NMR measurements of this nucleus difficult: (1) the natural abundance of ^{29}Si is rather low (4.67%); (2) the magnetogyric ratio, γ, is relatively small (-0.55477), so that the energy difference between spin states, and hence the Boltzmann population difference, is also small (both of these factors limit the intrinsic NMR sensitivity

[1] For reviews of ^{29}Si NMR see ref. *1*.

ISBN 0-12-031123-2

of silicon); (3) because the magnetogyric ratio is negative, nuclear Overhauser effects (NOE) can lead to reduced or nulled ^{29}Si signals; and (4) the spin–lattice relaxation times, T_1, for ^{29}Si are characteristically quite long, so that the rate at which Fourier transform (FT) pulses can be repeated is slow, leading to long data acquisition times. Taken together, these factors make silicon-29 distinctly more difficult to observe than carbon-13 in NMR even though ^{29}Si is more abundant than ^{13}C.

Therefore, even with the power of modern high-field FT NMR, special techniques are necessary to increase the ^{29}Si signal. For silicon nuclei that are coupled to protons, the NMR spectrum can be greatly improved by transfer of polarization from ^{1}H to ^{29}Si. Earlier methods for bringing about polarization transfer are now being replaced by the multipulse NMR techniques known as INEPT (insensitive nuclei enhanced by polarization transfer) and DEPT (distortionless enhancement by polarization transfer) (2–4). Both of these pulse sequences and their variations transfer nuclear spin polarization from protons, which have large Boltzmann population differences and hence high NMR sensitivity, to other nuclei to which the protons are coupled.[2] The result for ^{29}Si is to increase the signal strength significantly (6). Equally or more important, with polarization transfer the short relaxation times of the protons govern the pulse repetition rate, so that data may be acquired much more rapidly. These two effects together lead to a time-saving factor of 30 to 300 in ^{29}Si-NMR experiments, compared with normal FT-NMR spectroscopy.

This article outlines several multipulse NMR polarization transfer methods (INEPT, INEPT+, DEPT, DEPT+, and DEPT++) and explains how they may be used in ^{29}Si-NMR spectroscopy. Section II,A briefly traces the development of polarization transfer methods; Section II,B introduces INEPT and DEPT pulse sequences and shows the advantages of INEPT and DEPT to previous polarization transfer techniques. Sections III,A–D provide examples of the application of INEPT and DEPT ^{29}Si-NMR methods to representative silicon compounds. In Section III,E the advantages and disadvantages of the various methods are described, and the optimal choice of pulse sequence for particular types of compounds is summarized. *Sections II,B and III,E together can be used as a brief instruction manual for ^{29}Si polarization transfer spectroscopy.* Finally, for those interested in the theory underlying INEPT and DEPT spectroscopy, an overview is presented in Sections IV,A and IV,B.

---

[2] INEPT and DEPT techniques have also been applied to nuclei other than ^{29}Si. See refs. 2–5.

II

POLARIZATION TRANSFER TECHNIQUES

A. Development of Polarization Transfer Spectroscopy

Although polarization transfer techniques have been available for over a decade, they have not been widely used to obtain ^{29}Si-NMR spectra. J-cross polarization (7) (JCP), which evolved from methods used to enhance the solid state spectra of rare spin nuclei, (8) has been applied to ^{29}Si-NMR spectroscopy (9). JCP suffers from several limitations: the proton and ^{29}Si pulses must be on resonance, and the Hartmann–Hahn condition (8) ($\gamma_H H_H = \gamma_{Si} H_{Si}$) must be established for full enhancement. Neither of these conditions is trivial to obtain, and the difficulty in establishing them has prohibited routine application of JCP methods to ^{29}Si-NMR spectroscopy.[3]

Selective polarization transfer (SPT) (11) is another NMR technique recently used to obtain enhanced ^{29}Si-NMR spectra. To obtain SPT spectra, one selectively irradiates a silicon satellite of a proton signal prior to a non-selective ^{29}Si pulse. The resulting population inversion of the ^{1}H–^{29}Si energy levels produces a coupled ^{29}Si-NMR spectrum in which only that specific silicon whose coupled protons have been irradiated will be enhanced. Obviously, this method is impractical for polysilanes and is also limited to molecules in which ^{29}Si satellites can be observed in the proton-NMR spectrum.

B. INEPT and DEPT Spectroscopy for ^{29}Si NMR

Two new polarization transfer techniques have recently been reported: INEPT (2) and DEPT (3, 4). These pulse sequences lack the limitations of previous polarization transfer methods, and allow the routine collection of ^{29}Si-NMR data. The principal virtues of both the INEPT and DEPT pulse sequences are that the polarization transfer enhancements are substantial (five- to ninefold) (12) and relatively nonselective and that they can easily be used by chemists familiar with normal FT-NMR spectroscopy on available commercial multinuclear FT-NMR instruments.

---

[3] It is, however, useful in obtaining solid state ^{29}Si-NMR spectra (10).

Several modified INEPT and DEPT pulse sequences have recently been introduced (13) (see Fig. 1). The new INEPT+, DEPT+, and DEPT++ sequences differ from the original INEPT and DEPT sequences only in that they employ additional refocusing and purging pulses. These serve to reduce or eliminate distortions inherent in the parent pulse sequences. The fundamental polarization transfer mechanism however remains unchanged.

The INEPT and DEPT pulse sequences shown in Fig. 1 (13, 14) are all multinuclear pulse sequences in which proton and/or silicon pulses are separated by free precession periods. However, INEPT and DEPT differ in both the number and duration of precession periods. In the INEPT pulse sequences, there are two precession periods of duration τ, and one of duration Δ [a refocusing pulse (15) bisects the Δ period]. Both τ and Δ are parameters set by the user to optimize enhancements, although τ is routinely set to a constant $(4J)^{-1}$ (where J is the $^1H-^{29}Si$ coupling constant). The Δ parameter can be set according to Eq. (5) (Section IV,A) to obtain optimal enhancement, or may be set to selectively invert or suppress specific silicon resonances as shown by Figs. 8–10 (Section III,D).

The DEPT pulse sequences contain three precession periods of duration τ, and a proton pulse parameter θ. Like τ and Δ in INEPT, τ and θ are parameters set by the user to obtain optimal enhancements. The τ parameter in DEPT sequences is usually set to $(2J)^{-1}$, exactly twice as long as the τ parameter in INEPT. (As a consequence, DEPT pulse sequences are approximately three times longer than INEPT sequences). In cases where the silicon spin–spin relaxation time T_2 is short, one observes significantly less enhancement using DEPT rather than INEPT because of relaxation of silicon magnetization during the longer DEPT pulse sequences. (This will be discussed in more detail in Section III,E.) The value of θ is set according to Eq. (10) (Section IV,B), and θ has a function similar to Δ in INEPT sequences. However, unlike Δ, θ is independent of J, making DEPT less sensitive than INEPT to variations in J within a sample.

In practice, the appearance of decoupled INEPT and DEPT ^{29}Si-NMR spectra are usually the same. However, coupled INEPT and DEPT spectra differ dramatically. Coupled DEPT spectra essentially appear as greatly enhanced standard acquisition spectra; the multiplicity, phase, and relative intensities of multiplets using DEPT are the same as those obtained from normal FT-NMR techniques. In contrast, coupled INEPT spectra contain several distinctive "distortions": (1) the outer lines of multiplets in INEPT spectra are much enhanced compared to relative multiplet intensities obtained using standard acquisition or DEPT-NMR techniques; (2) the central line of odd line multiplets in INEPT has zero intensity; and (3) the two halves of a multiplet in INEPT are 180° out of phase. Thus, a triplet and a quartet in INEPT would appear as $1:0:-1$ and $1:1:-1$ patterns, respectively, instead of the "normal" $1:2:1$ and $1:3:3:1$ patterns seen with DEPT (see Section IV,A).

DEPT

1H 90°_y – τ – 180°_x – τ – $\theta_{\pm x}$ – τ – decouple as required

29Si 90°_x – 180°_x – acquire

DEPT+

1H 90°_y – τ – 180°_x – τ – $\theta_{\pm x}$ – τ – 180_y

29Si 90°_x – 180°_x – acquire

DEPT++

1H 90°_y – τ – 180°_x – τ – $\theta_{\pm x}$ – τ/2 – 180°_x – τ/2 – 90°_x – τ/2 – 180_x – τ/2 – acquire

29Si 90°_x – 180°_x

Coupled INEPT

1H 90°_x – τ – 180°_x – τ – $90^\circ_{\pm y}$

29Si 180°_x – 90°_x – acquire

INEPT

1H 90°_x – τ – 180°_x – τ – $90^\circ_{\pm y}$ – Δ/2 – 180°_x – Δ/2 – decouple as required

29Si 180°_x – 90°_x – 180°_x – acquire

INEPT +

1H 90°_x – τ – 180°_x – τ – $90^\circ_{\pm y}$ – Δ/2 – 180°_x – Δ/2 – 90°_x

29Si 180°_x – $90^\circ_{\pm x}$ – 180°_x – acquire

Fig. 1. INEPT and DEPT pulse sequences.

III

APPLICATION OF INEPT AND DEPT TO ^{29}Si NMR

Although there are differences (noted above) between the INEPT and DEPT techniques, the fundamental polarization transfer mechanism is the same (3, 4, 13). Thus, for simplicity, INEPT will be used to demonstrate the use of polarization transfer in ^{29}Si-NMR, except in cases where a comparison with DEPT would be advantageous.

A. *Methyl-Substituted Polysilanes*

Dodecamethylcyclohexasilane (**1**) was chosen as a representative member of this important class of compounds upon which to test INEPT and DEPT methods. The decoupled ^{29}Si NMR of **1** consists of a single resonance

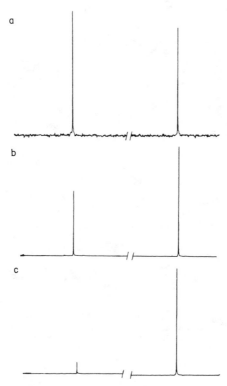

FIG. 2. ^{29}Si-NMR spectra of $SiMe_4$ (left) and Si_6Me_{12} (right) using (a) standard acquisition, (b) INEPT set for J^2_{Si-H}, and (c) INEPT set for J^3_{Si-H}.

corresponding to six equivalent $SiMe_2$ units, with $J^2_{Si-H} = 6.2$ Hz and $J^3_{Si-H} = 3.3$ Hz (16). (The superscript on J indicates the number of bonds linking the coupled nuclei.) Figure 2 shows ^{29}Si-NMR spectra of 0.2 M 1 and 1.25 M TMS in d_6-benzene, using INEPT with τ and Δ parameters set for both the two- and the three-bond ^{1}H–^{29}Si coupling constant ($J = 6.2$ Hz, $n = 6$: $\tau = 40.3$ msec, $\Delta = 21.6$ msec; $J = 3.3$ Hz, $n = 12$: $\tau = 71.2$ msec, $\Delta = 28.2$ msec).

The resulting INEPT enhancements were found to be 7.3 [theory 7.8, Eq. (3), Section IV,A] for 1 and 3.8 (theory 8.2) for TMS when the parameters were set for $J = 6.2$ Hz, and 5.3 (theory 10.8) and 0.55 (theory 0.85), respectively, for τ and Δ parameters set for $J = 3.3$ Hz. The spectra shown in Fig. 2 represent the same number of data acquisitions (four), but the pulse repetition rate for the INEPT spectra is 10 seconds, whereas for the standard acquisition spectrum (gated decoupling, 90° silicon pulse), the pulse repetition rate is 60 seconds. Not only are the INEPT spectra enhanced relative to the standard acquisition spectrum, but they were obtained with a sixfold time saving. It is also instructive to note the loss of enhancement observed when J is small, as when the J^3 coupling is used. The variables τ and Δ are inversely related to J; as J becomes smaller, τ and Δ increase and the pulse sequence becomes longer relative to silicon T_2 times. The spin–spin relaxation is thus more pronounced during the pulse sequence, resulting in loss of signal strength, hence enhancement.

Figure 3 shows coupled INEPT, DEPT, and standard acquisition spectra of 1 and TMS (parameters set for $J^2 = 6.2$ Hz). The most noticeable feature of the INEPT spectrum is the 180° phase inversion of the multiplets, and the additional enhancement of the outside lines of the multiplets. The TMS resonances are well resolved, but the multiplet for 1 is poorly resolved due to the complex nature of the two- and three-bond H–Si couplings.

B. Siloxanes

Siloxanes represent another large class of silicon compounds for which ^{29}Si-NMR data would be useful. The compound $HMe_2SiOSiMe_2H$ was chosen as a representative siloxane with a more complex spin system due to the Si–H moieties. The proton coupled spectrum (Fig. 4) consists of a major doublet ($J = 204$ Hz), for which INEPT and DEPT parameters were set, and a minor septet of doublets. In the INEPT spectrum (Fig. 4a) the major doublet shows the expected INEPT intensities ($+1:-1$), while the minor splittings show normal coupling intensities. A subtle but important distortion is present in the INEPT spectrum; the expected septet of doublets

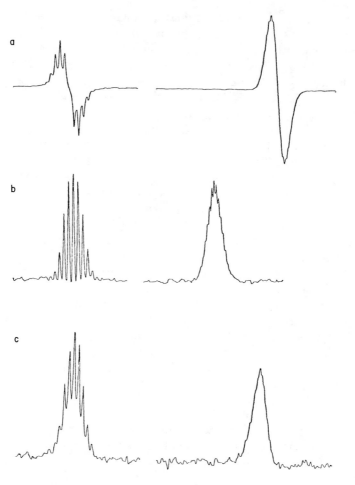

FIG. 3. Proton-coupled spectra of $SiMe_4$ (left) and Si_6Me_{12} (right) using (a) INEPT (500 seconds), (b) DEPT (500 seconds), and (c) standard acquisition (3000 seconds). Times in parentheses refer to the length of time required for data collection (50 scans).

appears as a *sextet* of doublets in INEPT, while the DEPT spectrum shows the correct multiplicity.

This extreme example highlights normally minor, but in this case important, differences between INEPT and DEPT. INEPT pulse sequences are more sensitive to J variations in a sample than DEPT since two INEPT parameters are J dependent, whereas only one DEPT parameter is. Also, the INEPT sequence contains more pulses than the DEPT sequence. Small instrumental pulse errors are thus more likely to combine to give noticeable

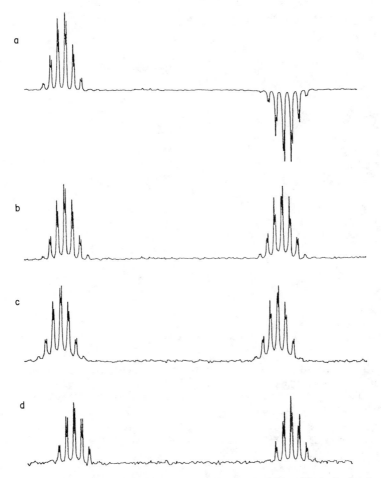

FIG. 4. Proton-coupled ^{29}Si-NMR spectrum of $HMe_2SiOSiMe_2H$ obtained using (a) INEPT (500 seconds), (b) DEPT (500 seconds), (c) standard acquisition (3000 seconds), and (d) standard acquisition (500 seconds). Times in parentheses refer to time required for data collection (a–c, 50 scans; d, 8 scans).

distortions in INEPT than in DEPT (*3, 4*). Figure 5 shows the decoupled INEPT, DEPT, and standard acquisition spectra of $HMe_2SiOSiMe_2H$, again with INEPT and DEPT parameters set to the major coupling constant. The difference in *J* dependence between INEPT and DEPT is again shown, as the TMS signal is totally suppressed in the INEPT spectrum but appears in the DEPT spectrum. One may anticipate that for complex spin systems containing large *J* variations, the DEPT experiment will be more useful than INEPT since it is less likely to suppress signals.

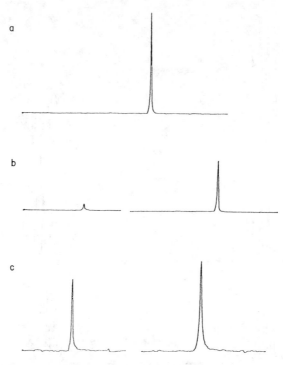

FIG. 5. ^{29}Si-NMR spectra of Me_4Si (left) and $HMe_2SiOSiMe_2H$ (TMS) (right) using (a) INEPT (40 seconds), (b) DEPT (40 seconds), and (c) standard acquisition (240 seconds). Times in parentheses refer to time required for data collection (4 scans). The TMS signal is not seen in the INEPT spectrum.

C. *Other Silanes*

As shown for dodecamethylcyclohexasilane (Section III,A), long-range couplings (provided they are resolvable) can be used for polarization transfer. From $(MeO)_4Si$, for example, one can obtain well-resolved proton coupled spectra (Fig. 6) and decoupled spectra (Fig. 7). The coupled spectra vividly show the alteration of the relative intensities of the $(MeO)_4Si$ multiplet in comparing the standard acquisition spectrum or DEPT spectrum to the INEPT spectrum. The relatively greater enhancement of the outer lines in coupled INEPT spectra is shown by the observation of 11 lines (including the zero-intensity central line) in the INEPT spectrum (Fig. 6a), compared to nine lines seen in the DEPT and standard acquisition spectra (Figs. 6b and c). Multiplet intensities in the DEPT and standard acquisition spectra are nearly identical, as expected.

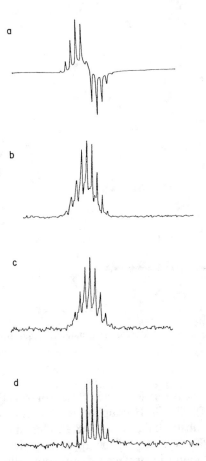

FIG. 6.   Proton coupled ^{29}Si-NMR spectra of Si(OMe)$_4$, (a) INEPT (500 seconds), (b) DEPT (500 seconds), (c) standard acquisition (3000 seconds), and (d) standard acquisition (500 seconds). Time in parentheses refers to time required for data collection (a–c, 50 scans; d, 8 scans).

A selection of silanes bearing representative substituents has been studied using the INEPT technique. Table I (6, 17) shows the signal enhancements found for silicon in simple silanes with Me, t-Bu, OMe, Ph, and H substituents. All of the compounds show enhancement, although some fall significantly below the predicted value. The compounds that show poor enhancement often contain quadrupolar nuclei such as Cl and ^{15}N whch shorten T_2 relaxation times. Compounds containing Si–H bonds have large Si–H couplings ($J = 185$–286 Hz) and generally give good enhancements due to the correspondingly short INEPT pulse sequences.

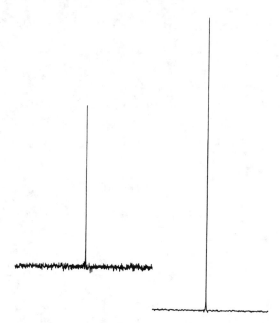

Fig. 7. ^{29}Si-NMR spectra of (MeO)$_4$Si using standard acquisition (left) and the INEPT pulse sequence (right).

Enhancement was measured for a wide variety of trimethylsilyl-substituted compounds (Table I). The Si–H coupling in all cases was 7 ± 0.5 Hz, and enhancements were substantial ($E_d = 4.4$–8.9). This suggests that INEPT or the equivalent DEPT sequence can be applied to virtually any trimethylsilylated compound found in organic or organometallic chemistry using a single set of standard parameters ($J = 7$ Hz, $n = 9$; INEPT: $\tau = 36$ msec, $\Delta = 15$ msec; DEPT: $\tau = 71$ msec, $\theta = 0.33$ radians).

D. *Other Applications*

The variation of enhancement, E_d, with Δ or θ has been used to distinguish CH$_2$ groups from CH and CH$_3$ groups in carbon spectra (3, 4, 14, 18). This type of experiment may also be valuable for polysilanes and polysiloxanes where silicon substitution varies from zero to three methyls. The method was tested using a mixture of TMS and decamethyltetrasiloxane in d_6-benzene. The three resonances Me$_4$Si, Me$_3$Si, and Me$_2$Si each gave a well-resolved multiplet in the proton coupled ^{29}Si-NMR spectrum, with

TABLE I

MEASURED AND THEORETICAL DECOUPLED ENHANCEMENTS FOR ^{29}Si-NMR OF VARIOUS
SILICON COMPOUNDS

Compound	Measured E_d	Theoretical $E_d{}^a$	δ (ppm)	J (Hz)
Me$_4$Si	9.3	10.8	0	6.6
Me$_3$SiCl	7.5	9.4	29.87	6.8
Me$_2$SiCl$_2$	4.8	7.8	31.66	7.3
MeSiCl$_3$	2.9	5.8	12.34	8.1
t-Bu$_2$SiCl$_2$	5.5	13.1	39.10	7.8
t-Bu$_2$SiF$_2$	5.8	13.1	−7.76	6.3 ($J_{Si-F} = 325.3$)
t-BuSiCl$_3$	3.6	9.4	17.75	10.4
t-BuMeSiCl$_2$	4.9	10.8	36.85	8.1
Ph$_2$SiCl$_2$	3.2	6.5	6.47	6.6 broad
Ph$_2$SiH$_2$	5.0	5.0	−33.19	198.3, 5.9 (ortho), 1.0 (meta)
Et$_3$SiH	3.6	5.0	0.28	185.3, 6.9
(EtO)$_3$SiH	4.7	5.0	−58.82	286.1, 3.4
(MeO)$_4$Si	5.5	10.8	−78.25	3.7
(HMe$_2$Si)$_2$O	5.1	5.0	−4.42	204.3, 7.0 (septet), 1.3 (doublet)
(Me$_3$SiOMe$_2$Si)$_2$O	6.5	9.4	7.17	6.7
(Me$_3$SiOMe$_2$Si)$_2$O	6.2	7.8	−21.66	7.3
(Me$_2$SiO)$_5$	7.1	7.8	−21.14	7.4
p-(Me$_3$Si)$_2$C$_6$H$_4$	7.3	9.4	−4.22	Unresolved
(Me$_3$Si)$_2$C$_2$	8.9	9.4	−19.13	7.1
(Me$_3$Si)$_2$NH	4.4	9.4	2.40	Broad unresolved
(Me$_3$Si)$_2$S	6.6	9.4	14.27	6.8
(Me$_3$Si)$_2$	7.3	9.4	−19.60	6.5, 2.7
(HMes$_2$Si)$_2$	4.7	5.0	−53.01	184.1
(t-BuMeSi)$_4$	4.2	10.8	−13.41	Unresolved
(Me$_2$Si)$_6$	7.3	7.8	−41.86	6.2, 3.3

^{a} Calculated from $E_d = (\nu_H/\nu_{Si})_n{}^{1/2}[1 - (1/n)]^{(1/2)(n-1)}$, according to ref. 12.

coupling constants of 6.58, 6.76, and 7.30 Hz, respectively. A series of decoupled INEPT and DEPT spectra was then obtained with τ set at an intermediate value (INEPT: 36 msec; DEPT: 71.4 msec); for INEPT, Δ was varied between 5 and 145 msec, and for DEPT, θ was varied between 5 and 120 μsec.

Graphs of INEPT signal enhancements of each silicon resonance versus Δ are shown in Fig. 8; graphs of DEPT enhancements versus θ are shown in Fig. 9. These can be compared to a graph of theoretical enhancements versus Δ shown in Fig. 14 (Section IV,A).

As expected, the silicon resonances drop to zero intensity at $\Delta = 1/(2J)$ ($\theta = \pi/2$). The Me$_4$Si and Me$_2$Si signals become negative as Δ or θ

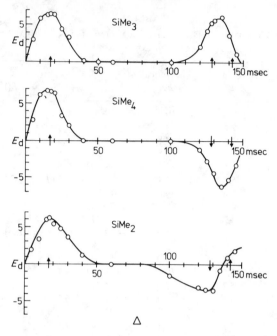

FIG. 8. Graphs of decoupled enhancement (E_d) of SiMe$_4$ and (Me$_3$SiOMe$_2$Si)$_2$O versus delay time Δ (INEPT). Arrows indicate spectra shown in Figs. 10b–d.

is increased, while the Me$_3$Si signal returns to a positive value. Representative INEPT spectra are shown in Fig. 10, in which Δ was varied in order to selectively invert silicon resonances. Both INEPT and DEPT show great sensitivity in producing resonances 180° out of phase from differences in J of only 0.5–0.7 Hz.

E. *Recommendations to Users*

From previous sections it is apparent that INEPT and DEPT are not always equivalent methods for every ^{29}Si-NMR application. This section outlines the factors that will enable users to choose the technique that will most improve the ^{29}Si-NMR spectrum.

The single most important consideration in choosing between INEPT and DEPT is the length of the pulse sequence relative to the silicon spin–spin relaxation time, T_2, which is fairly constant (about 160 msec) for most silanes. The DEPT pulse sequence is about three times longer than an equivalent INEPT pulse sequence (Section II,B). The length of INEPT or

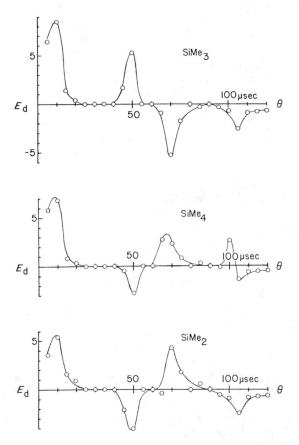

Fig. 9. Graph of decoupled enhancement E_d of $SiMe_4$ and $(Me_3SiOMe_2Si)_2O$ versus pulse angle θ (DEPT). 90° ¹H pulse is approximately 15 μsec.

DEPT pulse sequences is determined by the length of the τ precession times, which are inversely proportional to J. For large J values ($J > 50$ Hz) INEPT and DEPT sequences are both short relative to T_2 and give similar enhancements. For small J values ($J < 10$ Hz) the INEPT sequence is still short relative to T_2, but the DEPT sequence length becomes comparable to T_2. Consequently, the DEPT enhancements are poorer due to relaxation during the pulse sequence.

For ²⁹Si-NMR spectra in general, INEPT is superior to DEPT in cases where J is small or when quadrupolar nuclei such as ¹⁵N, ³⁵Cl, or ³⁷Cl are present and shorten T_2. For large J values, INEPT and DEPT are equivalent. [It is possible to increase T_2 somewhat by deoxygenating NMR samples.

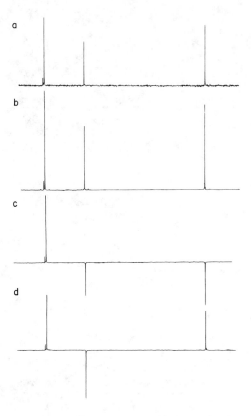

FIG. 10. ^{29}Si-NMR spectra of SiMe$_4$ and (Me$_3$SiOSiMe$_2$)$_2$O using standard acquisition (a) and INEPT (b)–(d) as described in the text.

Since oxygen is paramagnetic, it acts as a relaxation agent (*19*); degassed samples give DEPT enhancements that are nearly as great as INEPT enhancements, even for small *J* values.]

There are two cases in which the DEPT technique offers advantages over INEPT. First, in coupled spectra, DEPT gives normal multiplet phases and intensities, which may aid interpretation and are certainly more familiar in appearance than those obtained from INEPT. In addition, coupled DEPT spectra tend to be better resolved than the corresponding INEPT spectra, especially for complex spin systems (see Fig. 3). This is probably a consequence of DEPT's lesser *J* dependence compared to INEPT. Second, for samples in which large *J* variations are present or suspected, DEPT is the method of choice because it is less likely to suppress signals than INEPT (see Fig. 5).

IV

THEORY OF INEPT AND DEPT SPECTROSCOPY

Bendall, Pegg, and Doddrell (20) showed that the results of multipulse sequences, such as INEPT and DEPT, can be analyzed using the Heisenberg vector approach. The INEPT sequence is amenable to pictorial representation using familiar vector diagrams. However, the mechanism of the DEPT sequence is not well understood, and a more mathematical approach will be employed to describe it.

A. *INEPT Theory*

By employing a doubly rotating frame of reference, polarization of I and H in an IH system can be followed simultaneously (Fig. 11). It is important

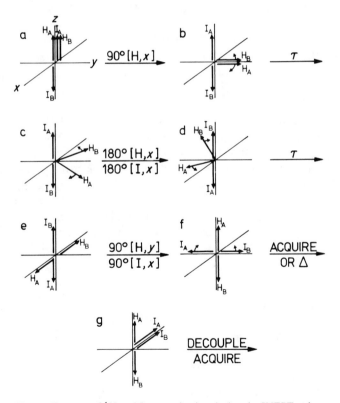

FIG. 11.   Vector diagrams of ^{1}H and I magnetization during the INEPT pulse sequence.

to recognize that $(H_A + H_B)$ represents the proton magnetization vector (excess population aligned along $+z$ axis) while I_A and I_B represent I atoms (half in the $+z$ direction and half in the $-z$ direction) that are coupled to H_A and H_B, respectively. For the time being, initial magnetization of I can be ignored.

The initial state in the presence of a magnetic field is shown in Fig. 11a. A $90°_x$ pulse rotates the H magnetization onto the y axis (Fig. 11b), where H_A and H_B begin to precess due to coupling to I_A and I_B. After a delay period $\tau = 1/(4J_{H-I})$, H_A and H_B have diverged 90° (Fig. 11c). At this point the $180°_x$ refocusing pulses are employed (Fig. 11d). The 180° proton pulse serves to refocus all H precessions, while the 180° I pulse reverses the directions of H_A and H_B thereby maintaining coupling information. After an additional delay time τ, H_A and H_B are aligned in opposite directions along the x axis (Fig. 11e). A $90°_y$ proton pulse then rotates the H magnetization back onto the z axis, while a simultaneous $90°_x$ I pulse aligns I_A and I_B with the $-y$ and $+y$ axes (Fig. 11f). I_A and I_B then represent enhanced I magnetization, coupled to H (therefore a doublet) with the two halves of the doublet 180° out of phase. Acquisition of data at this point gives a doublet with relative intensities $+1:-1$, while decoupling gives no net signal.

If a decoupled spectrum is desired an additional delay time, $\Delta = 1/2J$, is employed to allow the two halves of the doublet to come into phase (Fig. 11g). As before, a refocusing pulse is applied halfway through this delay period.

The key to polarization transfer is the manipulation of H magnetization. The H magnetization in Fig. 11f differs from that in Fig. 11a in that H_A has returned to its equilibrium position while H_B has an inverted population. The net effect can be seen from energy level diagrams that show the initial population distribution of the coupled system (Fig. 12a) and the population distribution following the pulse sequence (Fig. 12b). Since the coupled nuclei have common energy levels, the original larger population difference of the ^1H transition becomes the population difference for the I transition. The population difference for the I transition is thereby increased by a factor of γ_H/γ_I.

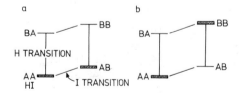

FIG. 12.   Energy level diagrams showing population distribution (a) before pulse sequence (see Fig. 11a) and (b) after pulse sequence (see Fig. 11f).

Although the natural I magnetization will be small compared to that produced by polarization transfer, it can affect the I-NMR INEPT spectrum. The natural magnetization is out of phase with the polarization transfer magnetization and also has different intensities associated with IH_n multiplets (see below) and therefore is best deleted. Alternation of the $90°_y$ proton pulse between $90°_{-y}$ and $90°_{+y}$ has no net effect on the natural magnetization, but reverses the precession of I_A and I_B. Subtraction of alternate signals then gives cancellation of natural magnetization but summation of polarization transfer magnetization. This provides the added result of eliminating all resonances where there is no I–H coupling, for example solvent and quaternary carbon resonances in ^{13}C spectra. (The same effect can be obtained in DEPT by alternating θ_x between θ_{-x} and θ_{+x}.)

The interpretation of pulse sequences is somewhat more complex in IH_n systems where $n > 1$. For $n = 2$ the nuclei I_A and I_B receive polarization from the surplus population of protons in the ↑↑ state over the ↓↓ state. There is, however, no net proton magnetization for I nuclei coupled to protons in opposite eigenstates (↑↓). Hence the expected triplet has lines of intensities $+1:0:-1$. Such a zero-intensity center line is found in all cases where n is even and allows easy distinction between even and odd multiline patterns.

In the case of $n = 3$, all four lines of the quartet are observed, however the intensities are distorted. As stated previously I nuclei receive polarization from the excess of protons aligned with the field over those opposed to the field. In this case two such proton conditions exist, ↑↑↓ versus ↓↓↑ and ↑↑↑ versus ↓↓↓. The first can be formed by three different combinations of spins while the second has only one form. This provides the usual 1:3:3:1 pattern. The second however, has three times as great an energy gap as the first and hence a threefold greater Boltzmann population difference. When these two effects are combined the result is a $1:1:-1:-1$ spectrum.

The expected intensities for any multiplet from a proton coupled INEPT sequence can be determined similarly by multiplying the terms of the appropriate binomial expansion (intensities expected for a normal H-coupled spectrum) by $n, n - 2, \ldots, 2, 0, -2, \ldots, 2 - n, -n$ for $n = $ even, and $n, n - 2, \ldots, 1, -1, \ldots, 2 - n, -n$ for $n = $ odd. From this it is seen that INEPT enhances the weak outer lines $[E = n(\gamma_H/\gamma_I)]$ by more than the center lines $(E = \gamma_H/\gamma_I)$, and thereby allows observation of many more lines in a multiline pattern.

Doddrell et al. (12) derived the expression for the theoretical enhancement of decoupled INEPT spectra, E_d [Eq. (1), for $\tau = \tau_{opt}$]. Furthermore, Burum

$$E_d = n(\gamma_H/\gamma_I)\sin(\pi J\Delta)\cos^{n-1}(\pi J\Delta) \qquad (1)$$

and Ernst (14) have shown that E_d depends on the initial delay period τ [Eq. (2)]. The two expressions can be combined to give Eq. (3), which shows

$$E_d \propto \sin 2\pi J\tau \tag{2}$$

$$E_d = n(\gamma_H/\gamma_I)\sin(\pi J\Delta)\cos^{n-1}(\pi J\Delta)\sin(2\pi J\tau) \tag{3}$$

that the enhancement depends on the four variables γ_I, n, τ, and Δ.

The variable γ_I is the magnetogyric ratio of nucleus I and contributes to the enhancement in the term γ_H/γ_I. This ratio varies from 4.0 for ^{13}C to 5.0 for ^{29}Si to 9.9 for ^{15}N. This factor represents the ratio of the ^1H population difference, which becomes the new I population difference, to the natural I population difference.

As before, n is the number of protons to which I is coupled. The effect of coupling to additional protons is not totally additive, as is shown by the $\cos^{n-1}$ term which reduces E_d. The derived relationship between enhancement and n is shown in Eq. (4). The theoretical enhancement of ^{29}Si coupled

$$E_d \propto n[1 - (1/n)]^{(n-1)/2} \tag{4}$$

to n protons is shown in Table II.

The only two parameters to be manipulated in obtaining an INEPT spectrum are the delay times τ and Δ (Fig. 11). As indicated above the optimal value of τ is $1/(4J)$, which will give a 90° divergence between H_A and H_B. From Eq. (2) the graph in Fig. 13 was obtained. This sinusoidal relationship makes E_d insensitive to minor variations in τ, which allows all systems with similar coupling constants $(J \pm 25\%)$ to be significantly enhanced (85% E_d) from a properly chosen τ. One useful feature of τ is that if it is set for $1/(4X)$ the INEPT sequence will give full enhancement of a nuclei where $J_{H-I} = X$, 71% enhancement when $J_{H-I} = X/2$, but zero

TABLE II

Optimal Values of E_d and Δ as a Function of Number of Coupled Protons, n[a]

| | n | | | | | | |
	1	2	3	6	9	12	18
E_d[b]	5.04	5.04	5.82	7.83	9.44	10.82	13.15
Δ_{opt}[c]	0.5	0.25	0.196	0.134	0.108	0.093	0.076

[a] From Ref. 12.
[b] E_d for ^{29}Si, where $\gamma_H/\gamma_{Si} = 5.04$.
[c] In units of J^{-1}.

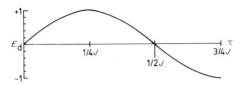

FIG. 13. Plot showing the dependence of decoupled enhancement E_d on delay time τ.

enhancement[4] when $J_{H-I} = 2X$. This means that if τ is set for $J = 10$ Hz it will give 71% enhancement of a system where $J = 5$ Hz, but if it is set for $J = 5$ Hz it will give a zero enhancement of a system where $J = 10$ Hz.

The final parameter, Δ, is the most useful variable in INEPT. Its optimal value depends on both the number of protons and the coupling constant [Eq. (5), see also Table II]. Figure 14 shows the relationship of Δ to E_d and

$$\Delta_{opt} = \frac{1}{\pi J} \arcsin n^{-1/2} \tag{5}$$

the dependence of that relationship on n. As n is increased, Δ_{opt} becomes smaller and E_d becomes more sensitive to variations in Δ. Furthermore if a larger value is chosen for Δ, the enhancements of nuclei coupled to an even number of protons become negative. This fact was used by Burum and Ernst (14) and by Doddrell and Pegg (18) to identify CH, CH$_2$, and CH$_3$ groups in ^{13}C NMR.

B. DEPT Theory

As mentioned above, DEPT theory is best described mathematically. Two different analyses have been employed: the Schrödinger approach (3, 4) and the Heisenberg approach (20). An overview of the Schrödinger approach is presented here; more detailed treatments of the Schrödinger and Heisenberg methods are contained in the references noted above.

For simplicity all pulses in the DEPT sequence are considered to be on resonance, and refocusing pulses are ignored. This results in the simplified DEPT sequence below:

$$90^\circ_y(H) - (2J)^{-1} - 90^\circ_x(I) - (2J)^{-1} - \theta_x(H) - (2J)^{-1} - \text{decouple, acquire (I)}$$

For a simple H–I system, the DEPT sequence in its initial stages resembles the INEPT sequence.

---

[4] Since enhancement is defined as $E = (S/N)_{INEPT}/(S/N)_{STD}$, $E = 1$ means no enhancement, while $E = 0$ means that no signal is observed in the ^{29}Si-NMR spectrum.

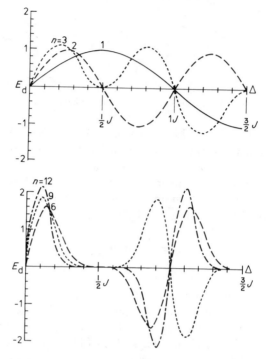

FIG. 14. Plot showing the relationship of Δ to E_d for ^{29}Si and the dependence of that relationship on n.

Again, we employ a doubly rotating reference frame and consider only the Boltzmann excess proton spin population and the I spin population (neglecting Boltzmann excess I spins). As in INEPT, an initial 90°_x (H) pulse rotates the Boltzmann excess proton spins to the x axis (Fig. 11b). The proton magnetization then splits into two groups depending on the spin state of the I nucleus to which the proton is coupled. Precession for $(2J)^{-1}$ aligns the protons of one group along the $+y$ axis, the other group along the $-y$ axis. The 90_x (I) pulse then rotates the I magnetizations to the x,y plane, into alignment with their respective coupled protons.

These states are described by the notation (3) $|+, +\rangle$ and $|-, -\rangle$, where the sign before the comma refers to the I states, and after the comma to the proton states. These states are coherent; a proton in the $|+\rangle$ state is coupled to a $|+\rangle$ state I. After the second $(2J)^{-1}$ period, the H and I states again evolve; the resulting states can then be described as superpositions of the original $|+, +\rangle$ and $|-, -\rangle$ states [Eq. (6)]. Note that

$$2^{-1/2}|+, +\rangle + i2^{-1/2}|-, -\rangle \text{ and } 2^{-1/2}|-, -\rangle + i2^{-1/2}|+, +\rangle \quad (6)$$

again, coherence is retained; there are no $|+, -\rangle$ or $|-, +\rangle$ states. This implies that the probability of single spin flips is zero. The effects of the final θ_x (H) pulse and $(2J)^{-1}$ precession period on the states of the H–I spin system are complex, but are described by Eq. (7).

$$2^{1/2}(\cos 1/2\phi|+\rangle - i \sin 1/2\phi|-\rangle)|\uparrow\rangle$$
$$+ 2^{-1/2}(\sin \phi 1/2\phi|+\rangle + i \cos 1/2\phi|-\rangle)|\downarrow\rangle \qquad (7)$$

and

$$2^{-1/2}(i \cos 1/2\phi|+\rangle - \sin 1/2\phi|-\rangle)|\uparrow\rangle$$
$$+ 2^{-1/2}(i \sin 1/2\phi|+\rangle + \cos 1/2\phi|-\rangle)|\downarrow\rangle$$

In Eq. (7), $\phi = 90 - \theta$, while $|\uparrow\rangle$ and $|\downarrow\rangle$ describe proton spins that are parallel or antiparallel to an axis at angle ϕ to the y axis; $|+\rangle$ and $|-\rangle$ again describe I spin states. It can be shown from the above equations (3, 4) that the upfield and downfield components of the signal vector are of opposite sign but proportional to $\sin \theta$. The final $(2J)^{-1}$ period serves to bring these two signals into phase, giving total enhancement of $(\gamma_H/\gamma_I) \sin \theta$. This signal appears as an enhanced doublet if no decoupling is used.

The more complex spin systems, IH_n, $n > 1$, follow the same logic as above (4) although the equations describing the spin states are correspondingly more complex. In each case, the maximal enhancement is γ_H/γ_I, with a sinusoidal θ dependence, which varies with n. It can also be shown (4) that decoupled DEPT spectra give "normal" multiplet intensities, i.e., for $n = 2$ the intensities are 1:2:1.

Enhancements and optimal settings for the τ and θ parameters of DEPT can easily be obtained from equations developed for INEPT using the relation

$$\theta = \pi J \Delta \qquad (8)$$

where J is the coupling constant and Δ is the INEPT parameter. Using Eq. (8), theoretical enhancements for DEPT can be derived from Eq. (3):

$$E_d = n(\gamma_H/\gamma_I) \sin \theta \cos^{n-1} \theta \sin(2\pi J \tau) \qquad (9)$$

where $\tau = (4J)^{-1}$. Similarly θ_{opt} can be derived from Eq. (5) and Eq. (8):

$$\theta_{opt} = \arcsin(n)^{-1/2} \quad \text{(radians)} \qquad (10)$$

One should especially note that Eq. (10) is independent of J thereby decreasing the sensitivity of the DEPT sequence to variations in J within a molecule (4).

Signal enhancement from INEPT and DEPT pulse sequences differs substantially from nuclear Overhauser enhancement. While polarization transfer enhancements are dependent on the number of protons coupled to

I (see above), NOE is independent of that number and has a theoretical limit (21) of $1 + \frac{1}{2}\gamma_H/\gamma_I$. For a $^{13}CH_2$ group the theoretical enhancement factors are 4.0 for INEPT and 3.0 for NOE. The difference is much more important for ^{29}Si which has a negative magnetogyric ratio. Me_3Si has a theoretical INEPT enhancement of 9.4 while its theoretical NOE is $1 + \frac{1}{2}(-5.0)$ or -1.5. Other advantages of INEPT and DEPT over NOE include faster polarization transfer, less sample heating, and independence of cross-relaxation effects (14).

In addition to signal enhancement the INEPT and DEPT sequences also allow a faster pulse repetition rate, thereby further decreasing the time required to obtain a spectrum. When polarization transfer is not employed, the pulse repetition rate is dependent on the spin–lattice relaxation times, T_1, which can be very long for some observed nuclei, I (Table III). However when the INEPT or DEPT sequences are employed the enhanced population

TABLE III

SPIN–LATTICE RELAXATION TIMES (T_1) OF ^1H, ^{13}C, ^{15}N, AND
^{29}Si ATOMS IN SELECTED COMPOUNDS[a]

^1H			^{13}C		
H_2O		3.5	C_6H_{12}		20
C_6H_{12}		7.1	CH_3COCH_3	CH_3	11
CH_3COCH_3		15.6		CO	36
C_6H_6		19	C_6H_6		28
CH_3CN		16.1	CH_3CN	CH_3	13
$C_6H_5CH_3$	CH	16		CN	5
	CH_3	9	C_6H_5CH	CH	15
$C_6H_5OCH_3$	CH	16		CH_3	16
	CH_3	5.8		C	38
$CHCl_3$		78	$CHCl_3$		32

^{15}N		^{29}Si	
N_2	13	$SiMe_4$	19
$HCONH_2$	21	$FSiMe_3$	19
CN^-	21	$HSiMe_3$	16
Pyrrole	40	$PhSiMe_3$	42
Pyrrolidine	58	$Me_2Si(CH_2)_4$	52
$BuNH_2$	70	$Me_3SiOSiMe_3$	40
Pyridine	85	$Me_3SiSiMe_3$	45
CH_3CN	90	$(MeO)_4Si$	73
$PhNO_2$	400	$(Me_2SiO)_4$	100

[a] Values in the table are given in seconds. From Refs. 1, 5, and 17.

differences are "pumped" from ^{1}H to I during each multipulse cycle. Therefore only the protons, and not the enhanced nuclei, need be allowed to relax.

INEPT and DEPT provide many advantages over standard heteronuclear NMR methods. These include signal enhancement, faster pulse repetition, enhanced outer lines of multiline patterns (for INEPT), clear distinction between even- and odd-lined patterns, distinction between IH$_n$ groups where n is odd or even, and elimination of signals from solvent or uncoupled I nuclei. Furthermore, the variable parameters (Δ, θ, and τ) are easily predicted and fairly insensitive to differences between systems, especially for the DEPT sequence.

The limitations of these methods are that (1) correct values of Δ, θ, and τ must be selected for a given I–H interaction; (2) the spin–spin relaxation times, T_2, must be long relative to the length of the pulse sequence, so that there is still observable signal when acquisition is begun, especially for DEPT; and (3) while I and H can be separated by any number of bonds, the resulting coupling J_{I-H} must be resolvable on the instrument used. Difficulties in any of these areas will cause the observed enhancement to be less than the theoretical value.

REFERENCES

1. B. E. Mann, *in* "Spectroscopic Properties of Inorganic and Organometallic Compounds," Vol. 1–14. The Chemical Society, London, 1968–1981; G. C. Levy and J. D. Cargioli, *in* "Nuclear Magnetic Resonance Spectroscopy of Nuclei Other than Protons" (T. Axenrod and G. A. Webb, eds.), Ch. 17. Wiley (Interscience), New York, 1974; J. Schraml and J. M. Bellama, *in* "Determination of Organic Structures by Physical Methods" (F. C. Nachod, J. J. Zuckerman, and E. W. Randall, eds.), Vol. 6, p. 203. Academic Press, New York, 1976; E. A. Williams and J. D. Cargioli, *in* "Annual Reports on NMR Spectroscopy" (G. A. Webb, ed.), Vol. 9, p. 221. Academic Press, New York, 1979; H. Marsmann, *in* "NMR Basic Principles and Progress" (P. Diehl, E. Fluck, and R. Kosfeld, eds.), Vol. 17, p. 65. Springer-Verlag, Berlin and New York, 1981; D. Corey, A. Wong, and W. M. Ritchey, *J. Organomet. Chem.* **235**, 277 (1982); G. A. Olah and L. D. Field, *Organometallics* **1**, 1485 (1982).
2. G. A. Morris and R. Freeman, *J. Am. Chem. Soc.* **101**, 760 (1979).
3. D. M. Doddrell, D. T. Pegg, and M. R. Bendall, *J. Magn. Reson.* **48**, 323 (1982).
4. D. T. Pegg, D. M. Doddrell, and M. R. Bendall, *J. Chem. Phys.* **77**, 2745 (1982).
5. G. A. Morris, *J. Am. Chem. Soc.* **102**, 428 (1980); D. T. Pegg, D. M. Doddrell, W. M. Brooks, and M. R. Bendall, *J. Magn. Reson.* **44**, 32 (1981); C. Brevard, G. C. van Stein, and G. van Koten, *J. Am. Chem. Soc.* **103**, 6746 (1981); C. Brevard and R. Schimpf, *J. Magn. Reson.* **47**, 528 (1982); P. L. Rinaldi and N. J. Baldwin, *J. Am. Chem. Soc.* **104**, 5791 (1982).
6. B. J. Helmer and R. West, *Organometallics*, **1**, 877 (1982).
7. R. D. Bertrand, W. B. Moniz, A. N. Garroway, and G. C. Chingas, *J. Am. Chem. Soc.* **101**, 5227 (1979).
8. S. R. Hartmann and E. L. Hahn, *Phys. Rev.* **128**, 2042 (1962); A. Pines, M. G. Gibby, and J. S. Waugh, *J. Chem. Phys.* **59**, 569 (1973).
9. P. D. Murphy, T. Taki, T. Sagabe, R. Metzler, T. Squires, and B. Gerstein, *J. Am. Chem. Soc.* **101**, 4055 (1979).

10. E. Lippma, M. Mägi, A. Samoson, G. Engelhardt, and A.-R. Grimmer, *J. Am. Chem. Soc.* **102**, 4889 (1980); G. E. Maciel and D. W. Sindorf, *ibid.* **102**, 7606 (1980); D. W. Sindorf and G. E. Maciel, *ibid.* **103**, 4263 (1981); P. F. Barron, M. A. Wilson, A. S. Campbell, and R. L. Frost, *Nature (London)* **299**, 616 (1982).

11. S. A. Linde, H. J. Jakobsen, and B. J. Kimber, *J. Am. Chem. Soc.* **97**, 3219 (1975); S. Li, D. L. Johnson, J. A. Gladysz, and K. L. Servis, *J. Organomet. Chem.* **166**, 317 (1979).

12. D. M. Doddrell, D. T. Pegg, W. M. Brooks, and M. R. Bendall, *J. Am. Chem. Soc.* **103**, 727 (1981).

13. O. W. Sørensen and R. R. Ernst, *J. Magn. Reson.* **51**, 477 (1983).

14. D. P. Burum and R. R. Ernst, *J. Magn. Reson.* **39**, 163 (1980); G. A. Morris, *ibid.* **41**, 185 (1980).

15. See for example: D. L. Rabenstein and T. T. Nakashima, *Anal. Chem.* **51**, 1465A (1979).

16. D. A. Stanislawski, Ph.D. Thesis, University of Wisconsin, Madison, 1978.

17. M. L. Martin, G. J. Martin, and J. Delpuech, *in* "Practical NMR Spectroscopy." Heyden, London, 1980; G. Martin, M. Martin, and J. Gouesnard, *in* "NMR Basic Principles and Progress" (P. Diehl, E. Fluck, and R. Kosfeld, eds.), Vol. 19. Springer-Verlag, Berlin and New York, 1981.

18. D. M. Doddrell and D. T. Pegg, *J. Am. Chem. Soc.* **102**, 6388 (1980).

19. E. D. Becker, *in* "High Resolution NMR," 2nd ed., p. 50. Academic Press, New York, 1980.

20. D. T. Pegg, M. R. Bendall, and D. M. Doddrell, *J. Magn. Reson.* **44**, 238 (1981); M. R. Bendall, D. T. Pegg, and D. M. Doddrell, *ibid.* **45**, 8 (1981); D. T. Pegg and M. R. Bendall, *ibid.* **53**, 229 (1983).

21. K. F. Kuhlmann, D. M. Grant, and R. K. Harris, *J. Chem. Phys.* **52**, 3439 (1970).

C- and O-Bonded Metal Carbonyls: Formation, Structures, and Reactions

COLIN P. HORWITZ

and

DUWARD F. SHRIVER

Department of Chemistry
Northwestern University
Evanston, Illinois

I

INTRODUCTION

Metal carbonyls, which were among the earliest discovered organometallic compounds, continue to play a central role in organometallic chemistry. One recently emerging aspect of carbonyl chemistry is the synthesis and chemistry of compounds having carbonyl ligands bonded simultaneously through C and O.

In the mid-1960s infrared spectroscopic evidence was obtained for ion pairing between alkali metal cations and metal carbonyl anions in solvents of moderate polarity (*1*). In light of subsequent research (see Section II,A,2), the infrared data in the CO stretching region indicate that the primary interaction in these tight ion pairs occurs between the cation and the oxygen of a CO ligand. The first explicit evidence for C- and O-bonded carbon monoxide was obtained in studies on the interaction of molecular Lewis acids with polynuclear carbonyl compounds (**1**) reported in 1969 (*2*). The compound contains bridging carbonyl groups bonded to Lewis acids through the carbonyl oxygen, illustrating the pronounced Lewis basicity of bridging carbonyl ligands. In a simultaneous and independent investigation it was demonstrated that an isolable compound could be produced by the formation of a —CO— interaction between a metal carbonyl anion (actually a zwitterion in this case) and a molecular Lewis acid (3). Since these first studies there have been hundreds of examples in which C and O bonding is implicated in compound formation or in transient reactive species.

Most of the structurally characterized C- and O-bonded metal carbonyls fall into one of two classes: those involving an essentially end-on array **1** and those in which CO is bound in dihapto fashion with respect to a metal atom in a metal carbonyl cluster **2**. However structures **1** and **2** do not

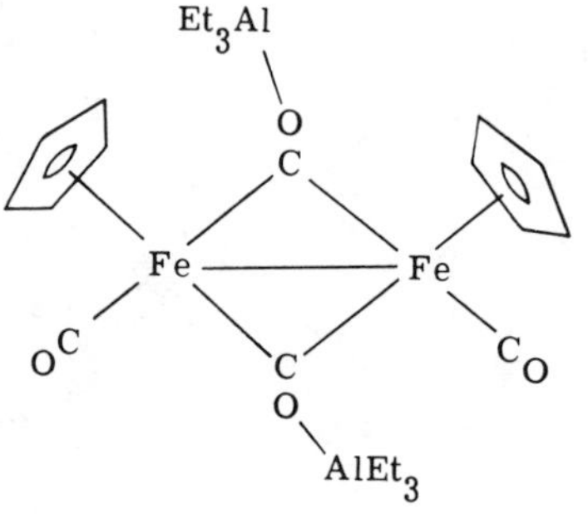

(1) 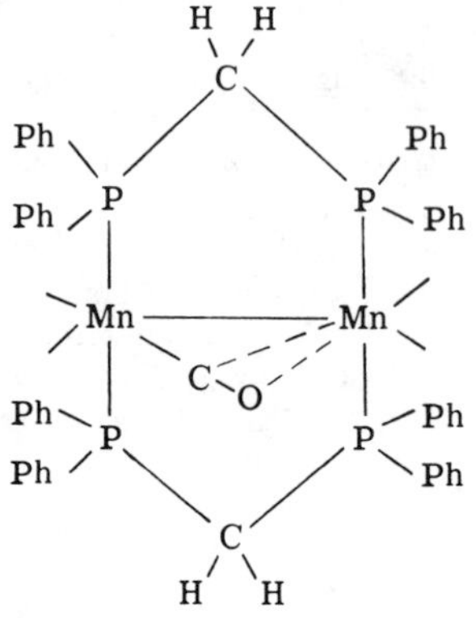   **(2)**

exhaust the possibilities for C and O bonding. Table I presents a more comprehensive view of the various idealized modes of bonding of CO in bimetallic and cluster compounds. The bottom row of Table I lists the well-known edge-, M_2CO, and face-, M_3CO, bridging carbonyl arrays as well as the hypothetical square face-bridging CO. For these three entries the carbonyl ligand is bonded exclusively through carbon. All other entries in the table represent some degree of oxygen bonding as well, and compounds are known that fit most of these remaining categories.

Σ —CO— *Bonding.* Compounds in the first row in Table I are variously referred to as end-on or isocarbonyls. We have adopted the Σ —CO— terminology to imply the essentially end-on nature of these species, although it must be recognized that these compounds rarely, if ever, contain strictly linear COM' arrays. In general Σ —CO— compounds are formed from the interaction of a basic carbonyl moiety with a Lewis or proton acid, repre-

TABLE I

THE DISPOSITION OF CO IN POLYMETAL SYSTEMS

CO disposition	Metal nuclearity		
Σ—CO—	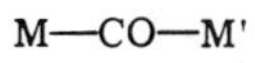 M—CO—M'	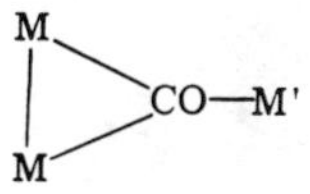 M, M / CO—M' (b)	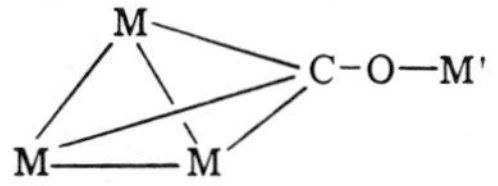 M / M——M C-O—M' (c)
	(a)	(b)	(c)
Π—CO	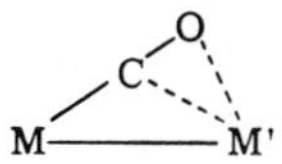 M——M' with C=O bridging (d)	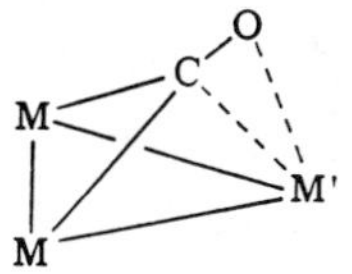 M, M / M' with C=O (e)	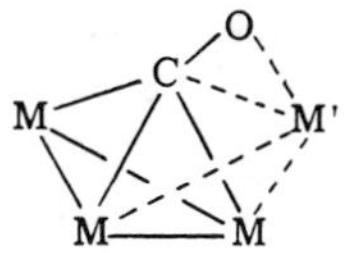 M, M / M'——M with C=O (f)
	(d)	(e)	(f)
		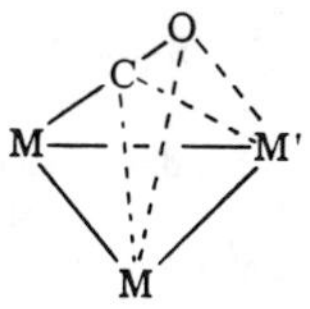 M——M' with C=O and M below (e')	
		(e')	
—CO	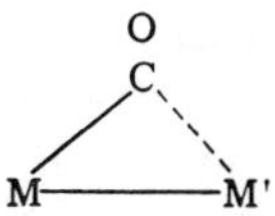 M——M' with O=C bridge (g)	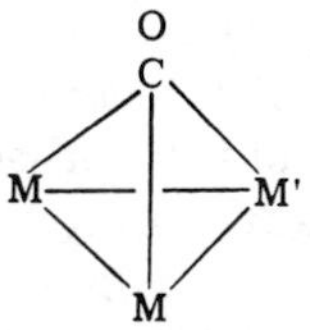 M——M' / M with O=C (h)	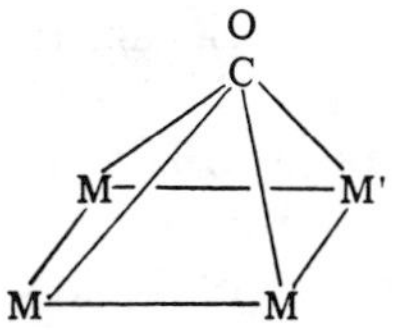 M'——M / M'——M with O=C (i)
	(g)	(h)	(i)

sented as M' in Table I. For example, anionic and donor-substituted metal carbonyls (Section II,A) as well as C-bridging carbonyls (Section II,B) are basic and susceptible to interaction with acceptor ions or molecules.

For similar compounds, the order of carbonyl basicity is face-bridging CO > edge-bridging CO > terminal CO (Section II). These trends in carbonyl basicity reflect the degree of delocalization of electron density onto the CO ligand. Carbonyl stretching frequencies provide a guide to the extent

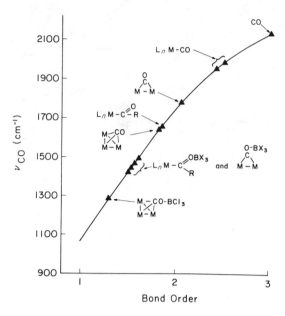

FIG. 1. CO stretching frequency versus bond order. The curve was determined from data on organic compounds. Organometallic compounds are entered on the curve based on observed CO stretching frequencies. Formation of isolable adducts generally does not occur when all CO stretching frequencies in the parent carbonyl are above 1900 cm^{-1}, which corresponds to bond orders greater than 2.

of this charge delocalization (Fig. 1), and therefore one can employ infrared spectra of metal carbonyl compounds to assess their potential basicity at the carbonyl oxygen. The formation of a donor–acceptor bond is dependent on the concentration of the reactants and on properties such as acceptor strength and steric bulk of the Lewis acid. With this caveat in mind, a rough general rule is that formation of isolable C- and O-bonded compounds of the donor–acceptor type is unlikely if the CO stretching frequencies in the parent carbonyl are all above 1900 cm^{-1}.

Infrared spectroscopy is a convenient tool for the identification of the Σ —CO— compounds because carbonyl stretching frequencies are greatly reduced upon the formation of these adducts. This effect can be attributed to an increase in back π bonding induced by the Lewis acid, which delocalizes electrons from the transition metal to CO π-antibonding orbitals and thereby lowers the CO stretching force constant. These perturbations are illustrated schematically by the difference between **3a** and **3b**. The magnitude of this reduction in CO stretching frequency varies with the nature of the parent metal carbonyl and the acceptor, but it often ranges between 100 and

$$\text{O} \Equiv \text{C} ==\!\!=\!\!= \text{M} ==\!\!=\!\!= \text{C} \Equiv \text{O} \quad \xrightarrow{\text{AlCl}_3} \quad \overset{\text{increase } \nu_{CO}}{\text{O} \equiv \text{C} - \text{M}} = \overset{\text{decrease } \nu_{CO}}{\text{C} = \text{O}}$$

increase ν_{CO} decrease ν_{CO}

shortened   lengthened

AlCl$_3$

(3a)   **(3b)**

200 cm^{-1}. Some examples are given in Fig. 1. In addition, a moderate but distinct increase usually is observed in the CO stretching frequencies for all other CO ligands in the molecule. As illustrated schematically in **3a** and **3b**, these changes can be attributed to changes in bond order in the conjugated $M(CO)_n$ system owing to competitive M—C back π-bonding. As with simple metal carbonyls, selection rules for infrared activity based on the molecular symmetry provide a useful means of characterizing the C- and O-bonded carbonyls. Figure 2 illustrates the systematic shifts in CO stretching frequencies upon interaction of a carbonyl with a Lewis acid. Many other examples are given in Section II.

Carbon-13 NMR provides another convenient spectroscopic technique that is diagnostic of Σ —CO— bonding (4, 5). The ^{13}CO resonance is shifted downfield upon attachment of an acceptor to the oxygen end of the CO ligand, and in general this shift is so great that interpretation of the spectra is unambiguous. Specific examples are given in Section II.

The electronic factors that influence the CO stretching frequencies of Σ —CO— compounds also are manifested in bond lengths. When precise structural data are available, the C—O bond distance is lengthened and the M—C distance usually is shortened upon adding an acceptor to a CO. These changes, which are illustrated graphically in **3a** and **3b**, can be substantial. An increase of 0.1Å in C—O and a decrease of this same magnitude in M—C is fairly common when strong acceptors are involved.

Π —*CO Bonding.* This class of C- and O-bonded carbonyls is displayed in the second row of Table I. In contrast with the simple Lewis acid–base interactions discussed above, the Π —CO bonded systems, such as **2**, generally involve less polar bonding, and are known primarily in dinuclear M—M-bonded compounds (6, 7). A few examples of Π —CO ligands have been observed recently in polynuclear carbonyls (Section III). A continuous gradation appears to exist between the Π —CO structure illustrated in **2** and the conventional C-bridged carbonyls such as that illustrated in **4**.

Unlike the straightforward identification of Σ —CO— ligands by spectroscopic methods, the spectroscopic identification of Π —CO ligands is less readily achieved because CO stretching frequencies overlap conventional

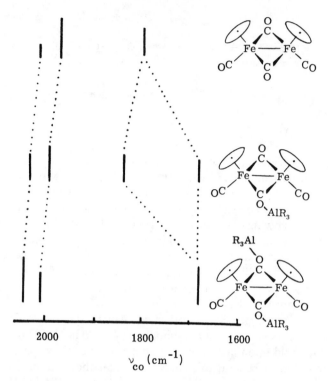

2000   1800   1600

$$\nu_{co} \, (cm^{-1})$$

FIG. 2. CO stretching frequencies for the parent $Cp_2Fe_2(CO)_4$ and its 1:1 and 1:2 adducts with $AlEt_3$. Small increases in terminal stretching frequencies are evident in the 2000 cm^{-1} region. Much larger decreases are evident in the lower frequency bridging CO stretching frequencies. For reasons of symmetry the parent and the 1:2 adduct each display only one strong bridging CO stretch.

Σ —CO— species such as a and b (Table I) and ^{13}CO chemical shifts often occur in the region for conventional terminal or bridging carbonyls, such as g and h in Table I. In a simple C-bonded CO bridge, the CO ligand is perpendicular to the metal–metal axis whereas in **2** there is one nearly linear M—C—O array with the CO tilted toward the other metal to give an apparent Π —CO ligand–metal interaction. The distinction is somewhat less

(4)

clear between **2** and CO in a semibridging position **4**; however, geometric criteria are sometimes helpful in making distinctions between the various types of CO dispositions.

Figure 3 presents a plot of the M—C—O angle, θ, against a distance function, $(D_2 - D_1)/D_1$, which separates the carbon-bonded bridging CO and the Π —CO arrays into two different curves (7). The two become indistinguishable only in the case of a nearly linear M—C—O array having a large difference between the two C—M separations. This corresponds to a slightly tipped terminal CO.

The above geometric criterion was not designed to distinguish nearly linear Σ —CO— compounds from the rest. We propose a different empirical correlation to differentiate Σ —CO—, Π —CO, and conventional C-bridging carbonyls. The function chosen, $\Omega = \exp[D(M'—C)/D(M'—O)]$, provides a measure of the extent of interaction of M' with the C and O ends of the CO ligand. A high value of Ω is indicative of Σ —CO— bonding, and intermediate values, indicating comparable M'—O and M'—C interaction, correspond to Π —CO. Conventional C-bonded bridging carbonyls have

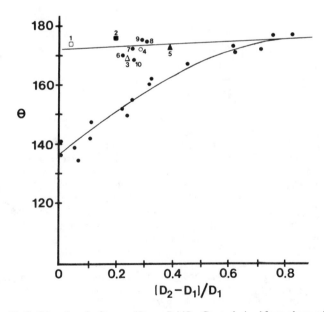

FIG. 3. M–C–O bond angle, θ versus $(D_2 - D_1)/D_1$. Curve derived from observed structural data. The suggested Π —CO adducts are on the linear portion of the upper curve starting at $\theta \cong 170$. Conventional C-bonded carbonyls fall on the lower curve. 1, $Mn_2(CO)_5(dppm)_2$; 2, $Cp_2Mo_2(CO)_4$; 3, $Cp_2V_2(CO)_5$; 4, $Cp_2Cr_2(CO)_4$; 5, $Cp^*_2Cr_2(CO)_4$; 6, $[Cp_2Mo_2(CO)_4CN]^-$; 7, $CpW(CO)_3GaMe_2$; 8, $Cp_2Mo(CO)_3ZnBr \cdot 2\,THF$; 9, $[CpMo(CO)_3 \cdot ZnCl \cdot Et_2O]_2$; 10, $CpW(CO)_3AuPPh_3$. (Provided by M. D. Curtis. See Ref. 7 for an earlier version.)

low values of Ω. This correlation is illustrated for some representative compounds in Fig. 4. It will be noted that the Ω values provide clear differentiation of the known Σ —CO— and Π —CO arrays but that there is no obvious break in Ω values between conventional —C-bridging carbonyls and Π —CO.

Reactions. The focus of much current research in the area of C- and O-bonded carbon monoxide is on the great changes in reactivity which this type of interaction exerts on the carbon monoxide ligand. Perhaps the simplest is the influence on tautomeric equilibria involving the CO ligands in metal carbonyl clusters. The principle involved here is that bridging carbonyl ligands are more basic than terminal carbonyls and therefore carbonyl bridging can be induced by interaction with a Lewis acid (Sections II,B,4 and

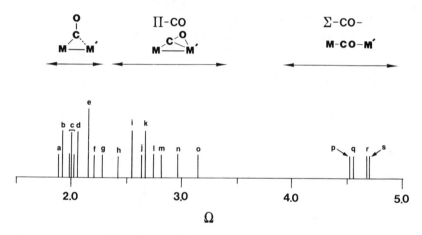

FIG. 4. The Ω function showing the clear demarcation between Σ —CO— compounds (high Ω values) and Π —CO compounds as well as the gradual progression of conventional C bridged carbonyls to Π —CO. a, $[Cp_2Co_2(CO)_2]^-$ ($\Omega = 1.89$), L. M. Cirjak *et al.*, *Inorg. Chem.* **21**, 940 (1982); b, $Cp_2Ni_2(CO)_2$ ($\Omega = 1.92$), L. R. Byers and L. F. Dahl, *Inorg. Chem.* **19**, 680 (1980); and $Co_2(CO)_8$ ($\Omega = 1.92$), G. G. Sumner *et al.*, *Acta Crystallogr.* **17**, 732 (1964); c, $Cp_2V_2(CO)_5$ ($\Omega = 1.99$) Ref. *179*; bridging CO in $[Fe_3Cr(CO)_{14}]^{2-}$ ($\Omega = 2.00$) Ref. *189*; bridging CO in $CpCo(CO)_2ZrCp^*_2$ ($\Omega = 2.00$) Ref. *167*; and semibridging CO in $[Fe_4(CO)_{13}]^{2-}$ ($\Omega = 2.02$) Ref. *153*; d, $[CpMo(CO)_3ZnCl \cdot Et_2O]_2$ ($\Omega = 2.06$) Ref. *70*; and semibridging CO in $[Fe_3Cr(CO)_{14}]^{2-}$ ($\Omega = 2.06$) Ref. *189*; e, $Cp_2Mo_2(CO)_4$ ($\Omega = 2.16$) Ref. *7*; $Cp_2Cr_2(CO)_4$ ($\Omega = 2.16$) Ref. *174*; and semibridging CO in $Cp_2NbMo(CO)_3Cp$ ($\Omega = 2.16$) Ref. *171*; f, $[CpMo(CO)_3]_2Zn$ ($\Omega = 2.20$) Ref. *70*; g, $CpMo(CO)_3ZnBr \cdot 2THF$ ($\Omega = 2.28$) Ref. *178*; h, $Mn_2(CO)_5(dppm)_2$ ($\Omega = 2.41$) Ref. *6*; i, Π —CO in $CpCo(CO)_2ZrCp^*_2$ ($\Omega = 2.55$) Ref. *167*; and $Cp_2MoW_2(\mu CR)(CO)_6$ ($\Omega = 2.55$) Ref. *169*; j, $Cp_3TiW(\mu CR)(CO)_2$ ($\Omega = 2.63$) Ref. *168*; k, Π —CO in $Cp_2NbMo(CO)_3Cp$ ($\Omega = 2.65$) Ref. *171*; and $Cp_3TiW(\mu CR:CH_2)(CO)_2$ ($\Omega = 2.65$) Ref. *170*; l, $Cp_2Zr(\eta^2-Ac)Mo(CO)_2Cp$ ($\Omega = 2.81$) Ref. *60*; m, $Cp_3Nb_3(CO)_7$ ($\Omega = 2.75$) Ref. *182*; n, $[HFe_4(CO)_{13}]^-$ ($\Omega = 2.96$) Ref. *184*; o, $Cp_2Cp^*_2Mo_2Co_2(CO)_4$ ($\Omega = 3.16$) Ref. *185*; p, $Cp^*_2(THF)YbCo(CO)_4$ ($\Omega = 4.53$) Ref. *55*; q, $Cp^*_2(CH_3)TiMo(CO)_3Cp$ ($\Omega = 4.56$) Ref. *58*; r, $(py)_4Mg[CpMo(CO)_3]_2$ ($\Omega = 4.69$) Ref. *27*; s, $V(B)_4[V(CO)_6]_2$ ($\Omega = 4.70$) Ref. *63*.

II,C,4). More drastic transformations of the CO ligand are discussed in Section IV. Of these, one of the most interesting is the very large acceleration of the CO migratory insertion into metal alkyl bonds by alkali metal ions and molecular Lewis acids. Similarly, cleavage and reduction of CO in metal clusters are greatly facilitated by alkylation or protonation of the carbonyl oxygen. This controlled CO cleavage reaction is opening up an extensive carbide chemistry for metal cluster carbonyls. Two areas in which initial observations have been made are the possibility that the carbonyl oxygen may be the general site of initial attack by electrophiles on metal carbonyls and that redox reactions of metal carbonyls may be mediated by a CO bridge between oxidant and reductant.

II

Σ —CO— COMPOUNDS:

FORMATION, STRUCTURES, AND SPECTRA

As mentioned in Section I, Σ —CO— interaction generally occurs between a basic carbonyl oxygen and an electron acceptor. The Ω values characteristic of Σ —CO— linkages range from 4 to 5.5. Carbonyl basicity is found for C-bridging CO ligands in metal–metal bonded compounds, CO ligands in anionic metal carbonyls, or CO ligands in donor-substituted metal carbonyls. Metal centers are known to be basic, and the latter two situations are known to enhance metal basicity (8, 9). Thus basic metal and ligand sites will often coexist. To a rough first approximation the ultimate site of attachment of an acceptor is sterically controlled. Thus, the most stable protonated species generally contain M—H bonds whereas the much bulkier molecular Lewis acids generally attach to carbonyl oxygens. Simple M^+ and M^{2+} ions may attach to either site depending in part on the bulk of other ligands attached to these cations. Independent of the ultimate structure of the product, it is physically reasonable that carbonyl oxygens, which are much more exposed than the metal centers, may be the sites of initial attack by electrophiles.

A. Terminal CO Donors

1. Proton and Hydrogen Bonding Acceptors

The interaction of protons with mononuclear metal carbonyls readily occurs at the metal center rather than the CO oxygen [Eq. (1)]; no stable or

$$HX + Co(CO)_4^- \longrightarrow HCo(CO)_4 + X^- \tag{1}$$

transient MCO—H species appear to have been observed. Similarly, hydrogen bonding to metal centers appears to occur in preference to hydrogen bonding with the carbonyl oxygen. For example, a recent X-ray crystallographic study of $[Me_3NH]^+$ and $[Et_3NH]^+$ salts of $[Co(CO)_4]^-$ reveals a hydrogen bond between nitrogen and the cobalt **5** (*10*).

$$
\begin{array}{c}
\text{O} \\
\text{C} \\
| \\
\text{Et}_3\text{N}-\text{H}----\text{Co}----\text{CO} \\
\text{O}^{\text{C}}\ \text{C} \\
\text{O}
\end{array}
$$

(**5**)

One example of hydrogen bonding to a terminal carbonyl has been indicated by crystallographic data for $[(i\text{-}Pr)_2NH_2][HFe_3(CO)_{11}]$ (*11*). The nitrogen to terminal oxygen distance is 3.11 Å, which places the hydrogen around 2.09 Å from the oxygen, assuming van der Waals radii. The presence of a hydrogen bond is also apparent from the IR absorption at 1880 cm^{-1}, which is approximately 50 cm^{-1} lower than the lowest frequency absorption for terminal carbonyls of salts which do not exhibit —C—O—H interactions.

2. *Alkali Metal Cation Acceptors*

In solvents of moderate to low polarity alkali metal cations frequently form ion pairs with a variety of anions, including metal carbonylates (*12*). A review of ion pairing in carbonyl chemistry has recently appeared (*13*), so the emphasis here will be placed on general phenomena and on more recent results.

Infrared spectroscopy in the CO stretching region provides a sensitive probe for detecting contact ion pair formation between cations and carbonylates (*12, 13*). Good examples are provided by the early work on the interaction of cations with tetracarbonylcobaltate (*12*) and by more recent work on a variety of metal carbonyl anions (*13, 14*). In solvents such as THF, alkali metal carbonylates appear to be present in two or three different forms: tight (contact) ion pairs, looser ion pairs, and ion triplets. For example, Fig. 5 shows the IR spectrum of $Na[Mn(CO)_5]$ in THF and its deconvolution into two different sets of bands, which are associated with two different species in solution (*14*). For one of these the evidence points to a contact ion pair of structure **6**.

A second, less perturbed species is generally described as a solvent-separated ion pair. Recently, this species was attributed to a second-nearest neighbor, $[Mn(CO)_5]^-$, protruding into the solvent shell of the Na^+ (*14*).

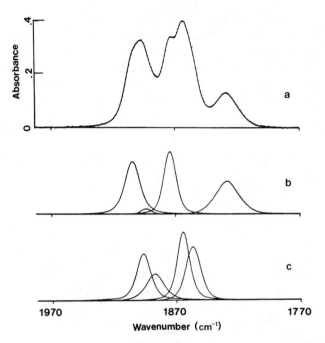

FIG. 5.   Experimental IR spectrum of Na[Mn(CO)$_5$] in THF in the CO stretching region (a); components due to the contact ion pair (6) (b); components due to the next-nearest neighbor in pair 7 (c). (Courtesy of W. F. Edgell, see also Ref. *14*.)

In this structural model, solvent molecules occupy the first coordination sphere of the Na$^+$, and the next-nearest interaction is with the carbonyl oxygen atoms of [Mn(CO)$_5$]$^-$, leading to the weak bidentate configuration as illustrated by 7.

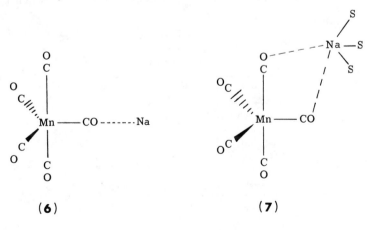

(6)   (7)

Similarly, vibrational spectroscopic evidence exists for the presence of a contact ion pair and a next-nearest neighbor ion pair for $Na[Co(CO)_4]$ in some polar organic solvents (15). In solutions of suitable concentration, the existence of ion triplets is possible (13, 16). The presence of these triple ions, shown schematically in **8** (13), can be detected by changes in electrical conductivity over the concentration range being investigated.

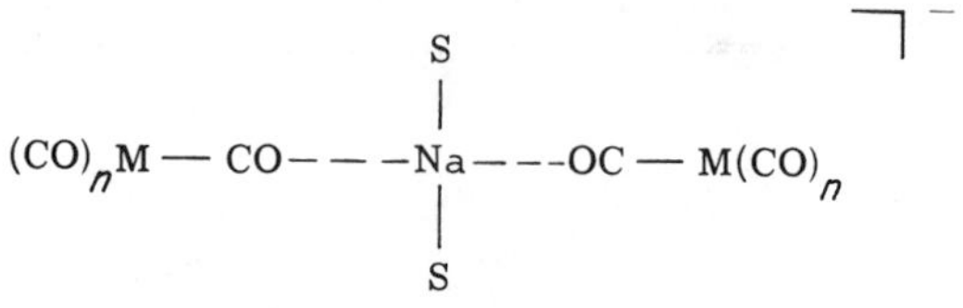

$$(CO)_nM - CO- - -Na- - -OC - M(CO)_n$$

with S above and below Na, overall charge $^-$

(8a)

$$S - Na- - -OC - M - CO- - -Na - S$$

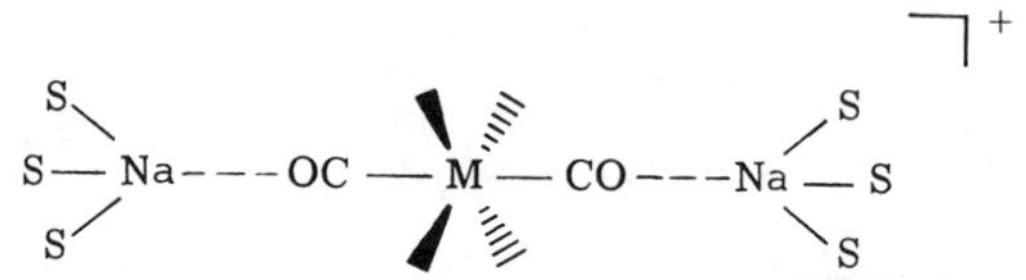

(8b)

Some ion pair dissociation constants obtained from conductivity data are given in Table II (13). As might be expected, $Na_2[Fe(CO)_4]$ has a very low dissociation constant in THF solution around room temperature. Under similar conditions, $Li[Mn(CO)_5]$ has a larger dissociation constant than $Na[Mn(CO)_5]$. Apparently THF competes more favorably than $Mn(CO)_5^-$ for coordination with the small Li^+ ion than with the larger Na^+.

Closely related to the association of alkali metal ions with metal carbonylates is the ion association, revealed by IR evidence, of

TABLE II

ION PAIR DISSOCIATION CONSTANTS
FROM CONDUCTIVITY DATA[a]

Compound	$10^5 K_D$
$Na_2[Fe(CO)_4]$	~0.05
$Na[Mn(CO)_5]$	1.98
$Li[Mn(CO)_5]$	4.30
$[Na \cdot HMPA][Mn(CO)_5]$	3.16
$[Na(15\text{-crown-}5)][Mn(CO)_5]$	3.90
$Na[Mn(CO)_4P(OPh)_3]$	0.93

[a] See Ref. 13.

$Mg[CpFe(CO)_2]_2$, $Mg[Co(CO)_3L]_2$, and $Tl[Co(CO)_4]$ in THF and similar solvents (13). Similarly, the anion $[Cr_2(CO)_{10}]^{2-}$, which contains only terminal carbonyls, forms contact ion pairs with Na^+ in THF (17, 18). In strong donor solvents such as dimethyl sulfoxide or hexamethylphosphoramide, the contact ion pair is broken up. As will be described in Section II,B,2, contact ion pairs are seen also in polynuclear carbonyls containing bridging carbonyls, in which case the cation has an affinity for the bridging CO ligand.

In addition to the C- and O-bridged contact ion pairs, it has been proposed that Na^+ and Mg^{2+} also interact directly with the iron in $[CpFe(CO)_2]^-$ in ether solvents (19–21). This proposition bears careful reinvestigation (13).

X-Ray crystallographic data on the interaction of alkali metal ions with mononuclear metal carbonyls are not very extensive. No data appear to be available for lithium salts of mononuclear carbonylates. The findings on sodium and potassium salts reveal interesting details in the variety of modes of interaction. The situation may be summarized as follows. An essentially end-on interaction (e.g., FeCO—Na) is seen in the majority of instances, but if the alkali metal ion is unencumbered with large ligands, such as crown ethers, an interaction also is seen with the carbonyl carbon atoms and with the metal. Thus the solid state species contain some of the same structural features which were discussed above for ion pairs in solution. Unfortunately, there are no detailed analyses of the vibrational spectra of the solids for which structures are available. This might provide one avenue to the refinement of our knowledge of the species present in solution.

The crystal structure of $Na_2[Fe(CO)_4]\cdot\frac{3}{2}$ dioxane (22a) reveals one set of Na^+--OC interactions, at 2.32 Å, to four CO oxygens from four different $[Fe(CO)_4]^{2-}$ units. These Na^+ ions also are coordinated to two dioxane molecules (Fig. 6). [A similar type of coordination about Na^+ is observed in $[Na(thf)_2]_2[Zn(Fe(CO)_4)_2]$, (22b).] A second set of Na^+ ions protrude into the distorted tetrahedron (Fig. 7). In this case the Na^+ has rather long contacts with two carbon atoms (3.05 and 2.86 Å) and the iron (3.09 Å). The Fe—C—O angle (171°) appears to be somewhat distorted, and the C—Fe—C angles (129.7°) deviate markedly from the tetrahedral angle. The first set of Fe—C—O—Na interactions found in this structure are of the Σ—CO— type and are analogous to those discussed above for the contact ion pairs in solution. The second type of interaction with the carbonyl carbon and metal may be analogous to the direct interaction between an alkali or alkaline earth metal ion and a transition metal which has been invoked as a possible solution species.

The salt $K_2[Fe(CO)_4]$ displays smaller distortions of the carbonylate than those described above for the sodium salt (23). The pattern of nearest neighbor interactions is similar to that seen for the sodium salt, however

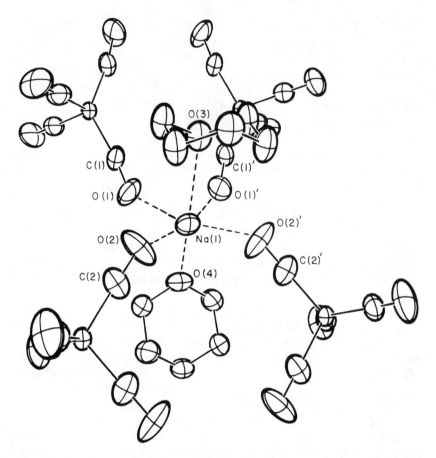

FIG. 6.   Coordination about one, (Na-1)$^+$, of two independent Na cations of Na$_2$Fe(CO)$_4$ · $\frac{3}{2}$C$_4$H$_8$O$_2$, emphasizing the Σ —CO— interaction of four [Fe(CO)$_4$]$^{2-}$ units with the (Na-1)$^+$ center. Two C$_4$H$_8$O$_2$ molecules complete the pseudooctahedral coordination about the cation. [By permission from the American Chemical Society, H. B. Chin and R. Bau, *J. Am. Chem. Soc.* **98**, 2434 (1976).]

there is only one type of K$^+$ present. The closest interactions with the potassium ion are K—O distances at 2.71 Å, in a Σ —CO— interaction. There is also a weaker interaction of the K$^+$ with another set of CO ligands where the closest approach is K—C (see Fig. 8). This array involves one K—C distance of 3.28 Å and the other of 3.40 Å. The K—Fe distance is 3.617 Å, which is approximately 0.4 Å longer than the sum of the covalent radii.

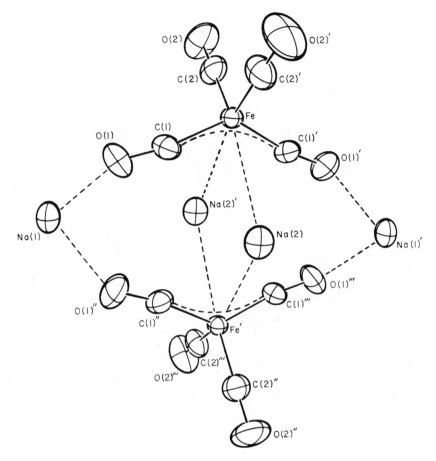

FIG. 7. Coordination about (Na-2)$^+$, the second type of Na cation in Na$_2$Fe(CO)$_4$ · $\frac{3}{2}$C$_4$H$_8$O$_2$. This view emphasizes the interaction of (Na-2)$^+$ with the C—Fe—C array. Two molecules of C$_4$H$_8$O$_2$ complete the approximate trigonal prism coordination about (Na-2)$^+$. [By permission from the American Chemical Society, H. B. Chin and R. Bau, *J. Am. Chem. Soc.* **98**, 2434 (1976).]

When Na$^+$ is encased in a cryptand ligand, [Fe(CO)$_4$]$^{2-}$ has small distortions from a tetrahedral geometry (*23*). In this case the least linear Fe—C—O group has an internal angle of 174°, and the largest C—Fe—C angle is 112°. The oxygens of the metal carbonyl are well separated from the sodium. In the case of the potassium cryptand, there appears to be a clear interaction (2.97 Å) between one of the carbonyl oxygens of [HCr$_2$(CO)$_{10}$]$^-$

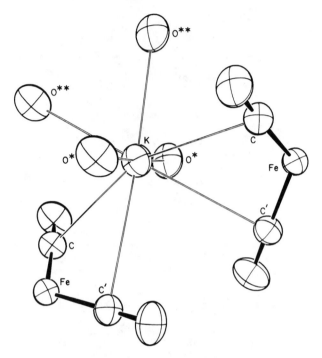

FIG. 8.   Coordination about K^+ in $K_2[Fe(CO)_4]$. Oxygen atoms involved in Σ —CO—K interactions are denoted with single or double asterisks. [By permission from the American Chemical Society, R. G. Teller *et al.*, *J. Am. Chem. Soc.* **99**, 1104 (1977).]

and K^+ (*24*). By comparison, the distances to the cryptand oxygen atoms range from 2.80 to 2.89 Å. Thus the interaction with the carbonyl anion is analogous to the next-nearest neighbor interaction which was postulated for solution species on the basis of infrared spectra.

Other cases in which an alkali metal ion is partially sterically encumbered are $[Na(18\text{-crown-}6)][W(CO)_5SH]$ (Fig. 9), $[Na(18\text{-crown-}6)][W_2(CO)_{10}\text{-}(\mu\text{-SH})]$, and $[Na(thf)(Co(salen)_2)][Co(CO)_4]$ (*25, 26*). The formulas are written here in a way which emphasizes the coordination of the alkali metal ion by large ligands. The structure of $[Na(18\text{-crown-}6)][W(CO)_5SH]$ shows an interaction between Na^+ and the oxygen of a carbonyl trans to the sulfur of a pseudo-octahedral tungsten. The Na—O distance is 2.41(2) Å and the Na—O—C angle is 140°. Coordination of Na to the oxygen has no apparent affect on the C—O distance.

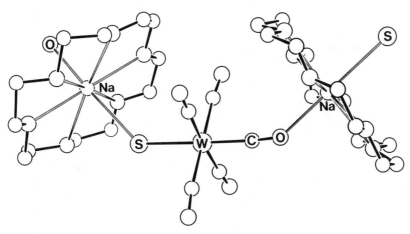

Fig. 9. Structure of [Na(18-crown-6)][W(CO)₅SH] showing the Σ—CO— inter-action with the Na(18-crown-6) cation. The structure is polymeric with the Na(18-crown-6) cation being coordinated axially by a Σ —CO— and a sulfur from a second [W(CO)₅SH]⁻ unit. [By permission from the Royal Society of Chemistry, M. K. Cooper *et al.*, *J. Chem. Soc., Dalton Trans.*, p. 2357 (1981).]

The reaction of NaCo(salen) with CO in THF–Et₂O results in the forma-tion of [Na(thf)Co(salen)₂][Co(CO)₄] [Eq. (2)] (*26*). The coordination

$$3\,NaCo(salen) + 4\,CO \xrightarrow[\text{Et}_2\text{O}]{\text{THF}} [Na(thf)[Co(salen)]_2][Co(CO)_4] + (salen)Na_2 \quad (2)$$

sphere of the sodium cation is pseudo-octahedral. It is chelated by two bi-dentate Co(salen) groups, and the remaining sites are occupied by the oxygen of a THF molecule and a carbonyl oxygen from a Co(CO)₄ unit. The carbonyl oxygen is 2.44 Å from the sodium and the Na—O—C angle is 142.1°. Neither the Co—C nor the C—O bond lengths are significantly perturbed by this interaction. The coordination about the Co(CO)₄ unit is thus similar to those discussed for tight ion pairs in solution.

3. *Main Group II Cation Acceptors*

The examples are limited to magnesium ion. A general synthesis for main group II cation acceptors is shown in Eq. (3) (*19, 27, 28*). Spectroscopic data

$$HgM_2 + Mg \xrightarrow[\text{Et}_2\text{O}]{\text{THF}} MgM_2 \cdot XTHF \xrightarrow{\text{pyridine}} MgM_2(py)_4 \quad (3)$$

$$M = C_5H_5Mo(CO)_3, C_5H_5Fe(CO)_2, Co(CO)_4, \text{ or } Mn(CO)_5$$

for these compounds, given in Table III, are consistent with coordination of oxygen to Mg since each compound displays low v_{CO} values not seen in the

TABLE III

Σ —CO— Stretching Frequencies for $M_2M'(B)_4$ Compounds

M	M'	B	$\nu_{\Sigma-C-O}(cm^{-1})(\Omega)$	Reference
$CpMo(CO)_3$	Mg	thf	1674	19
$CpMo(CO)_3$	Mg	py	1667 (4.69)	27[a]
$CpFe(CO)_2$	Mg	thf	1713	19
$CpFe(CO)_2$	Mg	py	1711	27
$Co(CO)_4$	Mg	py	1751	27
$Mn(CO)_5$	Mg	py	1721	27
$CpCr(CO)_3$	Mn	py	1652	66
$CpMo(CO)_3$	Mn	py	1650	66
$CpW(CO)_3$	Mn	py	1647	66

[a] See also text and Fig. 10.

parent. The structure of the Mo complex has been determined by single crystal X-ray diffraction and is shown in Fig. 10. The compound has an approximately octahedrally coordinated magnesium cation with the four pyridines occupying the equatorial sites and axial ligation provided by carbonyl oxygens from each molybdenum. Significant features include the shorter Mo—C bond of the coordinated carbonyl, 1.886(2), as well as the lengthened C—O bond, 1.189(3) Å, of this moiety. The noncomplexed carbonyls have average Mo—C distances of 1.947(3) Å and average C—O distances of 1.157(3) Å. The unique Mo—C—O bond angle is also close to linear, 177.2(1)°, while the Mg—O—C angle is significantly bent, 155°(2).

For the cobalt analog in which THF replaces py (29), $Mg(thf)_6 \cdot (Co-(CO)_4)_2$, an X-ray structure displays discrete $[Mg(thf)_6]^{2+}$ cations and pseudo-tetrahedral $[Co(CO)_4]^-$ anions with no close Mg carbonyl oxygen contacts. Two THF molecules are relatively easily lost, and judging from IR data, their place is occupied by two carbonyl oxygens, one from each $[Co(CO)_4]^-$ unit. The formation of these tetrakis base adducts of Mg^{2+} is dependent on both the transition metal and solvent basicity (19, 30). When the transition metal anion is only weakly basic, such as $[CpMo(CO)_3]^-$, the integrity of the tetrakis base is maintained in both THF as well as benzene solvents. Increasing the nucleophilicity of the anion by replacement of CO with a phosphine ligand results in partial loss of the base, THF or py, as evidenced by IR spectroscopy and molecular weights. For the highly basic $CpFe(CO)_2$ anion, IR spectra suggest that the species present in benzene solution contains an Mg—Fe rather than a Σ —CO— bond, and only two THF molecules are coordinated to Mg^{2+}. In THF solution this metal–metal

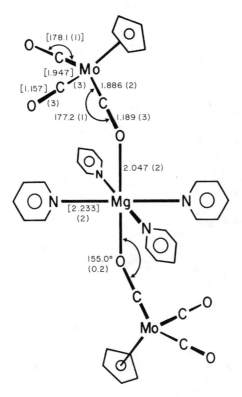

FIG. 10. Molecular structure of $(py)_4Mg[(OC)Mo(CO)_2Cp]_2$ showing the axial coordination of Mg^{2+} by two Σ —CO— groups. Note the nearly linear Mo—C—O array and the bent Mg—O—C array as well as the shortening of the Mo—C distance and lengthening of the C—O distance of the Σ —CO—. [By permission from the American Chemical Society, S. W. Ulmer *et al., J. Am. Chem. Soc.* **95**, 4469 (1973).]

bonded species is in equilibrium with a tetrakis THF species in which an Mg—O—C—Fe linkage is present [Eq. (4)].

$$(thf)_2Mg[CpFe(CO)_2]_2 + 2THF \rightleftharpoons (thf)_4Mg[(OC)Fe(CO)Cp]_2 \qquad (4)$$

4. *Main Group III and Rare Earth Acceptors*

Group III acceptors are present in the majority of compounds containing Σ —CO— carbonyls (*3, 8, 31–41*). Rare earth acceptors interact similarly with terminal oxygen atoms of carbonyls (*9, 53–56*). Compounds of these types are presented in Table IV. As with previous examples cited in this article, one of the most diagnostic features of these adducts is the low frequency carbonyl stretching modes characteristic of the Σ —CO— array (*42*).

TABLE IV

Σ —CO— Stretching Frequencies for Main Group III and Lanthanide
Adducts of Terminal Carbonyls

Compound (Ω)	$\nu_{\Sigma-CO-}$ $(cm^{-1})^a$	Reference
$(Ph_3PC_5H_4)Mo(CO)_3 \cdot AlMe_3$	1665	*3*
$[(n\text{-}Bu)_4N][CpW(CO)_3 \cdot AlPh_3]$	1660	*31, 38*
$(thf)_3Al[(OC)W(CO)_2Cp]_3$	1591; 1567; 1548	*32^b*
$trans\text{-}ReCl(PMe_2Ph)_4(CO) \cdot AlY_3,$		
$Y = Me; Ph; Cl$	1772; 1682; 1634	*33, 35*
$trans\text{-}ReCl(diphos)_2CO \cdot AlY_3,$		
$Y = Me; Ph; Cl$	1738; 1653; 1630	*33, 35*
$trans\text{-}ReCl(P)_4CO \cdot AlMe_3,$		
$P = PPh_3; PPh_2Et$	1728; 1738	*33, 35*
$cis\text{-}W(diphos)_2(CO)_2 \cdot 2AlMe_3$	1682	*33, 35*
$cis\text{-}Mo(diphos)_2(CO)_2 \cdot 2AlEt_3$	1660	*36*
$cis\text{-}Mo(PMe_2Ph)_4(CO)_2 \cdot 2AlEt_3$	1640	*36*
$WPR_3(CO)_5 \cdot AlBr_3; R = Ph; n\text{-}Bu$	$1665;^b 1630^c$	*37*
$[R_2Al(OC)W(CO)_2Cp]_2 (5.24);^d$		
$(5.09)^e; R = Me; Et$	1650; 1659	*39, 40^f*
$[Me_2Al(OC)W(CO)_2Cp] \cdot B;$		
$B = NMe_3; Et_2O$	1569; 1562	*40*
$Mo(phen)(PPh_3)_2(CO)_2 \cdot AlR_3;^g$		
$R = Et; i\text{-}Pr$	1731; 1633; 1727; 1633	*41*
$Mo(phen)_2(CO)_2 \cdot 2AlEt_3$	1684; 1584; 1565	*41*
$Mo(5,6\text{-}dmphen)(PPh_3)_2(CO)_2 \cdot 2AlEt_3^h$	1727; 1627	*41*
$BrMn(CO)_5 \cdot Cp_3Sm$	1936	*9*
$BrMn(P(OPh)_3)(CO)_4 \cdot Cp_3Sm$	1900	*9*
$(MeC_5H_4)Mn(CO)_3 \cdot n(MeC_5H_4)_3Sm$	1868	*53, 54*
$Me(C_5H_4)Mn(CO)_3 \cdot n(MeC_5H_4)_3Nd$	1865	*53, 54*
$(MeC_5H_4)Mn(CO)_3 \cdot Cp_3M; M = Er; Yb$	1868; 1868	*53, 54*
$Cp*_2(thf)Yb(OC)Co(CO)_3 (4.53)$	1761	*55^i*
$Cp*_2YbM(CO)_5 \cdot \frac{1}{4}PhMe; M = Mn; Re$	1775; 1750	*56^f*

^{a} A semicolon separates the $\nu_{\Sigma-CO-}$ for respective derivatives listed in the first column.

^{b} Before loss of PR_3.

^{c} After loss of PR_3.

^{d} Ω For "linear" Al—O—C.

^{e} Ω For "bent" Al—O—C.

^{f} X-Ray crystal structure available in reference.

^{g} Phen = 1,10-phenanthroline.

^{h} 5,6-Dmphen = 5,6-dimethyl-1,10-phenanthroline.

^{i} See also text and Fig. 12.

The somewhat smaller increase in CO stretching frequency of all other CO ligands also is characteristic (*42, 43*) (**3b**).

Despite the extensive information on the perturbation of CO stretching frequencies and ^{13}C-NMR chemical shifts by Σ —CO— formation, very little has been reported on the influence of Σ —CO— on electronic spectra. The sole study in this area demonstrated that the electronic spectrum of $Mo(phen)(PPh_3)_2(CO)_2$ changes dramatically in the presence of Lewis acids (*41*). The starting compound exhibits an intense blue-green color due to a charge transfer transition which can be described roughly as an Mo → phen π^* transition with an absorption maximum at 693 nm. The addition of excess $AlEt_3$ to a solution of the substituted carbonyl shifts this band across the visible spectrum to 499 nm.

A molecular orbital model, which fits the observed spectra, identifies the filled orbital from which the charge transfer occurs as a π system involving Mo d_{xz} and C p_x and O p_x of the two CO ligands. The transition occurs from this $Mo(CO)_2$ π orbital into a vacant π^* orbital on the 1,10-phenanthroline ligand. As illustrated in Fig. 11, attachment of an acceptor to the oxygen of the CO ligands increases the back π bonding in the $Mo(CO)_2$ system and lowers the energy of the π orbital [labeled Mo(d) in the figure], but leaves the phen π^* orbital unchanged. The net result is an increase in the energy of the charge transfer band. This explanation for the origins of the electronic spectral changes is closely related to the changes in electronic structure used to describe the influence of Σ —CO— formation on shifts in CO stretching frequencies and CO distances (**3b**).

The majority of simple Lewis acid adducts of terminal carbonyls is formed with anionic metal carbonyls. But the above example demonstrates that

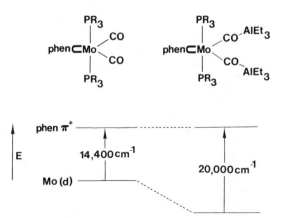

FIG. 11.   Energy level scheme and charge transfer energies for $Mo(phen)(PPh_3)_2(CO)_2$ and $Mo(phen)_2(PPh_3)_2(COAlEt_3)_2$.

carbonyl ligands may be quite basic if a neutral carbonyl is heavily sub-stituted by electron donor ligands.

Most of the remainder of this section will be organized around the methods of preparation of Σ —CO— complexes containing main group III and lanthanide acceptors. These preparative methods are simple adduct forma-tion, protolysis, and redox processes. Simple adduct formation is by far the most common mode of forming these compounds. One example has just been given in which aluminum alkyls were employed. Also to be noted (Table IV) are several complexes formed between transition metal carbonyls and tris(cyclopentadiene)lanthanide acceptors (9, 53, 54).

Simple adduct formation also occurs between a metal ion such as Al^{3+} and the carbonylate anions, usually with the formation of Σ —CO— bridges. For example, in THF solution, $(thf)_3Al[(OC)W(CO)_2Cp]_3$ forms readily. As with the analogous Mg complex discussed in the last section, Al(III) is pseudo-octahedrally coordinated but in this case there is a meridional arrangement of THF and carbonyl oxygens (32). The Al—O distances are longer to THF [1.94(2) Å] than to the carbonyl [1.827(9) Å]. This pattern has been observed in several other compounds. As commonly observed, the Σ —CO— carbonyls have shorter W—C distances than the strictly terminal carbonyls, 1.85 versus 1.95 Å, and the respective CO distances are longer: 1.25(2) versus 1.16(2) Å.

Aluminum alkyls and lanthanide organometallics are highly carbanionic and thus susceptible to protolysis by slightly acidic transition metal carbonyl hydrides. An example of this preparative route is provided by the interaction of acidic $CpW(CO)_3H$ with Al_2Me_6 to yield $[CpW(CO)_3 \cdot AlMe_2]_2$. This product occurs as a 12-membered ring containing W—CO—Al bridges (39, 40). The ring is slightly puckered due to the bending of one of the Al—OC—W linkages about the oxygen (149°) whereas the other related linkage is close to linear (172–176°).

As shown in Eqs. (5) and (6), reaction between $Cp^*_2Yb(OEt_2)$ and neutral metal carbonyl compounds occurs by a redox reaction to form a Yb(III)

$$Cp^*_2Yb(OEt_2) + Co_2(CO)_8 \xrightarrow[\text{2. THF}]{\text{1. toluene}} Cp^*_2(THF)Yb(OC)Co(CO)_3 \qquad (5)$$

$$Cp^*_2Yb(OEt_2) + Mn_2(CO)_{10} \xrightarrow{\text{toluene}} Cp^*_2YbMn(CO)_5 \cdot \tfrac{1}{2}C_6H_5CH_3 \qquad (6)$$

center, which has significant Lewis acidity, and a metal carbonylate which can display carbonyl basicity (55, 56). The product of such a reaction is illustrated in Fig. 12. The Yb(III)—O distance to the carbonyl oxygen is 2.258(2) Å, which is shorter than that to the THF oxygen, 2.412(5) Å. For the Σ —CO— group the typical lengthening of the CO bond is observed (0.05 Å) as well as shortening of the Co—C bond (0.07 Å).

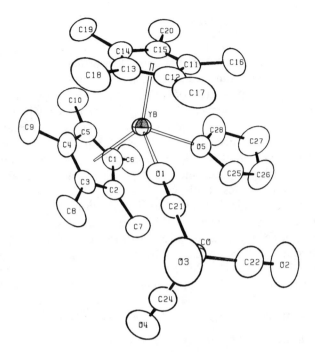

FIG. 12. Structure of $Cp^*_2(thf)Yb(OC)Co(CO)_3$ showing the Σ —CO— interaction of the Yb(III) with a CO from $[Co(CO)_4]^-$. [By permission from the Royal Society of Chemistry, T. D. Tilley and R. A. Anderson, *J. Chem. Soc., Chem. Commun.*, p. 985 (1981).]

Tungsten and molybdenum carbonyls in conjunction with molecular or solid state Lewis acids are effective olefin metathesis catalysts (44–47), but the nature of the catalytic species is in general poorly understood. An IR band at $1745 \, cm^{-1}$ for W(CO)L [L = CO, P(OPh)$_3$, PPh$_3$, and P(n-Bu)$_3$] complexes adsorbed on activated η-Al$_2$O$_3$ is taken as evidence for the formation of a Σ —CO— interaction with the surface Al^{3+} (48, 49). It is also thought that the metal carbonyl loses a ligand, L, to the surface. Similarly, in chlorobenzene solution, AlBr$_3$ initially forms a 1:1 adduct with CO stretching frequencies at approximately 1900 and $1665 \, cm^{-1}$. The lower peak indicates that a Σ —CO— may be present, and ^{31}P-NMR spectra confirm that the P—W bond is intact in the initial compound. Upon standing the P—W coupling is lost but the Σ —CO— persists as indicated by IR spectral evidence. Aluminum alkyl reagents with Mo and W carbonyls are metathesis catalysts, providing some O$_2$ is introduced to the mixture (50–52). Low frequency CO stretching frequencies indicate that Σ —CO— may be present in these systems as well (50).

5. *Transition Metal Acceptors*

An interesting variety of methods has been used for the preparation of transition metal Σ —CO— compounds. The principal methods to be emphasized here are (1) simple adduct formation with molecular Lewis acids or ionic species, (2) protolysis of carbanionic reagents, (3) redox processes, (4) solid state stereochemical control, and (5) molecular stereochemical control. The early transition metals in high oxidation states are relatively strong Lewis acids, especially toward hard donors such as oxygen. Thus, it is not surprising that the bulk of the known Σ —CO— compounds are formed with early transition metal acceptors such as Ti, Zr, and Hf in $+3$ and $+4$ oxidation states (Table V) (57–62).

An example of simple adduct formation with early transition metal acceptors is the reaction between $TiCl_4$ and *trans*-$ReCl(PMe_2Ph)_4(CO)$ to produce a species formulated as $[ReCl(PMe_2Ph)_4(CO)]_2TiCl_4$, which has a low frequency CO stretch at $1610\ cm^{-1}$ indicating the presence of Σ —CO— [Eq. (7)] (57). Similarly, a simple metathetical displacement reaction between Cp_2HfCl_2 and $Na_2Fe(CO)_4$ yields $[Cp_2HfFe(CO)_4]_2$ (62).

$$2\ \textit{trans-}ReCl(PMe_2Ph)_4(CO) + TiCl_4 \xrightarrow{CH_2Cl_2} [\textit{trans-}ReCl(PMe_2Ph)_4(CO)]_2TiCl_4 \quad (7)$$

A characteristic low-frequency CO stretching frequency is observed at $1683\ cm^{-1}$, and a cyclic Σ —CO— bridged structure is proposed. Another example of simple adduct formation with metal carbonylates is provided by a series of Mn(II) complexes (66), analogous to the previously discussed

TABLE V

Σ —CO— Stretching Frequencies for Transition Metal Adducts of Terminal Carbonyls

Compound (Ω)	$v_{\Sigma-CO-}\ (cm^{-1})$	Reference
$[\textit{trans-}ReCl(PMe_2Ph)_4CO]_2TiCl_4$	1610	57
$Cp^*_2(CH_3)Ti(CO)Mo(CO)_2Cp$ (4.56)	1623	58[a]
$Cp_2(thf)Ti(OC)Mo(CO)_2Cp$	1650	59[a]
$Cp_2(CH_3)Zr(OC)Mo(CO)_2Cp$	1545	60, 61
$Cp_2(\eta^2\text{-Ac})Zr(OC)Mo(CO)_2Cp$ (4.23)	1590[b]	60[a]
$[Cp_2HfFe(CO)_4]_2$	1683	62
$(thf)_4V[(OC)V(CO)_5]_2$ (4.70)	1684	63[c]
$Co_3[Fe(CN)_5CO]_2 \cdot 5.7\,H_2O$	1950	67
$[ClZnMo(CO)_3Cp]_n$	1750	70[d]

[a] Broad absorption probably with component due to η^2-Ac.

[b] X-Ray crystal structure available in reference.

[c] Text structure **9**.

[d] See text, Scheme 2, for proposed polymeric structure.

[CpM(CO)$_3$]$_2$Mg(py)$_4$ (Table III). The CO stretching frequencies of these Mn complexes (Table III) are close to those of the Mg compounds.

Protolysis also is a useful way of preparing Σ—CO— adducts with the early transition metal centers. Typically, the early transition metal alkyls have carbanionic character and readily react with the acidic metal carbonyl hydrides. In the reaction of Cp$_2$ZrMe$_2$ with CpMo(CO)$_3$H, methane is evolved and Cp$_2$ZrMe$_2$(OC)Mo(CO)$_2$Cp forms (60, 61). A low-frequency Σ—CO— stretching frequency is observed at 1545 cm^{-1}, and the formulation is confirmed by an X-ray structure determination. When this complex is placed under a CO atmosphere, a migratory insertion reaction occurs (Scheme 1) to produce a species having both η^2-acetyl and Σ—CO— ligands.

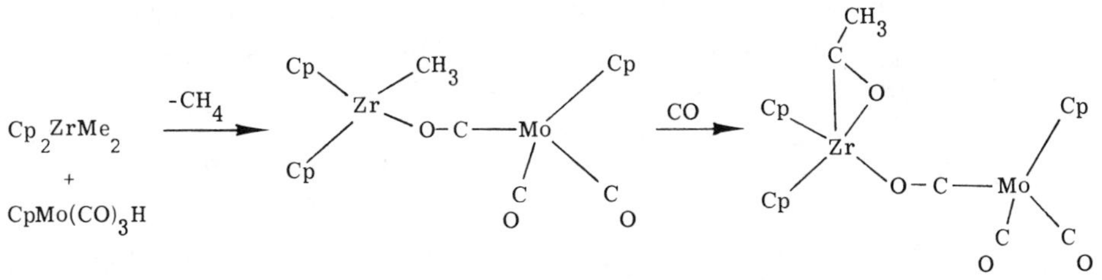

SCHEME 1

Zinc alkyls also have significant carbanionic character, which forms the basis of the reaction of CpMo(CO)$_3$H with C$_2$H$_5$ZnCl in ether solution to yield ethane and [Cp(CO)$_3$MoZnCl·Et$_2$O]$_2$ (70). Removal of the ether causes the dimer to polymerize. The IR spectrum of this polymer contains a CO stretching frequency at ~1750 cm^{-1}, which is attributed to the Zn—O—C—Mo linkage, as depicted in Scheme 2.

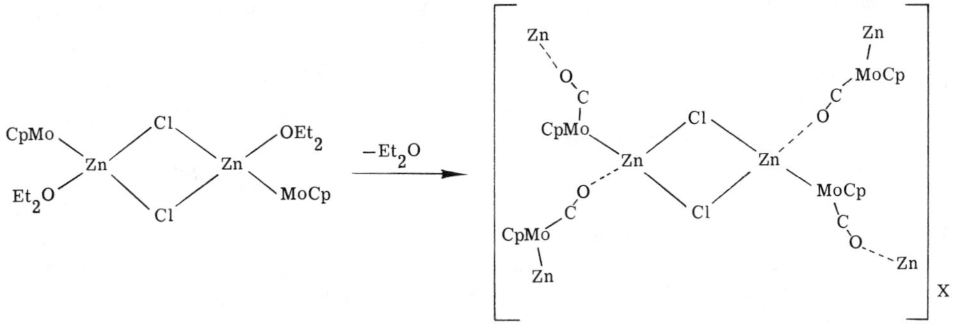

SCHEME 2

A novel reaction based on proton transfer is given in Eq. (8) *(58)*. In this instance, protonation occurs on a $C_5Me_4=CH_2$ ring *(71, 72)* to form C_5Me_5 and no alkane is evolved. Formation of a Σ —CO— in this reaction was established by a crystal structure determination, and the characteristic low frequency IR signature of the CO bridge was observed at 1623 cm^{-1}.

$$CpMo(CO)_3H + Cp^*[C_5(CH_3)_4=CH_2]Ti(CH_3) \xrightarrow{\text{toluene}} Cp^*_2(CH_3)Ti(OC)Mo(CO)_2Cp \quad (8)$$

Redox processes are fairly common in the formation of Σ —CO— complexes of transition metals, and an example is given in Eq. (9). In this reaction, titanium is oxidized from the $+2$ to the $+3$ state, thus becoming a better Lewis acid, and the molybdenum dimer is reductively cleaved, thus developing Σ —CO— donor character *(59)*. A characteristic low-frequency Σ —CO— band is observed in the IR spectrum, and a crystal structure is available. A proposed mechanism for the redox process, based on CO mediated electron transfer, is discussed in Section IV,C.

$$Cp_2Ti(CO)_2 + Cp_2Mo_2(CO)_4 \xrightarrow{\text{THF}} Cp_2(thf)Ti(OC)Mo(CO)_2Cp \quad (9)$$

Another quite interesting case in which a redox process leads to a Σ —CO— adduct is provided by the redox disproportionation of $V(CO)_6$, which is induced by Lewis bases. If the base is relatively weak, such as THF or Et_2O, a product formulated as $V(B)_4[(OC)V(CO)_5]_2$ can be isolated *(63–65)*. The X-ray crystal structure of the product when B = THF contains a central V^{2+} coordinated approximately octahedrally by four equitorial THF molecules and two axial Σ —CO— carbonyl oxygens **(9)** *(63)*. Thus the disproportionation reaction produces a Lewis acid center, V^{2+}, and basic carbonyls in the anion $[V(CO)_6]^-$, which interact via a Σ —CO— bridge. The mechanistic details of this redox disproportionation are discussed in Section IV,C.

A clever example of the way in which solid state stereochemistry can be employed to force CO into a bridging environment which it might not otherwise assume is provided by the reaction of $Na_3Fe(CN)_5(CO)$ with $CoCl_2$ in aqueous solution *(67)*. The solid which precipitates from this reaction mixture, $Co_3[Fe(CN)_5(CO)]_2 \cdot 1.5H_2O$, appears to have a Prussian blue structure, in which case the CO ligand is forced into a Σ —CO— bridging position. In keeping with this interpretation, the CO stretching frequency, which is above 2015 cm^{-1} in the parent compound, shifts down to 1950 cm^{-1} in the product.

Two cases in which molecular stereochemical control may promote Σ —CO— formation are provided by the two dicopper species shown in **10** and **11** *(68, 69a)*. It is thought that the ligands constrain the Cu(I) centers to lie 3.5–6 Å apart in these complexes. In both instances one CO is absorbed

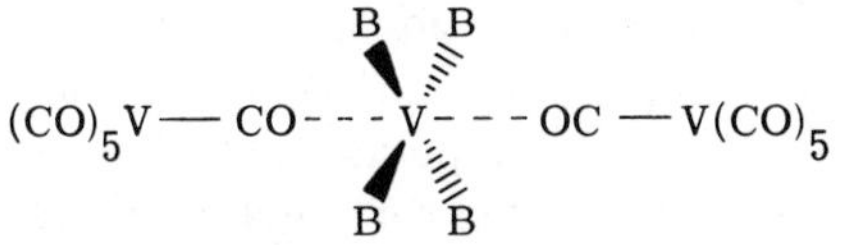

$$(CO)_5V - CO - - - \overset{\displaystyle B \quad B}{\underset{\displaystyle B \quad B}{V}} - - OC - V(CO)_5$$

(**9**)

$$\left(\begin{array}{c}O\\S\\S\\N\end{array}Cu - C \equiv O - Cu\begin{array}{c}O\\S\\S\\N\end{array}\right)$$
$$- Y - X - Y -$$

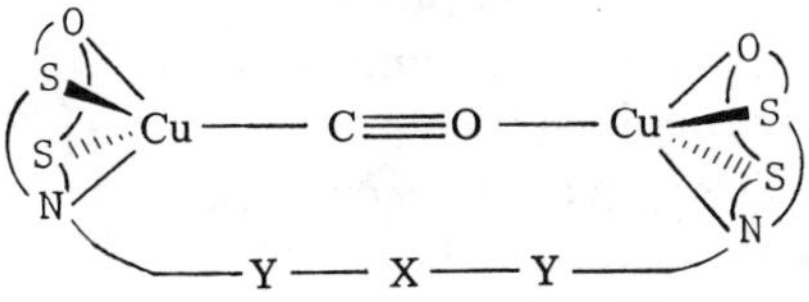

(**10**) $X = C_6H_4$, $Y = CH_2$

(**11**) $X = Rh(CO)Cl$, $Y = PPh_2$
or $Rh(CO)Cl$, $Y = PMe_2$
or $Ir(CO)_2Cl$, $Y = PPh_2$

per bimetallic complex, and it is claimed on the basis of this stoichiometry and the CO stretching frequency ($\sim 2070 \text{ cm}^{-1}$) that the CO forms a $\Sigma - CO -$ bridge between the copper centers. This proposition deserves further investigation because the quoted frequency is in the same region as that for mononuclear Cu(I) CO complexes (*69b*), and, in addition, 2070 cm^{-1} is an unusually high frequency for a $\Sigma - CO -$.

B. *Edge-Bridging CO Donors*

The greater basicity of bridging CO than terminal CO leads to a strikingly richer $\Sigma - CO -$ chemistry for polynuclear carbonyls than for their mononuclear counterparts. For example, simple neutral carbonyls such as $Fe(CO)_5$ and $Cr(CO)_6$ do not form stable adducts with boron or aluminum halides, whereas neutral polynuclear carbonyls containing bridging CO ligands readily form such adducts.

1. *Proton and Hydrogen Bonding Acceptors*

Hydrogen has been identified as an acceptor toward a bridging carbonyl oxygen in both fully protonated $\Sigma - CO -$ (*73, 74*) and hydrogen-bonded

Σ —CO— (11, 75, 76). The two known compounds in which an edge-bridging carbonyl is fully protonated (73, 74) are thermally unstable and have been detected only by low-temperature NMR spectroscopy.

The first example of a protonated carbonyl ligand was realized from the reaction of $[HFe_3(CO)_{11}]^-$ with HSO_3F at $-90°C$ in CD_2Cl_2. The Σ —CO—H was identified by resonance at $\delta = 15.0$ ppm in the ^1H-NMR spectrum, a region typical for oxygen-bound protons of carboxylic acids. ^{13}C-NMR results corroborate the protonation of carbonyl oxygen by the appearance of a signal shifted significantly downfield ($\delta = 358.78$ ppm) from the unprotonated cluster anion $[PPN][HFe_3(CO)_{11}]$ ($\delta = 285.7$ ppm) (75). A downfield shift of this magnitude is consistent with studies of molecular Lewis acids coordinated to carbonyl oxygens. Decomposition of the adduct to $Fe_3(CO)_{12}$ occurs at temperatures above $-30°C$. In this connection, one of the classic methods for the synthesis of $Fe_3(CO)_{12}$ is the reaction of $[HFe_3(CO)_{11}]^-$ with acid (77). A reaction scheme for these transformations is proposed in Eq. (10).

$$[HFe_3(CO)_{11}]^- + H^+ \xrightarrow{-90°C} HFe_3(CO)_{10}COH \xrightarrow{-30°C} Fe_3(CO)_{12} + ? \quad (10)$$

The ruthenium analog of $[HFe_3(CO)_{11}]^-$, $[HRu_3(CO)_{11}]^-$, also reacts with acid (74). At $-60°C$ the proton NMR shows two signals: $\delta = 16.11$ and -14.60 ppm. The downfield resonance is again consistent with formation of an oxygen–hydrogen interaction. An interesting transformation occurs upon warming the solution, in which the two low-temperature signals are replaced by two new resonances in the typical metal hydride region. Thus the proton initially bound to a carbonyl oxygen has migrated to the metal framework. Further increase in temperature results in decomposition of the dihydride. A general outline of this reaction sequence is given in Eq. (11).

$$[HRu_3(CO)_{11}]^- + H^+ \longrightarrow HRu_3(CO)_{10}COH \longrightarrow$$
$$H_2Ru_3(CO)_{11} \longrightarrow Ru_3(CO)_{12} + H_2 + \cdots \quad (11)$$

These examples demonstrate that the oxygen end of a carbonyl can serve as the initial site of attack for reactions which ultimately take place on the metal center.

Hydrogen bonding has been detected in solution both by IR and NMR spectroscopy for $[Et_3NH][HFe_3(CO)_{11}]$ (75). In solvents of low polarity, such as C_6H_6 and CH_2Cl_2, the bridging carbonyl displays stretching frequencies of 1639 and 1650 cm^{-1}, respectively (**12a**). This is a shift to lower energy of approximately 80 to 100 cm^{-1} from the comparable species in more polar solvents such as CH_3CN or CH_3NO_2 (**12b**). This relatively

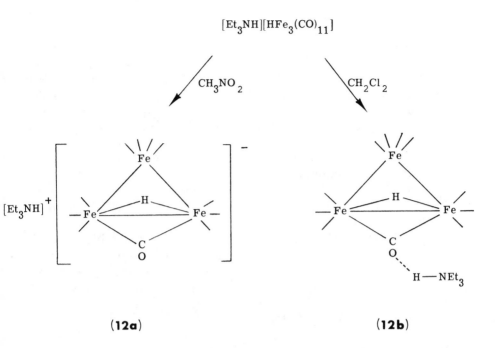

(12a)   **(12b)**

strong donor–acceptor interaction is reflected also in the position of the resonance due to the bridging carbonyl in the ^{13}C-NMR spectrum (301.3 ppm in $CHCl_2F$–CD_2Cl_2 at $-107°C$ versus 285.7 ppm for the anion in the presence of a non-hydrogen-bonding cation, $[PPN]^+$).

A few examples of hydrogen bonding in the solid state are known. The compound $[(PPh_3)_2Rh(CO)_2]_2 \cdot 2CH_2Cl_2$ (78), which contains two bridging carbonyl groups, shows the carbon atom of the solvent molecule within the van der Waals distance (79) of the carbonyl oxygen. This places the hydrogens within hydrogen bonding distance. The solid state structures of both $[Et_3NH][HFe_3(CO)_{11}]$ (80) and $[(i\text{-}Pr)_2NH_2][HFe_3(CO)_{11}]$ (11) display a distinct hydrogen bonding interaction between the bridging carbonyl oxygen and the hydrogen on the nitrogen cation. For the $[(i\text{-}Pr)_2NH_2]^+$ salt, the oxygen–hydrogen distance is approximately 1.89 Å.

Finally, Σ —CO— hydrogen bonding has been observed with hydroxylic solvents (76). It also is claimed to occur with the carbon-bound hydrogens in CH_3NO_2 and CH_3CN (11). Results of the former IR study indicate that two forms of each metal carbonylate molecule can exist in solution: the normal bridged form and the hydrogen-bonded bridged form. The position of this equilibrium is controlled by the pK_a of the solvent.

2. Alkali Metal Cation Acceptors

As with the terminal carbonyls, alkali metal cations have been identified as acceptors toward bridging carbonyls, both in solution and in the solid state. Analysis of these weak interactions by IR spectroscopy is hampered by the general spectral complexity of polynuclear carbonyl species. Despite this difficulty, some recent reports have appeared for ion pairing in tri- (*11, 81*) and tetranuclear (*82–84*) anionic metal clusters. Contact ion pairing in these systems is usually evidenced by a splitting of the bridging carbonyl band from the unperturbed ion. Little or no change is observed for the terminal carbonyl absorptions, presumably because these secondary perturbations are diminished by charge distribution in the polynuclear species. For example, $[PPN][HRu_3(CO)_{11}]$, the structure of which is probably similar to $[HFe_3(CO)_{11}]^-$ (*85*), has a stretching frequency for the bridging CO at 1733 cm^{-1} in THF, while the Na salt displays two absorptions at 1733 cm^{-1} and 1678 cm^{-1} in the same solvent. This shift of approximately 50 cm^{-1} in CO frequency is of the same general magnitude as that seen for terminal carbonyl contact ion pairs (see Section II,A,2). The ^{13}C-NMR spectrum also reflects the presence of contact ion pairs. Comparison of the K^+ (THF-d_8) and PPN$^+$ (CD$_2$Cl$_2$) salts of $[H_2Ru_4(CO)_{12}]^{2-}$ (*83*) shows eight and seven resonances, respectively, at $-80°C$. The eight-band pattern is consistent with the presumed structure (**13**).

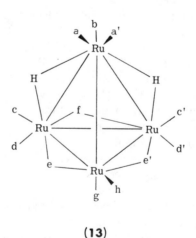

(**13**)

X-Ray structures provide two examples of alkali metal cations interacting with bridging carbonyl oxygens (*86, 87*). The first is an unusual species formulated as $[Cs_9(18\text{-crown-}6)_{14}]^{9+}[Rh_{22}(CO)_{35}H_x]^{5-}[Rh_{22}(CO)_{35^-}H_{x+1}]^{4-}$. This compound is composed of two virtually identical Rh$_{22}$

clusters differing only by one hydrogen and three distinctly different types of Cs(18-crown-6) cations. One of the cations displays close contact (3.51 Å) of the cesium with two symmetry-related bridging carbonyls on one of the metal cluster anions.

The Nujol mull IR spectrum of $Li[Co_3(CO)_{10}] \cdot i\text{-}Pr_2O$, with a CO stretching frequency at 1800 cm^{-1}, is suggestive of a bridging carbonyl (87). The X-ray crystal structure confirms the presence of three bridging carbonyls about the triangle of cobalt atoms. Two of the three bridging carbonyls participate in an ion-pairing interaction with two different lithium cations. The oxygen–lithium distances are 1.989(10) Å and 2.049(10) Å.

3. Carbocation Electrophiles

Unlike mononuclear carbonylates, reaction of carbocation reagents with anionic polynuclear carbonylates often results in coordination at the bridging carbonyl oxygen rather than at the metal center. This type of interaction was first observed for the reaction of $[HFe_3(CO)_{11}]^-$ with CH_3SO_3F (88) [Eq. (12)], and a higher yield route is indicated in Eqs. (13) and (14). The X-ray crystal structure of this molecule is shown in Fig. 13.

$$[HFe_3(CO)_{11}]^- + CH_3SO_3F \xrightarrow{CH_2Cl_2} HFe_3(CO)_{10}COCH_3 \qquad (12)$$

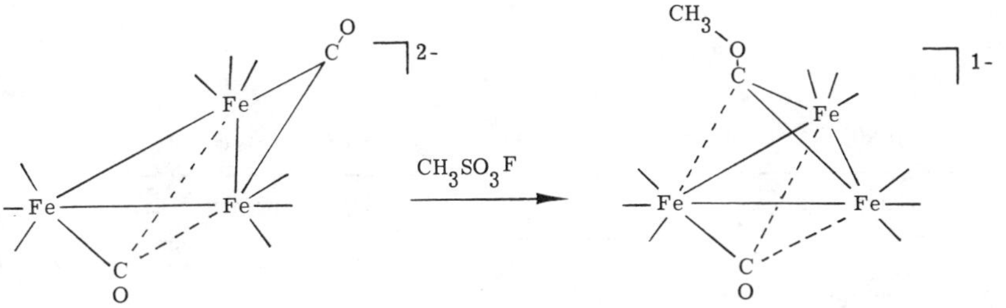

$$(13)$$

$$[Fe_3(CO)_{10}(COCH_3)]^- + HSO_3F \longrightarrow HFe_3(CO)_{10}(COCH_3) \qquad (14)$$

Coordination at the carbonyl oxygen causes a lengthening of the bridging carbonyl C—O bond by 0.1 Å when compared to the anion, and the C—O—Me angle of 118° shows approximate sp^2 hybridization of the oxygen. Alkylation has also been successfully performed on both the ruthenium and osmium analogs of the triiron cluster (89–91). A number of other alkylated species have been isolated and their spectroscopic data are presented in Table VI (5, 88–92).

Reaction of the anionic metal carbonylates with alkyl carbocations causes an increase in stretching frequency of approximately 50 to 100 cm^{-1} in the

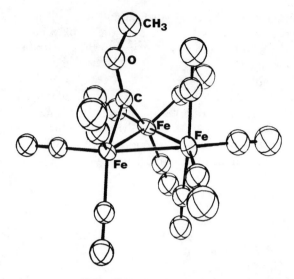

FIG. 13.   Molecular structure of $HFe_3(CO)_{10}(\mu\text{-}COCH_3)$ (hydrogens not shown). Note that the H_3C—O—C angle is significantly bent. [By permission from the American Chemical Society, D. F. Shriver *et al.*, *J. Am. Chem. Soc.* **97**, 1594 (1975).]

TABLE VI

IR AND ^{13}C-NMR SPECTRAL DATA FOR ALKYLATED Σ—CO— COMPOUNDS

Compound	$\nu_{\Sigma-CO-}$ (cm^{-1})	$\delta_{\Sigma-CO-}$ (ppm)	Reference
[PPN][Fe$_3$(CO)$_{10}$(COMe)]	1360	336	5
HFe$_3$(CO)$_{10}$(COMe)	1452	356	5, 88[a]
[PPN][Fe$_3$(CO)$_{10}$(COC$_2$H$_5$)]	—	—	5
HFe$_3$(CO)$_{10}$(COC$_2$H$_5$)	1472, 1398, 1378	—	5
[PPN][Fe$_3$(CO)$_{10}$(COC(=O)CH$_3$)]	1365	292	5
HRu$_3$(CO)$_{10}$(COMe)	1450	366.5, (328)[b]	89[c], 91
HOs$_3$(CO)$_{10}$(COMe)	—	346	90, 91
HOs$_3$(CO)$_{10}$(COC$_2$H$_5$)	—	—	90
HOs$_3$(CO)$_9$(CN-t-Bu)(COMe)	—	—	90
PPh$_3$AuFe$_3$(CO)$_{10}$(COMe)	—	361	92
PPh$_3$AuRu$_3$(CO)$_{10}$(COMe)	—	381.8	92[c]

[a] Text Fig. 13.

[b] Value in parentheses from ref. *91*.

[c] X-Ray crystal structure available in reference.

IR bands of the unreacted carbonyl ligands. The dominant shift is observed in the O-alkylated carbonyl which displays a decrease in stretching frequency ranging from about 100 to 350 cm^{-1}. The assignment of a band for the CO stretch in the Σ —CO—R group is often difficult owing to the weakness of the absorption as well as the low frequency, which overlaps absorptions due to many other vibrational modes. In the absence of interfering or overlapping bands, the 1300 to 1500 cm^{-1} spectral region is reasonably diagnostic for the Σ —CO—R functionality.

As with other Σ —CO— compounds, ^{13}C-NMR spectroscopy can be used as a diagnostic tool for the identification of an O-alkylated carbonyl. As previously mentioned, edge-bridging carbonyls, if they are observed, have resonances which appear approximately 30–60 ppm downfield of the corresponding terminal carbonyls. Upon coordination to the oxygen, this resonance shifts an additional 60–100 ppm downfield. This region of the spectrum is generally free of any other resonances due to organic molecules and can therefore be used to identify Σ —CO—R species. The resonances for the protonated, methylated, and BF$_3$ Σ —CO— products of [HFe$_3$(CO)$_{11}$]$^-$ occur in the narrow range of 355 to 359 ppm, which shows the relative insensitivity of ^{13}C-NMR spectroscopy to the type of acceptor.

4. Main Group III and Rare Earth Acceptors

Main group III Lewis acids form much stronger interactions with bridging carbonyls (93–97) than do the rare earths (53, 54, 98). Similarly, the group III Lewis acids bring about larger Σ —CO— frequency shifts (Table VII).

TABLE VII

Σ —CO— Stretching Frequencies for Main Group III and Lanthanide Adducts of Edge-Bridging Carbonyls

Compound	$v_{\Sigma-CO-}$ (cm^{-1})^a	Reference
Cp$_2$Fe$_2$(CO)$_4$ · 2 AlR$_3$, R = Et, i-Bu	1682, 1680	2^b, 93
Cp$_2$Ni$_2$(CO)$_2$ · AlR$_3$, R = Et, i-Bu	1761, 1761	93
Cp$_2$Ru$_2$(CO)$_4$ · [Al(i-Bu)$_3$]$_n$, n = 1, 2	1679, 1679	93
Cp$_2$Fe$_2$(CO)$_4$ · BX$_3$, X = Cl, Br	1463, 1438	95
Co$_2$(CO)$_8$ · AlX$_3$, X = Cl, Br	1618, 1600	95, 96
Fe$_2$(CO)$_9$ · AlBr$_3$	1557	95
M$_3$(CO)$_{12}$ · AlBr$_3$, M = Fe, Ru	1548, 1535	95
Cp$_2$Fe$_2$(CO)$_4$ · 2(RC$_5$H$_4$)$_3$Sm, R = H, Me	1700, 1700	53, 54
Co$_2$(CO)$_8$ · 2(MeC$_5$H$_4$)$_3$Sm	1781	53, 54

^{a} A comma separates the $v_{\Sigma-CO-}$ for respective derivatives listed in column 1.

^{b} See also text Fig. 14 (R = Et).

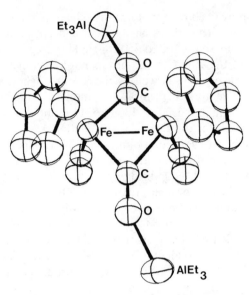

FIG. 14. X-Ray structure of $Cp_2Fe_2(CO)_4 \cdot 2AlEt_3$, illustrating the bent Al—O—C coordination of the bridging carbonyls and the cis configuration of Cp ligands. [By permission from the American Chemical Society, N. J. Nelson *et al.*, *J. Am. Chem. Soc.* **91**, 5173 (1969).]

Owing to the weak interaction between Σ —CO— donors and rare earth compounds the stoichiometry of the interaction is often not well defined.

As previously mentioned, the first species unambiguously identified as possessing both a C- and O-bonded carbon monoxide was Cp_2Fe_2-$(CO)_4 \cdot 2AlR_3$ (*2*) (see Fig. 14). The evolution of the IR spectrum from the parent $Cp_2Fe_2(CO)_4$ through the mono AlR_3 complex to Cp_2Fe_2-$(CO)_4 \cdot 2AlR_3$ is given in Fig. 2. The synthesis of $Co_2(CO)_8 \cdot AlBr_3$ (*99*) had been performed approximately a decade prior to the report of Cp_2Fe_2-$(CO)_4 \cdot 2AlR_3$. At that time it was proposed that a three-centered Co_2-to-Al donor–acceptor bond was present (**14a**) by analogy with similar bonding modes observed for protons with polynuclear carbonyl compounds. More recent IR studies of this adduct (*95, 96*) suggest that this complex contains a Σ —CO— carbonyl ligand (**14b**).

The $Cp_2Fe_2(CO)_4$ system normally exists as a mixture of both cis and trans bridged isomers in solution (*100, 101*). However, coordination with $AlEt_3$ results in exclusive formation of the cis isomer. A different type of Lewis acid-induced structural change involving the shift of CO from terminal to bridging positions has been observed for $Cp_2Ru_2(CO)_4$ (*93, 94*) and $Ru_3(CO)_{12}$ (*95*). The former exists in solution as a mixture of bridged and

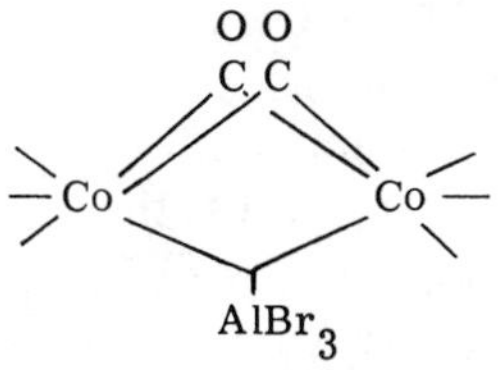

(14a) 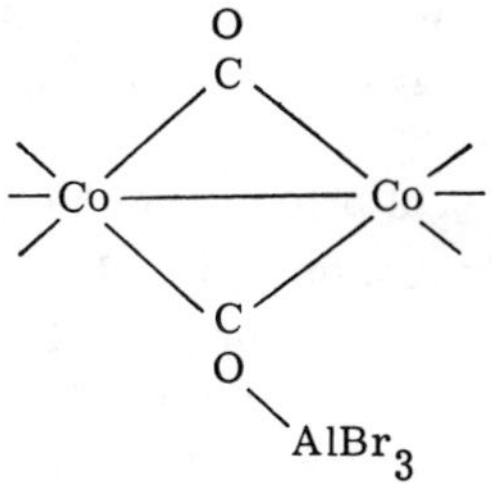   (14b)

nonbridged forms (102, 103). Addition of two equivalents of alkylaluminum results in formation of both cis- and trans-bridged forms only [Eq. (15)]. The

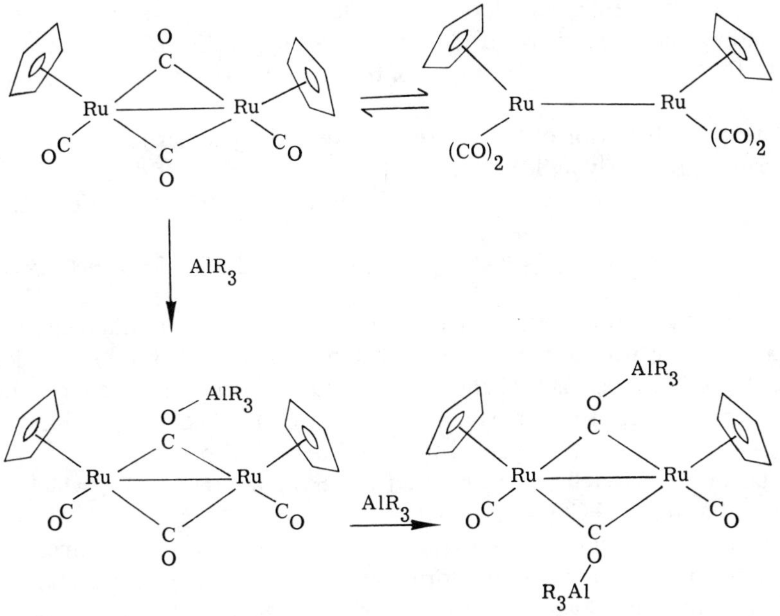

(15)

triruthenium cluster possesses only terminal carbonyls in the solid state as well as in solution (104, 105). Addition of one equivalent of $AlBr_3$ to a toluene solution of $Ru_3(CO)_{12}$ causes the appearance of a low-frequency ν_{CO} stretch at 1535 cm^{-1}. This is consistent with formation of a CO bridged adduct. The basis for the shift of CO groups from terminal to bridging positions upon coordination to the Lewis acid is the greater basicity of bridging CO ligands

(*107*). Attempts to induce CO bridging in $Mn_2(CO)_{10}$ by interaction with Lewis acids have been unsuccessful (*93*).

In molecules that have more than one C-bridging carbonyl, coordination of an acceptor to one of the bridging carbonyls can cause the uncoordinated bridging carbonyls to adopt terminal positions. The reaction of $Fe_2(CO)_9$, which contains three bridging carbonyls (*107*), with an excess of $AlBr_3$ results in the formation of a yellow powder for which analysis indicates the formula $Fe_2(CO)_9 \cdot AlBr_3$ (*95*). The IR spectrum of this adduct shows the presence of only one bridging carbonyl, in a region typical of $C—O—AlBr_3$ ($1557 \, cm^{-1}$). Infrared data indicate that the remainder of the carbonyls assume a terminal bonding mode, therefore the proposed structure of the adduct is similar to that envisioned for $Os_2(CO)_9$ (*107, 108*). The explanation offered for this transformation is that coordination of the $AlBr_3$ to one of the bridging CO ligands results in a significant withdrawal of electron density from the metal centers. With decreased electron density available for back π-bonding the noncomplexed C-bridging carbonyls shift to terminal positions. Further examples of this type of behavior will be presented in Section II,C,4.

The redox behavior of $Cp*_2Yb(OEt_2)$ toward metal carbonyls has been described previously in Section II,A,4 (*55, 56*). This reagent displays similar behavior toward $Fe_3(CO)_{12}$ [Eq. (16)]. Structurally, $(Cp*_2Yb)_2Fe_3(CO)_{11}$

$$2Cp*_2Yb(OEt_2) + Fe_3(CO)_{12} \longrightarrow [Cp*_2Yb]_2[(OC)_4Fe_3(CO)_7] \qquad (16)$$

(*109*) (see Fig. 15) is unusual. The compound has an approximately linear array of three iron atoms (Fe—Fe distances average 2.531 Å) with four bridging carbonyls. Each of the bridging carbonyls in turn is coordinated through the oxygen to an ytterbium. The average Yb—O distance, 2.243(5) Å, is similar to that in the cobalt compound described previously. Two low-frequency CO stretches are observed at 1667 and $1604 \, cm^{-1}$, which are assigned to the Σ —CO— carbonyls.

The number of cluster valence electrons (CVE) expected for a linear M_3 array is 50 (*110*), however the formulation from the X-ray and magnetic susceptibility data suggests a 48 CVE count for the $[Fe_3(CO)_{11}]^{2-}$ moiety. It is possible that the $Fe_3(CO)_{11}$ unit is electron deficient, but another explanation is that it is a 50-electron system with two hydride ligands. No evidence for an M—H bond is present in the ^1H-NMR spectrum. Similarly, the disposition of the carbonyl ligands on the terminal iron atoms does not suggest the presence of hydrogen ligands.

The interaction of Lewis acid sites on metal oxide surfaces with bridging carbonyls also has been observed (*111, 112*). These metal carbonyl–surface interactions are of interest as intermediates in the formation of supported

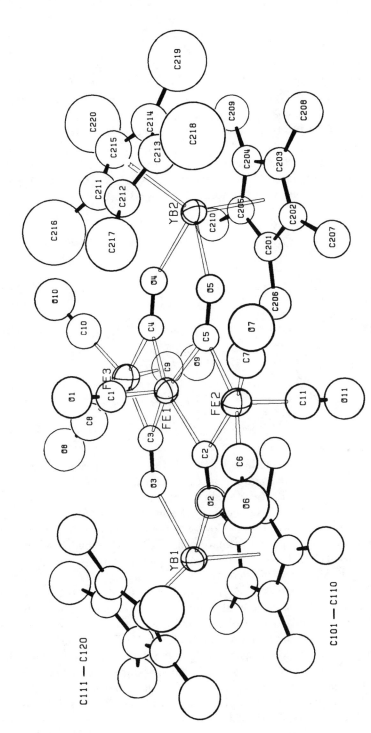

FIG. 15. Molecular structure of $(Cp^*_2Yb)_2[(OC)_4Fe_3(CO)_7]$ showing a nearly planar array of three linear Fe atoms and four edge-bridging carbonyls forming Σ —CO— interactions with two Yb(III) centers. [By permission from the American Chemical Society, T. D. Tilley and R. A. Andersen, *J. Am. Chem. Soc.* **104**, 1772 (1982).]

metal catalysts from organometallic precursors and as potential models for "support effects" which metal oxide supports exert on metal particles (113). When $Cp_2Ni_2(CO)_2$ is deposited from solution onto a dehydroxylated γ-Al_2O_3 surface, IR spectra suggest an initial Σ—CO— interaction with surface Al^{3+}, followed by break-up of the molecule (111).

Hydroxylated γ-Al_2O_3 and hydrated H—Y zeolite surfaces have been found to be reactive toward some iron carbonyls ($112, 114$). Infrared and NMR spectroscopy indicate that reaction of $Fe_3(CO)_{12}$ [or $Fe(CO)_5$] with hydroxylated alumina results in formation of $[HFe_3(CO)_{11}]^-$ (Scheme 3).

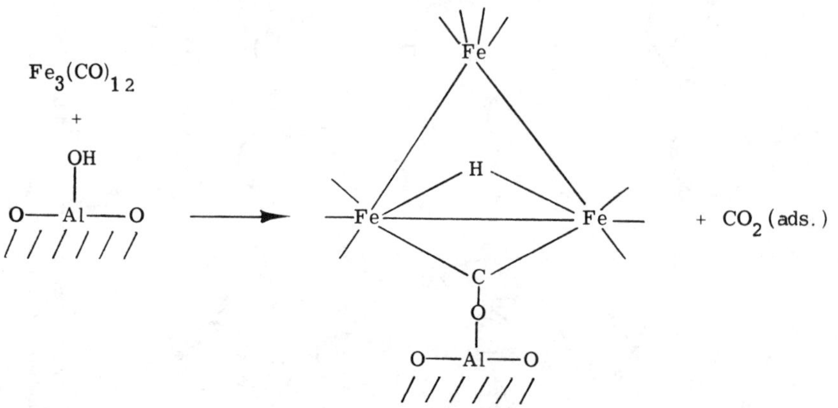

SCHEME 3

The initial step in this surface reaction, OH^--induced redox disproportionation, is well known in homogeneous chemistry ($115, 116$).

In contrast with hydrated alumina, silica gel does not have basic OH groups and instead the SiOH groups are mild proton acids. This is demonstrated by the adsorption of $Co_2(CO)_8$ on the SiO_2 surface (pretreated at 675 K) with the apparent formation of a Σ—CO— hydrogen bonded species, Si—$OH(OC)Co_2(CO)_7$ (117). Analogous results are obtained when $Fe_3(CO)_{12}$ is adsorbed on dehydrated H—Y zeolite (118). A decrease in intensity for both the OH hydroxyl groups of the supercage ($\nu_{OH} = 3640\ cm^{-1}$) and hexagonal prism ($\nu_{OH} = 3540\ cm^{-1}$) is accompanied by appearance of a broad absorption near 3530–$3520\ cm^{-1}$ upon adsorption of the iron carbonyl. Furthermore, new bridging CO absorptions at 1795 and $1760\ cm^{-1}$ are observed. It has been suggested that no chemical change has occurred and the low CO stretching frequencies arise from interaction of hydroxyl groups with bridging carbonyl oxygens. The magnitude of the shift to lower energy, $\sim 80\ cm^{-1}$, is similar to that for the hydrogen bonding interaction in $[Et_3NH][HFe_3(CO)_{11}]$ (75).

5. *Transition Metal Acceptors*

There appear to be no examples of transition metal acceptors attached to edge-bridging carbonyls in molecular compounds. This paucity of data undoubtedly arises from the lack of synthetic work rather than fundamental chemical reasons. There is, however, one example in surface chemistry which is likely to fit into this category.

In a study of the reactions of H_2–CO mixtures catalyzed by rhodium supported on silica gel, it was noted that Mn, Ti, and Zr additives have a large influence in the product distribution, and this observation prompted an IR spectroscopic investigation of these systems (*119*). In the absence of the early transition metal additives, Rh on silica gel exhibits an IR band around 2040 cm^{-1} attributed to terminal CO and another band in the vicinity of 1885 cm^{-1} assigned to conventional C-bonded edge-bridging CO. Addition of the dopants in an Rh:M ratio of 1:1 results in a decrease in the bridging CO stretching frequency but no change in the bands associated with terminal CO. In the case of the addition of manganese a broad and probably composite band is observed around 1650 cm^{-1}. Smaller shifts were observed for the Ti- and Zr-doped material. Under the conditions of this reaction, the early transition metals undoubtedly exist as oxides, and it appears probable from the IR spectra that the edge-bridging CO ligands are Σ —CO— bonded to these transition metal ions.

C. *Face-Bridging CO Donors*

The face-bridging carbonyl is more basic than terminal or edge-bridging CO, and it readily enters into Σ —CO— interactions. Furthermore, the interaction of a face-bridging CO with strong electrophiles leads to long C—O bond distances and low CO stretching frequencies, so the resulting ligand bears little resemblance to free carbon monoxide.

1. *Proton and Hydrogen Bonding Acceptors*

The lone example of hydrogen acting as an acceptor toward a face-bridging CO is provided by $Co_3(CO)_9COH$ (*120–123*). Synthesis of this compound is shown in Eq. (17). Proton NMR spectroscopy confirms the

$$Co_2(CO)_8 + LiCo(CO)_4 \xrightarrow[\text{low P, 15°C}]{\text{Bu}_2\text{O}}$$

$$Li[Co_3(CO)_{10}] + 2CO \xrightarrow[\text{hexane, } -20°C]{\text{HCl}} Co_3(CO)_9(COH) + LiCl \quad (17)$$

presence of the OH group with a resonance appearing at 11.25 ppm down-field of Me_4Si (*120*). This shift is similar to those previously described for $HFe_3(CO)_{10}COH$ ($\delta = 15.0$ ppm) (*73*), and $HRu_3(CO)_{10}COH$ ($\delta = 16.11$ ppm) (*74*).

This compound displays reasonable stability up to $-20°C$, unlike the protonated carbonyls previously discussed. Isolation of a crystalline solid was accomplished by performing all manipulations at reduced temperature, and the low-temperature structure of the crystal is shown in Fig. 16 (*122*). As expected, the structure shows a nearly regular triangular array of cobalt atoms [average Co—Co distance 2.478(5) Å], each with three terminally bound carbonyl ligands. The face-bridging CO is approximately equidistant from each of the cobalt atoms, $Co-C_{apical} = 1.928(8)$ Å. The hydrogen was not located directly in the solution of the structure, but the apical C—O bond length of 1.33(1) Å is significantly longer than in $LiCo_3(CO)_{10} \cdot (i\text{-Pr})_2O$, 1.190 Å (*124*) (see Section II,C,2). Hydrogen bonding also appears to be present between neighboring apical and terminal carbonyls.

In the absence of a CO atmosphere, this species decomposes at temperatures above $-20°C$ to form $HCo_3(CO)_9$ (*121*), providing another example of initial protonation on carbonyl oxygen followed by H^+ migrating to the metal framework. If decomposition is allowed to proceed at 40°C, the starting cluster is quantitatively converted to $HCo(CO)_4$ and $Co_4(CO)_{12}$ (*126*). Addition of $N(C_2H_5)_3$ to a solution containing $HCo(CO)_4$ and $Co_2(CO)_8$ induces the formation of $Co_3(CO)_9COH \cdot NEt_3$ (*123*), which may involve hydrogen bonding to CO such as that observed for $[NEt_3H]$-$[HFe_3(CO)_{11}]$.

2. Alkali Metal Cation Acceptors

Contact ion pairing between face-bridging carbonyls and alkali metals has been observed in solution as well as the solid state. The Li^+, Na^+, and K^+ salts of the $[Co_3(CO)_{10}]^-$ anion have been synthesized (*125, 126*) [Eqs. (18) and (19)], and of these the lithium salt is the most stable. The greater

$$LiCo(CO)_4 + Co_2(CO)_8 \xrightarrow{Bu_2O} Li[Co_3(CO)_{10}] \qquad (18)$$

$$2\,LiCo(CO)_4 + Co_4(CO)_{12} \longrightarrow 2\,Li[Co_3(CO)_{10}] \qquad (19)$$

stability of the lithium salt is interesting since large anions are usually stabilized by large cations.

The IR spectrum of the lithium salt shows a low frequency absorption both in solution and in the solid state at 1584 cm^{-1}. This is slightly lower than would be expected for a face-bridging carbonyl in the absence of

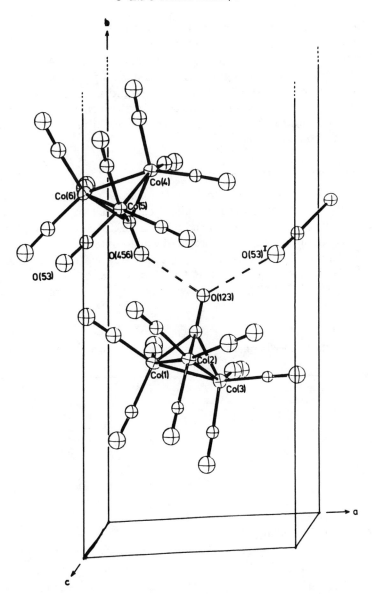

FIG. 16. Crystal structure of $Co_3(CO)_9COH$ emphasizing the hydrogen bonding interactions between the face-bridging COH group and carbonyl oxygens from neighboring clusters.
[By permission from Verlag Chemie, H. N. Adams *et al.*, *Angew. Chem. Int. Ed. Engl.* **20**,
125 (1981).]

perturbing forces. The structure of $Li[Co_3(CO)_{10}] \cdot (i\text{-}Pr)_2O$ confirms the presence of a contact ion pair between Li^+ and the oxygen of the face-bridging carbonyl (124), the Li—O distance being 1.859(10) Å. This distance is shorter than the edge-bridging CO—Li distances previously discussed for this molecule (Section II,B,2). The tight ion pair formed between the lithium and the oxygens is a possible explanation for the increased stability of this salt compared to the Na and K salts, which should interact more weakly with the carbonyl oxygens.

3. Main Group III Acceptors

The donor–acceptor interactions for the main group III elements described previously for terminal and edge-bridging carbonyls have been observed for face-bridging carbonyls as well. In addition, a class of compounds containing boron and aluminum have been prepared which contain the $Co_3(CO)_{10}$ core. These compounds are formulated in general as $Co_3(CO)_9COAX_2NR_3$, where A = B or Al, X can be H, F, Cl, Br, or I, and R = CH_3 or CH_3CH_2 (Table VIII) (127–133). The two synthetic approaches are illustrated in

TABLE VIII

Σ —CO— Stretching Frequencies for Main Group III Adducts of Face-Bridging Carbonyls

Compound	$v_{\Sigma-CO}$ (cm^{-1})[a]	Reference
$Cp_4Fe_4(CO)_4 \cdot (BX_3)_n$,		
$n = 1$, X = F, Cl, Br	1365; 1292; 1312	95
$n = 2$; X = F, Cl, Br	1435, 1405; 1360, 1327; 1365, 1320	
$Cp_4Fe_4(CO)_4 \cdot (AlBr_3)_n$		
$n = 1$	1392	95
$n = 2$	broad 1415, 1395, 1368	
$n = 3$	1470, 1439	
$n = 4$	1473	
$Cp_4Fe_4(CO)_4 \cdot 4AlEt_3$	1547	93
$Cp_3Ni_3(CO)_2 \cdot [Al(i\text{-}Bu)_3]_n$		
$n = 1, 2$	1637, 1637	93
$Co_3(CO)_9CO \cdot BX_2NR_3$		
X = H, F, Cl, Br;		
R = Me, Et	—	127[b,c], 129, 130
$Co_3(CO)_9CO \cdot AlX_2NEt_3$,		
X = Cl; Br	—	129, 130
$Co_3(CO)_9CO \cdot AlCl_2 \cdot AlCl_3$	—	129, 133

[a] A semicolon separates $v_{\Sigma-CO}$ for respective derivatives listed in column 1.
[b] Text structure **15**.
[c] X-Ray crystal structure for X = H available in reference.

Eqs. (20) and (21). The first of these species to be isolated was $Co_3(CO)_9$-$COBH_2N(C_2H_5)_3$ **(15)** *(127)*. The suggested formulation for this compound was based on the addition of an $-OBH_2N(C_2H_5)_3$ group to a $Co_3(CO)_9C$ core, and a large number of compounds are known in which the $Co_3(CO)_9C$

$$7Co_2(CO)_8 + 4X_3BNEt_3 \longrightarrow 4Co_3(CO)_9COBX_2NEt_3 + 16CO + 2CoX_2 \quad (20)$$

$$LiCo_3(CO)_{10} + X_3BNEt_3 \longrightarrow Co_3(CO)_9COBX_2NEt_3 + LiX \quad (21)$$

core is present with a variety of substituents other than oxygen attached to the apical carbon *(134)*. However, these compounds can also be synthesized from $[Co_3(CO)_{10}]^-$, so the $CO-BX_2NEt_3$ formalism also is an appropriate description.

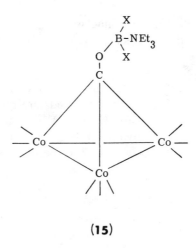

(15)

Studies of the interaction of the simple boron halides and aluminum alkyl or halides also have been performed (Table VIII). The two species that have been studied most extensively are $Cp_3Ni_3(CO)_2$ **(16a)** and $Cp_4Fe_4(CO)_4$ **(16b)** *(93, 95)*. The nickel trimer contains two face-bridging carbonyls *(135)* whereas four face-bridging CO's are present in the iron tetramer *(136)*.

The nickel trimer forms both 1:1 and 1:2 adducts with aluminum alkyls depending on the Lewis acid concentration. The formation constants differ only by the statistical factor so the electronic effects of one COAl on the other are negligible.

The cluster $Cp_4Fe_4(CO)_4$ has four potentially basic face-bridging CO ligands. The 1:4 (Fe$_4$:A) adduct as well as all intermediate adducts have been obtained for some of the aluminum alkyls *(93)* as well as BBr$_3$ and AlBr$_3$ *(95)*.

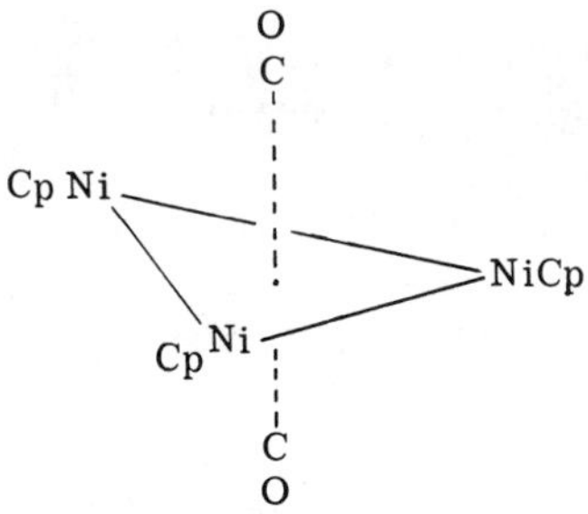

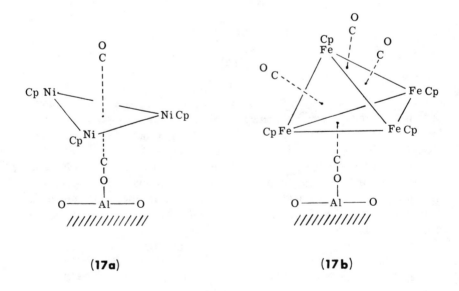

(16a) 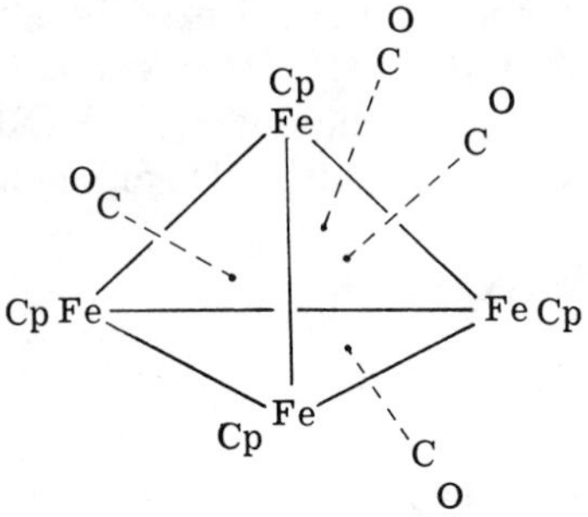   (16b)

Infrared spectra in the CO stretching region show the pattern of bands expected from selection rules for the series of $AlBr_3$ adducts ranging from 1:1 to 1:4 (Scheme 4).

In addition to the reaction of $Cp_3Ni_3(CO)_2$ and $Cp_4Fe_4(CO)_4$ with molecular Lewis acids in solution, a similar interaction is found with Lewis acid sites on γ-Al_2O_3 (*111*). These compounds are rapidly extracted from hydrocarbon solution by dehydroxylated γ-Al_2O_3, and IR spectra of the adsorbed species indicate an interaction between a surface aluminum acid site and a carbonyl oxygen (**17a** and **17b**). Infrared stretching frequencies

(17a)   (17b)

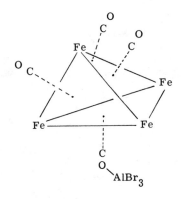

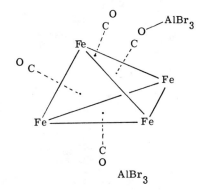

C_{3v} ; 1:1 adduct

ν_{CO} ; a_1, e

$\nu_{\Sigma-CO-}$; a_1

C_{2v} ; 1:2 adduct

ν_{CO} ; a_1, b_1

$\nu_{\Sigma-CO-}$; a_1, b_2

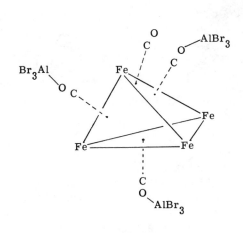

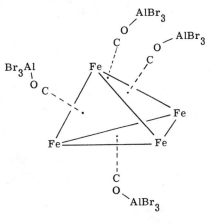

C_{3v} ; 1:3 adduct

ν_{CO} ; a_1

$\nu_{\Sigma-CO}$; a_1, e

T_d ; 1:4 adduct

$\nu_{\Sigma-CO-}$; t_2

SCHEME 4

TABLE IX

Σ —CO— Stretching Frequencies Following
Adduct Formation with Lewis Acids

	$\nu_{\Sigma-CO-}$ with indicated acceptor (cm^{-1})		
Compound	$AlBr_3$	γ-Al_2O_3	$AlEt_3$
$Cp_3Ni_3(CO)_2$	1531^a	1590^a	1637^b
$[CpFe(CO)]_4$	1392^c	1464^a	1547^b
$[CpNi(CO)]_2{}^d$	—	1707^a	1761^b

[a] Ref. 93.
[b] Ref. 111.
[c] Ref. 95.
[d] Edge-bridging carbonyl.

indicate that the strength of the surface interaction is intermediate between that with $AlBr_3$ and $Al(C_2H_5)_3$ (Table IX) (111).

4. Main Group IV Acceptors

Donor–acceptor interactions are known for both alkyl (92, 137–147) and silyl (128, 148–150) groups with face-bridging carbonyls. The presence of silyl groups is noteworthy since they have not yet been observed for either edge-bridging or terminal carbonyls. The synthesis of the silicon-containing compounds is outlined in Eqs. (22) and (23). The products from Eq. (22) are

$$Co_2(CO)_8 + R_2SiX_2 \longrightarrow Co_3(CO)_9CO \cdot SiR_2Co(CO)_4 \qquad (22)$$

$$X = H \text{ or } Cl, R = Et \text{ or } Ph$$

$$LiCo_3(CO)_{10} + R_3SiX \longrightarrow Co_3(CO)_9CO \cdot SiR_3 + LiX \qquad (23)$$

generally obtained in low yield, but some of the intermediates have either been isolated or observed spectroscopically. Among the intermediates identified in the reaction of Ph_2SiH_2 and $Co_2(CO)_8$ are $Ph_2SiHCo(CO)_4$ (151) and $Ph_2SiCo_2(CO)_7$ (152).

Alkyl- or acyl-containing face-bridging Σ —CO— species have been synthesized by two routes. The simplest is alkylation of a preexisting anionic μ_3-CO compound with an appropriate reagent such as CH_3SO_3F or $CH_3C(O)X$ (138, 143, 144). The second approach is addition of ligands to a species that initially contains an edge-bridging COR group, inducing a structural change with formation of a face-bridging COR moiety (91, 92, 141, 142).

The first instance in which a preformed face-bridging CO was directly methylated to form a μ_3-COMe ligand involved reaction of $[Fe_4(CO)_{13}]^{2-}$

with CH_3SO_3F in CH_2Cl_2 (*143*, *144*). A similar acylation reaction had been performed earlier on $[Co_3(CO)_{10}]^-$ (*138*). These two reactions are presented in Eqs. (24) and (25).

$$[Fe_4(CO)_{13}]^{2-} + CH_3SO_3F \longrightarrow [Fe_4(CO)_{12}(COMe)]^- \qquad (24)$$

$$[Co_3(CO)_{10}]^- + CH_3C(O)Br \longrightarrow Co_3(CO)_9COC(O)CH_3 \qquad (25)$$

Addition of the methyl group to $[Fe_4(CO)_{13}]^{2-}$ causes a reorganization of the carbonyl ligands. The parent compound was shown crystallographically to possess an approximately tetrahedral array of iron atoms with nine terminal carbonyls, three semibridging carbonyls about the base of the tetrahedron, and a face-bridging CO capping the base (*153*). Following Σ —CO— formation, only one of the semibridging CO groups remains (Fig. 17). This decrease in the number of ancillary bridging carbonyls is reminiscent of that discussed for the $AlBr_3$ adduct of $Fe_2(CO)_9$ (*95*). Coordination of the CH_3^+ group to the oxygen increases the C—O bond length from 1.20 Å to 1.36 Å, and the corresponding average iron–carbon bond for

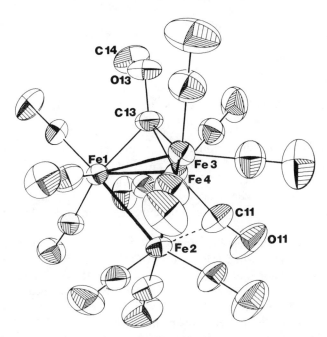

FIG. 17. Structure of $[Fe_4(CO)_{12}(\mu_3\text{-COCH}_3)]^-$ showing the bent H_3C—O—C array (118°), face-bridging Σ —CO—CH_3, C-13—O-13—C-14, as well as the presence of a single semibridging CO interaction between Fe-4—C-11—Fe-2. [By permission from the Royal Society of Chemistry, E. M. Holt *et al.*, *J. Chem. Soc., Chem. Commun.*, p. 778 (1980).]

the moiety is decreased by 0.07 Å. The ^{13}C-NMR spectrum of the parent compound shows a single resonance for all carbonyls at 227 ppm. Methylation of the oxygen causes appearance of a signal for the μ_3-C at 361.2 ppm, which is similar to the position for the Σ —CO— of edge-bridging carbonyls (Section II,B).

Reaction of $HM_3(\mu_2\text{-COMe})(CO)_{10}$ (M = Fe, Ru, or Os) with H_2 results in the loss of one CO and formal addition of one molecule of H_2 to the cluster, with a transformation of the μ_2-COMe group to a μ_3-COMe moiety (*91, 141, 147*) (**18**). The face-bridging CO—Me group was not identified by

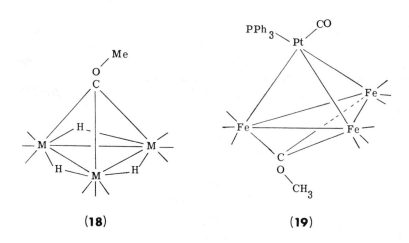

(**18**)   (**19**)

IR spectroscopy, but the terminal CO region appears to be similar to iso-structural clusters (*154, 155*). The ^{13}C-NMR spectrum of $H_3Os_3(CO)_9$-(COMe) has a resonance at 205.2 ppm which has been assigned to the μ_3-C. The NMR resonance characteristic of the μ_3-C in this molecule occurs at higher field than in analogous compounds but is similar to that for simple μ_3-CR ligands (*156*).

Conversion of an edge-bridging COR to a face-bridging COR functionality also occurs when $HFe_3(CO)_{10}COMe$ reacts with $Pt(C_2H_4)PPh_3$ (**19**) (*92*). Displacement of the labile C_2H_4 ligand is the driving force for this reaction. The ^{13}C-NMR spectrum of this species has a resonance at 342.6 ppm which has been assigned to the methylidyne carbon. The X-ray crystal structure confirms the presence of the face-bridging Σ —CO— and shows it to be slightly asymmetrically disposed between the three iron atoms. In addition, the C—O bond length of 1.34(1) Å in this moiety is very similar to that reported for $[Fe_4(CO)_{12}(COMe)]^-$ (*143, 144*).

5. Transition Metal and Actinide Acceptors

The Group IV transition metals and uranium form an interesting series of compounds with the $Co_3(CO)_{10}$ core (Table X) (157–161). Owing to the similarity of these compounds, only one will be discussed in detail. The reaction of $Li[Co_3(CO)_{10}]$ with one equivalent of Cp_2ZrCl_2 results in the formation of $Co_3(CO)_9CO-ZrCp_2Cl$ (160). The Zr—O distance is 2.036(4) Å and the C—O distance of the Σ —CO— is 1.28 Å. The slightly bent Σ —CO— (165°) in this compound compares with $Co_3(CO)_9COBH_2N-(C_2H_5)_3$ (127), in which the C—O—B angle is 135.4(30)°. This discrepancy is most probably due to unfavorable steric interactions between the $Co_3(CO)_{10}$ cluster and the bulky Cp_2ZrCl group.

If two equivalents of $Li[Co_3(CO)_{10}]$ are mixed with Cp_2ZrCl_2, $[Co_3(CO)_9CO]_2ZrCp_2$ is isolated. The zirconium is pseudo-tetrahedrally coordinated by two Cp rings and two apical oxygens from two $Co_3(CO)_{10}$ cluster units. The oxygen–zirconium bond distance averages 2.01 Å and the average C—O—Zr angle is 171.2°. Again the near linearity of the C—O—Zr bond is probably due to significant steric interactions between both the clusters and the cyclopentadiene rings. In general, C—O—X bond angles are highly variable, indicating a shallow potential for the angular deformation about the O—X bond.

TABLE X

Σ —CO— Stretching Frequencies for Transition Metal and Actinide Adducts of Face-Bridging Carbonyls

Compound	$v_{\Sigma-CO-}$ (cm^{-1})[a]	Reference
$Co_3(CO)_9COM(Cl)Cp_2$, M = Ti, Zr, Hf	1401; 1373; 1373	157[b], 160[c]
$[Co_3(CO)_9CO]_2MCp_2$, M = Ti, Zr, Hf	1403, 1337; 1372, 1307; 1378, 1319	160[c]
$[Co_3(CO)_9COMCp_2]_2O$, M = Ti, Zr, Hf	1413; 1377; 1379	160
$[Co_3(CO)_9CO]_2TiCpCo(CO)_4$	1268, 1233	159[d]
$Co_3(CO)_9COUCp_3$	1370	158[d]

[a] A semicolon separates the $v_{\Sigma-CO-}$ for respective derivatives listed in column 1.
[b] X-Ray crystal structure of M = Ti available in reference.
[c] X-Ray crystal structures of M = Zr and Hf available in reference.
[d] X-Ray crystal structure available in reference.

III

Π—CO COMPOUNDS:

FORMATION, STRUCTURES, AND SPECTRA

The chemical and structural characterization of molecular compounds containing Π —CO are of great interest because the Π —CO is implicated as a precursor to the cleavage of CO on metal surfaces, such as those employed in Fischer–Tropsch catalysis (*162*). Despite the physical appeal of this idea, there is only one case in which a Π —CO has been identified on a metal surface. In this one instance, photoelectron spectroscopy of CO adsorbed on the Cu (311) surface demonstrates that a significant fraction of the CO assumes an orientation which is nearly parallel to the surface (*163*) (Scheme 5).

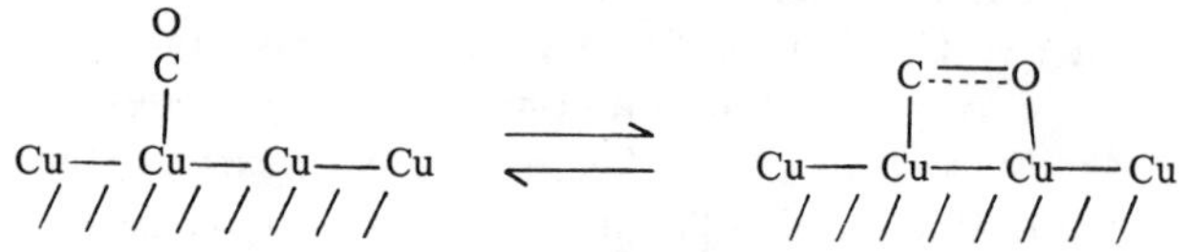

$$
\begin{array}{ccc}
\underset{\displaystyle \text{Cu}-\text{Cu}-\text{Cu}-\text{Cu}}{\overset{\displaystyle \overset{\text{O}}{\underset{\big|}{\text{C}}}}{}} \;\; /\,/\,/\,/\,/\,/\,/ & \rightleftharpoons & \underset{\displaystyle \text{Cu}-\text{Cu}-\text{Cu}-\text{Cu}}{\overset{\displaystyle \text{C}\text{-}\text{-}\text{-}\text{O}}{}} \;\; /\,/\,/\,/\,/\,/\,/
\end{array}
$$

Scheme 5

Although the discovery of Π —CO bonding in molecular chemistry is relatively recent, there exists a substantial number of structurally well-characterized examples, and the chemistry of Π —CO compounds is being developed. For example, a molecular analogy to Scheme 5 has been discovered (Section IV,B).

Another interesting aspect of Π —CO chemistry is the bonding. As will be described in this section, the most common situation is to regard Π —CO as a formal four-electron donor. However, some CO ligands that ostensibly have the Π —CO geometry are best described as two-electron donors and others as six-electron donors. The Ω parameter ranges from approximately 2.2 to 3.3 for Π —CO compounds (Fig. 4).

A. *Dinuclear Systems*

The first cluster containing a Π —CO was reported in 1975 (*6, 164, 165*) and was prepared according to the reaction given in Eq. (26). The X-ray

$$
\text{Mn}_2(\text{CO})_{10} + 2\,\text{dppm} \xrightarrow[p\text{-xylene}]{\Delta,\ 4\ \text{hours}} \text{Mn}_2(\text{CO})_5(\text{dppm})_2 \tag{26}
$$

crystal structure of this molecule (Fig. 18) shows a nearly linear Mn-1—C-1—O-1 array with the CO tipped toward Mn-2. The Mn-2—C-1 distance is

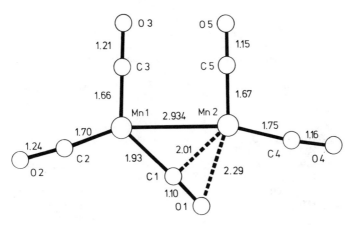

FIG. 18. Molecular structure of $Mn_2(CO)_5(dppm)_2$ showing the Π—CO interaction of C-1—O-1 with Mn-2. The Mn-1—C-1—O-1 array is nearly linear. The two dppm (PPH_2-CH_2PPh_2) ligands (not shown) would lie above and below the plane of the figure. [By permission from the Royal Society of Chemistry, R. Colton *et al.*, *J. Chem. Soc., Chem. Commun.*, p. 363 (1975).]

2.01 Å, and that between Mn-2—O-1 is 2.29 Å; indicating side-on bonding between Mn-2 and C-1—O-1. The Ω parameter is 2.41, which indicates significant C and O overlap with Mn-2. Thus it is reasonable to consider the Π—CO as a four-electron donor, which brings the electron count on each Mn to the expected 18. An IR band at 1645 cm^{-1} is assigned to the CO stretch of the Π—CO.

Since the report of the $Mn_2(CO)_5(dppm)_2$ structure, a number of additional binuclear complexes containing a Π—CO have been synthesized (*60, 61, 167–171*) (Table XI). Identification of Π—CO ligands is generally based on X-ray crystal structure data rather than spectroscopic methods, because the CO stretching frequencies are similar to those of Σ—CO— carbonyls and the ^{13}C-NMR signal of the Π—CO occurs in the same general region as that of C-bonded terminal and some edge-bridging carbonyls (*165b*).

A distinctive feature of the $Mn_2(CO)_5(PP)_2$ series of compounds (*166*) is their ability to reversibly add CO when heated in solution under a CO atmosphere. The IR spectrum of the product, for example, $Mn_2(CO)_6(dppm)_2$, suggests the presence of two terminal carbonyls on each manganese as well as two C-bonded edge-bridging carbonyl ligands. This compound decarbonylates to the starting $Mn_2(CO)_5(dppm)_2$ when it is allowed to stand in air for a few weeks. Systematic studies of the decarbonylation of $Mn_2(CO)_6(PP)_2$ complexes demonstrate that the tendency

TABLE XI

Π —CO Stretching Frequencies and Ω Values

Compound	$v_{\pi-CO-}$ (cm^{-1})	Ω	Reference
Mn$_2$(CO)$_5$(dppm)$_2$[a]	1648	2.41	6, 164[b]
Mn$_2$(CO)$_5$(depm)$_2$[c]	1640	—	166
Mn$_2$(CO)$_5$(dcpm)$_2$[d]	1643	—	166
CpCo(CO)$_2$ZrCp*$_2$	1683	2.55	167[e]
Cp$_3$TiW(μ-CR)(CO)$_2$[f]	1683	2.63	168[g]
Cp$_3$ZrW(μ-CR)(CO)$_2$[f]	1578	—	168
Cp$_2$MoW$_2$(μ-CR)(CO)$_6$[f]	1687	2.55	169[g]
Cp$_3$TiW[(μ-CR)=CH$_2$](CO)$_2$[f]	1649	2.65	170[g]
Cp$_3$NbMo(CO)$_3$	1560	2.65	171[g]
Cp$_3$ZrMo(CO)$_2$(η^2-Ac)	1534	2.81	60, 61[h]
Cp$_3$Nb$_3$(CO)$_7$	1330	2.75	182[i]
[Me$_3$NCH$_2$Ph][HFe$_4$(CO)$_{13}$]	—	2.96	184[j]
Cp$_2$Cp*$_2$Mo$_2$Co$_2$(CO)$_4$	1633	3.16	185[k]

[a] Dppm = Ph$_2$PCH$_2$PPh$_2$.
[b] Text, Fig. 18.
[c] Depm = Et$_2$PCH$_2$PPh$_2$.
[d] Dcpm = (c-C$_6$H$_{11}$)$_2$PCH$_2$PPh$_2$.
[e] Text, Fig. 19.
[f] R = p-C$_6$H$_4$Me.
[g] X-Ray crystal structure available in reference.
[h] Text, Fig. 20.
[i] Text, Fig. 22.
[j] Text, structure **22a**.
[k] Text, structure **22b**.

to decarbonylate increases with increasing bulk of the PP ligand (*166*). Thus the formation of Π —CO in this series of compounds is favored by steric repulsion in the precursor. This observation may provide a useful principle for the designed synthesis of new classes of metal carbonyls containing Π —CO ligands.

It is useful at this point to summarize some of the structural features which have been observed for Π —CO in bimetallic compounds. The M—C—O bond angle is slightly bent in Π —CO compounds, in which it ranges from 177° (*167*) to 169° (*169*). Some terminally bonded carbonyls have similar angles, so the slight bend is not unique to the Π —CO moiety. Second, the C—O bond distance is lengthened upon coordination. The degree of lengthening falls in the region typical for Σ —CO— Lewis acid systems (*27, 32, 40, 55, 58, 59*). Finally, the M—C distance is shortened in the Π —CO systems just as it is in the Σ —CO— Lewis acid cases.

The reaction involving equimolar quantities of CpCo(CO)$_2$ and Cp*$_2$ZrH$_2$ results in the evolution of 1 mol of H$_2$ and formation of CpCo(CO)$_2$ZrCp*$_2$

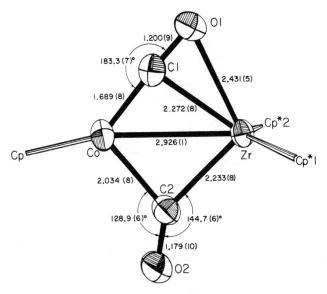

Fig. 19. Structure of the planar metal–carbonyl framework of CpCo(CO)$_2$ZrCp*$_2$ showing both the П —CO interaction as well as an edge-bridging CO group. Note the nearly linear Co—C—O array of the П —CO. [By permission from Elsevier Sequoia, S. A., P. T. Barger and J. E. Bercaw, *J. Organomet. Chem.* **201**, C39 (1980).]

according to Eq. (27) (*167*). The vibrational spectrum shows two absorptions

$$CpCo(CO)_2 + Cp^*_2ZrH_2 \longrightarrow CpCo(CO)_2ZrCp^*_2 + H_2 \qquad (27)$$

at 1683 and 1737 cm^{-1}. Figure 19 depicts the results of the crystal structure analysis of the molecule. A zirconium–cobalt bond is present (bond length 2.93 Å) as well as an edge-bridging carbonyl and a П —CO. It has been suggested (*167*) that presence of both C-bridging CO and П —CO permits both metal centers to achieve an inert gas electron configuration. It can be seen in Fig. 19 that the П —CO has the longer C—O distance.

A second example of a zirconium П —CO interaction is found in Cp$_2$Zr(η^2-Ac)Mo(CO)$_2$Cp (*60, 61*). After 2 days in inert atmosphere, one CO is lost from Cp$_2$Zr(η^2-Ac)Mo(CO)$_3$Cp (Section II,A,5), and Cp$_2$Zr(η^2-Ac)Mo(CO)$_2$Cp forms [Eq. (28)]. The structure is shown in Fig. 20. One of the carbonyls has a П —CO configuration, the carbon end of the

$$Cp_2Zr[\eta^2\text{-C(O)Me}]Mo(CO)_3Cp \xrightarrow{-CO} Cp_2Zr[\eta^2\text{-C(O)Me}]Mo(CO)_2Cp \qquad (28)$$

acetyl is now attached to the molybdenum, and a metal–metal bond (3.297 Å) is present. The П —CO moiety has an Mo—C—O angle of 172.3° whereas

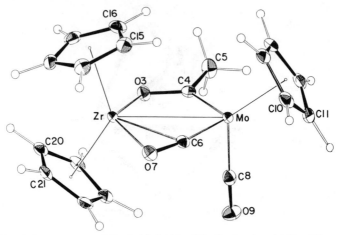

FIG. 20. X-Ray structure of $Cp_2Zr(\eta^2$-Ac)Mo(CO)$_2$Cp showing the Π—CO coordinated to the Zr as well as the η^2-CCH$_3$O interaction spanning the Mo—Zr bond. The Mo—C-6—O-7 group is slightly bent and the Zr—O-7 distance is shorter than Zr—C-6. [By permission from the American Chemical Society, B. Longato *et al., J. Am. Chem. Soc.* **103**, 209 (1981).]

the terminal carbonyl has an angle of 178°. Bond length differences in the two carbonyls are consistent with those previously described, i.e., Mo—C-8 = 1.957 Å versus Mn—C-6 = 1.876 Å, and C-8—O-9 = 1.147 Å versus C-6—O-7 = 1.241 Å. This molecule provides the only example to date in which the distance from the Π—CO oxygen to the metal is shorter than the corresponding carbon–metal distance in a bimetallic compound. Thus the structure is somewhat intermediate between a Σ—CO— and Π—CO, although the Ω value of 2.81 indicates that, of the two, the latter description is more accurate.

An additional example of Π—CO bonding involves the product of Eq. (29) (*171*). The infrared spectrum of Cp$_2$NbMo(CO)$_3$Cp exhibits three bands

$$Cp_2NbBH_4 + Et_3N + CH_3Mo(CO)_3Cp \xrightarrow{110° \text{ toluene}} Cp_2NbMo(CO)_3Cp \quad (29)$$

at 1870, 1750, and 1560 cm^{-1} corresponding to a terminal carbonyl and two different types of bridging carbonyl stretching modes. The results of the X-ray crystal structure (**20**) show that one of the two bridging carbonyls is C-bonded semibridging, where in this instance D_1 corresponds to the Mo—C distance of 2.02 Å, D_2 refers to the Nb—C distance of 2.53 Å, and the angle (θ) is 155° (**4**). The second bridging carbonyl is of the Π—CO configuration. No significant difference in the Mo—C distances for the Π—CO and the terminal carbonyl are observed. However, the C—O distances are significantly different, 1.22 Å and 1.16 Å for the Π—CO and terminal CO, respec-

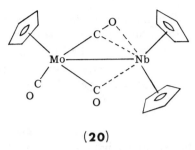

(20)

tively. The structural significance of this molecule is the presence in a single structure of a terminal carbonyl, a semibridging carbonyl, and a four-electron donating Π —CO.

An interesting feature of the heteronuclear Π —CO compounds studied to date is the consistent orientation of the carbon end of this CO group toward the late transition metal and the Π interaction with the early transition metal. Also the early transition metal has the same or higher formal oxidation state as the latter metal. In all of these cases, acceptable 18-electron structures on the individual metal centers result if the Π —CO is considered to be a formal 4-electron donor.

A series of compounds with the formula $Cp_2M_2(CO)_4$ (M = Cr, Mo, or W) which may be prepared by the thermal or photochemical decarbonylation of $Cp_2M_2(CO)_6$ (7, 172–177), presents a more difficult problem. As shown in Fig. 21, all four CO ligands are in a Π-type disposition. Although steric factors were originally invoked to explain the tilt of the CO groups in $Cp^*_2Cr_2(CO)_4$ (172, 173), this explanation is less tenable with the observation of the analogous structure for the much less sterically crowded

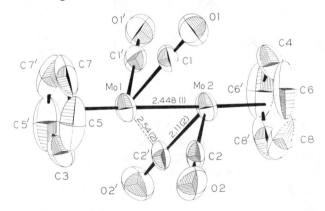

FIG. 21.   Structure of $Cp_2Mo_2(CO)_4$ illustrating the four Π —CO interactions and the linear Cp—Mo—Mo—Cp array. [By permission from the American Chemical Society, R. J. Klinger *et al., J. Am. Chem. Soc.* **100**, 5034 (1978).]

$Cp_2Mo_2(CO)_4$ (7). In all of the structural studies the M—M bond lengths suggest multiple M—M bonding, and an M–M triple bond would produce an 18-electron configuration on each of the metal centers, providing the CO groups are only formal 2-electron donors. This appears to be inconsistent with the Π—CO orientation, which suggests the presence of 4-electron donor CO ligands. The bonding in these compounds presents a clear breakdown of formal electron counting rules.

On closer inspection there are some other anomalies that suggest that the compounds may not be classifiable with other Π—CO compounds. One is the higher CO stretching frequency and the other is the low value of Ω (2.16), which suggests a rather higher carbon metal bonding than is found in the majority of Π—CO compounds. Extended Hückel molecular orbital calculations on $Cp_2M_2(CO)_4$ systems (180) suggest that in the Π—CO interaction the carbonyl is net electron accepting through carbon. Qualitatively this is the same situation that is encountered in a conventional C-bonded carbonyl bridge, as shown in Table I, g. Why then is the CO ligand tipped relative to the M—M axis in $Cp_2M_2(CO)_4$ rather than more nearly perpendicular, as in the conventional C-bonded bridging CO? A succinct answer to this question is illustrated by **21a** and **21b**, and described below (181).

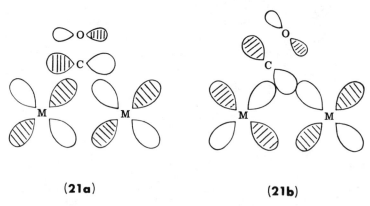

(21a)   **(21b)**

The overlap of the CO π* orbital with the highest filled metal molecular orbital in a conventional C-bonded carbonyl bridge is shown in **21a**. In this case, the metal atoms are held together by a single bond, and the highest filled d orbitals on the two metal centers are antibonding in relation to each other. This nodal character is in phase with the nodal properties of the p π* orbital on the carbonyl carbon, if the CO is perpendicular to the M—M axis (**21a**). By contrast, the highest filled orbital in $Cp_2M_2(CO)_4$ has M—M π bonding symmetry. This M—M π bond would give zero net overlap with the π* orbital of a perpendicular CO, but as illustrated in **21b**, positive overlap

TABLE XII

Ω Values and CO Stretching Frequencies at the Borderline
between Π —CO and C-Bonded CO

Compound	Ω	ν_{CO} (cm^{-1})[a]	Reference
$Cp_2Cr_2(CO)_4$	2.16	1881	174[a]
$Cp_2Mo_2(CO)_4$	2.16	1859	7[b]
$[CpMo(CO)_3]_2Zn$	2.20	1867	70[a]
$[CpMo(CO)_3ZnCl \cdot Et_2O]_2$	2.06	1758	70[a]
$CpMo(CO)_3ZnBr \cdot 2THF$	2.28	1849	178[a]
$Cp_2V_2(CO)_5$	1.99	1815	179[a]

[a] Lowest CO stretching frequency observed.
[b] X-Ray crystal structure available in reference. See also Fig. 21.

is achieved when the CO is tipped. It is interesting that the $Mn_2(CO)_5(PP)_2$ series and the $Cp_2M_2(CO)_4$ compounds appear to achieve the 18-electron count on each metal in quite different ways. For the $Mn_2(CO)_5(PP)_2$ compounds, the metal–metal bond order remains one and a Π —CO ligand is a 4-electron donor. By contrast, the $Cp_2M_2(CO)_4$ series of compounds appears to achieve the 18-electron count on each metal by multiple M—M bonding.

A group of zinc derivatives of transition metal carbonyls is known which appears to have Π —CO—Zn interactions (70, 178). The Ω values (Table XII) in this case are higher than for the $Cp_2M_2(CO)_4$ compounds. Although no molecular orbital calculations are available for the zinc compounds, a logical description of the bonding is that of CO π donation into vacant sp^n hybrid orbitals on Zn. Similarly, a vanadium compound, $Cp_2V_2(CO)_5$, also lies on the borderline between Π —CO and —C-bridging CO (179) (Table XII).

B. Tri- and Polynuclear Systems

Although there are a comparatively large number of binuclear systems which contain a Π —CO, a limited number of polynuclear species have been characterized. The only example of a trimetallic cluster with a Π —CO is $Cp_3Nb_3(CO)_7$ (182, 183). Equation (30) gives the synthesis for this species,

$$3\,CpNb(CO)_4 \xrightarrow[\text{hexane}]{h\nu} Cp_3Nb_3(CO)_7 + 5\,CO\uparrow \tag{30}$$

and the molecular structure is shown in Fig. 22. The cluster is composed of a triangular array of niobium atoms with six terminal carbonyls and an asymmetrically face-bridging carbonyl group. One Nb—C distance to this bridging CO is 1.97 Å while the other two distances are 2.28 and 2.23 Å. An

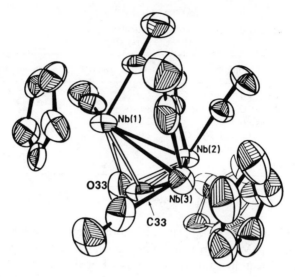

FIG. 22.   Structure of $Cp_3Nb_3(CO)_7$ showing the nearly linear Nb-3—C-33—O-33 group and the Π —CO interaction with Nb-1 and Nb-2. [By permission from Verlag Chemie, GmbH, W. A. Herrmann *et al.*, *Angew. Chem. Int. Ed. Engl.* **18**, 960 (1979).]

almost linear array is observed for Nb-3—C-33—O-33 (170°), and there are close contacts between Nb-2—O-33 (2.24 Å) and Nb-1—O-33 (2.21 Å). These metal–oxygen distances and the Ω values, 2.70 and 2.80, are similar to those in other systems containing Π —CO. The structural data indicate that this unique carbonyl functions as a 6-electron donor, and this also agrees with the resulting 48-electron count expected for three-metal tri-angular cluster. A significant lengthening of the C—O bond to 1.30 Å results, and this is reflected in the CO stretching frequency for this moiety ($1330 \, cm^{-1}$). Conversely, the Nb-3—C-33 bond is the shortest metal carbonyl bond in the structure, a behavior which has been discussed above for Π —CO groups in other molecules.

Four different Π —CO species containing four metal atoms have been structurally characterized (*184, 185, 190, 191*). The syntheses for two of these compounds are given in Eqs. (31) and (32), and structures are shown in **22a** and **22b**.

Some features are common to both clusters. The metal frameworks are rather open, allowing all four metals to be within bonding distance of the unique carbon. Unlike the other Π —CO containing species presented up to this point, none of the M—C—O angles for the Π —CO in these species approximates linearity, and in both instances the M'—O distance is signifi-cantly shorter than the M'—C distance. This results in the rather large Ω

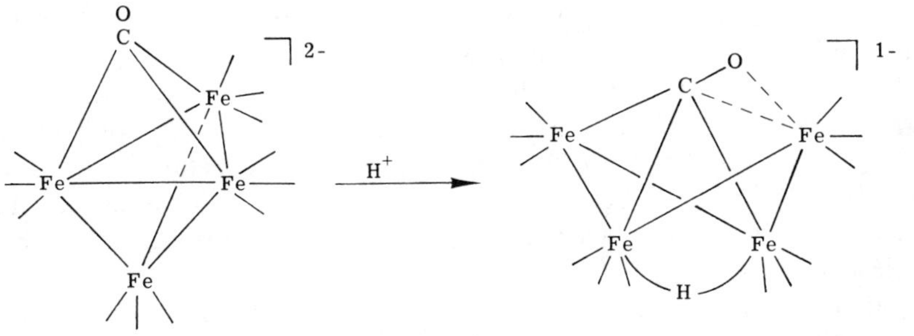

$$2Cp^*Co(C_2H_4)_2 + Cp_2Mo_2(CO)_4 \xrightarrow{h\nu} Cp_2Cp^*_2Mo_2Co_2(CO)_4 + 4C_2H_4 \quad (32)$$

values of 2.96 for the iron cluster and 3.16 for the mixed Mo—Co cluster, the latter is approaching the Ω values for Σ —CO—. Furthermore, the Π —C—O bond is significantly lengthened, 1.26 and 1.28 Å, for $[HFe_4(CO)_{13}]^-$ and $Cp_2Cp^*_2Mo_2Co_2(CO)_4$, respectively. The vibrational spectrum of the Mo—Co cluster contains an absorption at 1633 cm^{-1} from the Π —CO, and the ^{13}C-NMR signal associated with the Π —CO appears between 280 and 295 ppm downfield of TMS for these two compounds. However, these spectroscopic features cannot be used as diagnostic for these Π —CO containing species because Σ —CO— compounds have IR absorptions in this region (Section II) and conventional edge-bridging C-bonded carbonyls have ^{13}C-NMR resonances at similar positions (4).

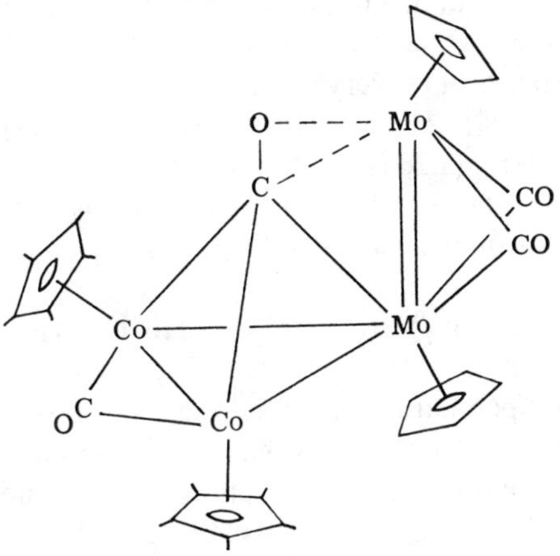

(22b)

The synthesis of $[HFe_4(CO)_{13}]^-$ from $[Fe_4(CO)_{13}]^{2-}$ (*186, 187*) is interesting because of the major structural rearrangement that occurs following protonation (*184*). The parent dianion has a tetrahedral metal framework (Section II,C,4) (*153*) which undergoes metal–metal bond scission and formation of the Π —CO upon protonation. The metal core geometry for $[HFe_4(CO)_{13}]^-$ is butterfly shaped, and the proper 62-electron count for a butterfly metal cluster is achieved if the Π —CO is assumed to be a 4-electron donor.

Steric arguments were used to rationalize the conversion of tetrahedral $[Fe_4(CO)_{13}]^{2-}$ to the butterfly $[HFe_4(CO)_{13}]^-$ (*184*). However, recent NMR studies (*188*) suggest that the species initially formed upon protonation has a closed tetrahedral metal framework and that in polar organic solvents an equilibrium exists between the open butterfly form and the closed tetrahedron. Furthermore, synthesis of the highly strained tetrahedral compound $[Fe_3Cr(CO)_{14}]^{2-}$ (*189*) suggests that the steric argument for adoption of the butterfly geometry is not compelling.

Unlike the case for other Π —CO containing species, reactivity at the oxygen has been observed in $[HFe_4(CO)_{13}]^-$ (*190*). The reaction of $[HFe_4(CO)_{13}]^-$ with HSO_3CF_3 at low temperature in an NMR tube results in the appearance of a resonance in the ^1H-NMR spectrum at $\delta = 13.2$ ppm in addition to a metal hydride resonance. Downfield resonances in this region have been observed previously for $H_2Fe_3(CO)_{11}$ (*73*), $H_2Ru_3(CO)_{11}$ (*74*) (Section II,B,1), and $Co_3(CO)_9(COH)$ (*121–124*) (Section II,C,1), in which C—O—H interactions were claimed. Furthermore, the ^{13}C-NMR spectrum displays a significant downfield shift, $\delta = 301$ ppm, for the Π —CO upon reaction. Although the protonated species could not be isolated due to thermal instability, the analogous methylated compound has been structured crystallographically (*144, 191*) (Fig. 23). Spectroscopically these species are very similar. The Π —CO bond length in $HFe_4(CO)_{12}(\eta\text{-}COCH_3)$, (1.39 Å) is substantially larger than that in $[HFe_4(CO)_{12}(\eta\text{-}CO)]^-$ (1.26 Å). The CO in this O-alkylated ligand is a formal six-electron donor so the CO bond order is substantially reduced. Similarly the O-protonated CO in $HFe_4(CO)_{12}(\eta\text{-}COH)$ is a six-electron donor.

The structure of $Cp_2Cp^*_2Mo_2Co_2(CO)_4$ (**22b**) (*185*) is more open than for $[HFe_4(CO)_{13}]^-$, and as a result the Mo_2Co_2 cluster is formally electron deficient (*110*). The open nature of this species, which probably results from the bulk of the Cp and Cp* ligands, has been likened to a stepped metal surface (*192*), approximating the intersection of (111) and (001) planes of a face-centered cubic lattice. Consequently, the C—O vector of the Π —CO is nearly perpendicular to the plane defined by the triangle of two cobalts and one molybdenum and parallel with the two molybdenum centers.

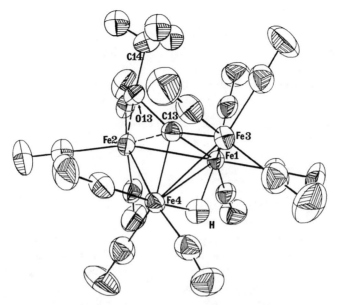

Fig. 23. Molecular structure of $HFe_4(CO)_{12}(COMe)$ illustrating the butterfly metal framework with the Π —CO, C-1—O-1 to Fe-2, and the CH_3 group on the Π —CO oxygen. [By permission of the Royal Society of Chemistry, K. H. Whitmire *et al.*, *J. Chem. Soc., Chem. Commun.*, p. 780 (1980).]

IV

INFLUENCE OF —CO— BONDING ON REACTIVITY

In previous sections numerous examples were given for large changes in C—O bond length and CO stretching frequencies which occur upon C and O bond formation. In some cases, for example, the C—O bond length approaches that of a single bond. In view of these large physical changes it perhaps is not surprising that C and O bonding also engenders large changes in CO reactivity (*193a*).

A. *CO Migratory Insertion*

Carbon monoxide insertion into a metal–alkyl bond [Eq. (33)] is a fundamental reaction in organometallic chemistry as well as an important

$$L_nM \overset{\displaystyle R}{\underset{\displaystyle CO}{<}} + L' \longrightarrow L_nM \overset{\displaystyle L'}{\underset{\displaystyle \underset{\displaystyle R}{\overset{|}{C}=O}}{<}} \tag{33}$$

step in the commercial production of oxygen-containing organics such as aldehydes and acetic acid (*193–195*). Extensive mechanistic studies of CO insertion into $Mn(CO)_5R$ (R = alkyl) and related compounds reveal the rate law given in Eq. (34). A mechanism consistent with the rate law and nonkinetic observations is illustrated in Eq. (35) (*194*). The detailed stereochemistry of the products of CO insertion reactions has been investigated to

$$\text{rate} = \frac{k_1 k_2 [L'][L_n MR(CO)]}{k_{-1} + k_2 [L']} \tag{34}$$

$$L_nM{\overset{R}{\underset{CO}{\big<}}} \underset{k_{-1}}{\overset{k_1}{\rightleftharpoons}} \left[L_nM{-}C{\overset{O}{\underset{R}{\big<}}} \right] \xrightarrow[k]{+\,L^1} L_nM{\overset{L'}{\underset{\underset{O}{\overset{\|}{C}}-R}{\big<}}} \tag{35}$$

identify the group that undergoes migration. Spectroscopic studies using isotope tracers indicate that in the case of $Mn(CO)_5R$ the R group migrates onto the CO and the incoming ligand takes the place of this migrating group in the coordination sphere of the metal (*196, 197*). Similarly, the stereochemistry of the product of the insertion reaction for CpFe(CO)LEt [L = PPh_3 and $P(OCH_2)_3CCH_3$] indicates greater than 95% formal alkyl migration in some solvents (*198*). For these cases and presumably many others which have not been studied in detail, the name CO insertion is inappropriate, but this terminology is still prevalent.

An extended Hückel study of the alkyl migration reaction indicates the cyclic transition state depicted in **23**. A small change in this transition state

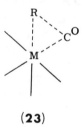

(23)

could change the nature of the reaction from an apparent alkyl migration to an apparent CO migration (*199*).

It is often found that the migratory insertion reaction is strongly solvent dependent. In addition, electron acceptors bring about great rate enhancements. As will be described in this section, the mechanism for this rate enhancement appears to proceed via initial attack of the electron acceptor on the oxygen of the carbon monoxide ligand involved in the insertion process.

1. *Proton-Assisted Migratory Insertion*

An example of the promotion of CO migratory insertion by a Brønsted acid is found for the enhanced rate of CO uptake by $Mn(CO)_5(CH_3)$ in the presence of dichloroacetic acid. Judging from the kinetic data, CO uptake is enhanced by both the monomer and the dimer of the acid (*200*). It was postulated that the basis for this enhancement in the migratory insertion reaction is hydrogen bonding with a carbonyl oxygen (**24**). In this connection,

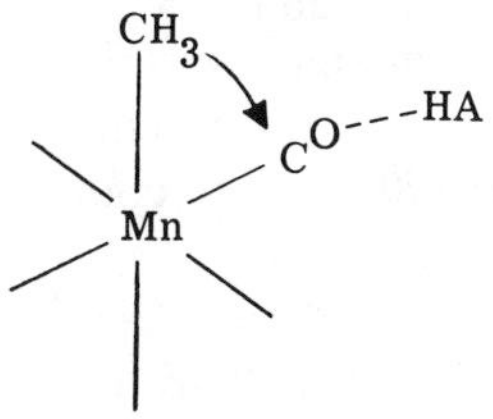

(24)

extended Hückel results indicate that the transition state (**23**) is stabilized relative to the ground state by 2–3 kcal/mol. Experimentally it has not been possible to carry out the reaction under conditions that might correspond to the complete protonation of the transition state, because strong acids cause protolysis of the $Mn-CH_3$ linkage. A correlation is observed between acidity and the degree of rate enhancement. Thus 0.4 M CCl_2HCOOH (pK_a 2.86), increases the rate by a factor of 2.4, whereas at the same concentration of CF_3COOH (pK_a 0.23) a 9.4-fold increase was observed. In addition to insertion, trifluoroacetic acid caused appreciable $Mn-CH_3$ protolysis, and with HBr only protolysis occurs.

2. *Cation-Assisted Migratory Insertion*

The nature of the cation has been shown to greatly influence the rate at which alkyl migration occurs in the alkyl tetracarbonyl ferrates, $[Fe(CO)_4R]^-$ (*201*). The small Li^+ increases the rate of insertion by a factor of nearly three over that with Na^+ and a factor of 10^3 over that when a large noncoordinating cation is present. In tetrahydrofuran solvent, this migratory insertion reaction is first order in ligand L and in the ion pair $[M^+Fe(CO)_4R^-]$ [Eqs. (36) and (37)] (*202*). The indication is that the

$$[M^+(OC)Fe(CO)_3R^-] + L \longrightarrow [M^+(ORC)Fe(CO)_3L^-] \qquad (36)$$

$$\text{rate} = k[M^+(OC)Fe(CO)_3R^-][L] \quad L = PPh_3, P(OMe)_3, PMe_3, \text{etc.} \qquad (37)$$

increased rate of insertion in the presence of small cations is brought about by the more favorable equilibrium constant $[K = k_1/k_{-1}$, Eq. (34)] for the formation of the coordinatively unsaturated acyl intermediate, $Fe(CO)_3(CRO)$. The greater basicity of acyl ligands than the CO ligand is well documented with other acceptors (203, 204). Detailed conductivity and spectroscopic data confirm the higher amount of tight ion pairs in the case of the acyl iron complex than the alkyl complex and also indicate that the site of ion binding is the acyl ligand.

Alkali metal ions have never been found to promote hydride migration to form metal formyls in a manner analogous to cation-induced alkyl migration In fact small cations have been shown to enhance the rate of decomposition of transition metal formates (205) and to coordinate to and abstract hydride from transition metal hydrides (206).

3. Main Group Lewis Acid-Assisted Migratory Insertion

Molecular Lewis acids can greatly enhance the rate and the position of equilibrium in the migratory insertion reaction. For example, $Mn(CO)_5(CH_3)$ reacts on the time of mixing with $AlBr_3$ to give an interesting product [Eq. (38)] (207). In contrast with the migratory insertion reactions discussed to this point, a simple donor ligand is not added to this system to effect the reaction; instead a Br atom bridged from the Al to Mn fills the vacant coordination site created on the Mn following migration. A stopped-flow kinetic study demonstrated that the reaction depicted in Eq. (38) is over on the time of mixing of $AlCl_3$ and $Mn(CO)_5(CH_3)$ in dilute solution (208). To explain this enhanced reaction rate one simple possibility to be considered is that the Lewis acid captures the transient intermediate $[(OC)_4Mn(CRO)]$ each time it is formed in the first step of Eq. (35). However, it can be shown that the rate of the reaction in Eq. (38) is at least 10^8 faster than would be

$$
\begin{array}{ccc}
\text{CH}_3 & & \text{Br}\!-\!\!-\text{AlBr}_2 \\
| \quad \text{CO} & \xrightarrow{\text{AlBr}_3} & | \quad \text{CO} \;\backslash \\
\text{OC}\!-\!\text{Mn}\!-\!\text{CO} & & \text{OC}\!-\!\text{Mn} \qquad\quad \text{O} \\
\text{OC}^{\diagup}\;\underset{\text{O}}{\overset{|}{\text{C}}} & & \text{OC}^{\diagup}\;\underset{\text{O}}{\overset{|}{\text{C}}}\;\;\underset{\diagdown \text{CH}_3}{\overset{\diagup}{\text{C}}} \\
\end{array}
\qquad (38)
$$

predicted for this possibility. Thus, the influence of the Lewis acid must be to enhance the rate of the primary insertion step. The rates of reaction also are far greater than those for a simple substitution on the metal center of a coordinatively saturated 18-electron octahedral complex; therefore the most probable mechanism involves initial Lewis acid coordination to the oxygen of a CO ligand followed by alkyl migration. Although compounds with CO stretching frequencies as high as those in $Mn(CO)_5(CH_3)$ are not known to form stable adducts with Lewis acids such as the aluminum and boron halides (Fig. 1), the transient existence of an O-coordinated carbonyl would be adequate to explain this rate enhancement.

The cyclic product of Eq. (38) is susceptible to further reaction if a suitable ligand is present. When carbon monoxide is this ligand, it displaces Br from the coordination sphere of Mn in the follow-up reaction, which obeys the rate law given in Eq. (39). Presumably this reaction proceeds by prior

$$\text{rate} = k[L][Mn(CO)_4(CCH_3)OAlBrBr_2] \tag{39}$$

dissociation of the Br from the coordination sphere of Mn [Eq. (40a)] as is usual for substitution reactions of octahedral 18-electron compounds. This would then be followed by the addition of CO [Eq. (40b)].

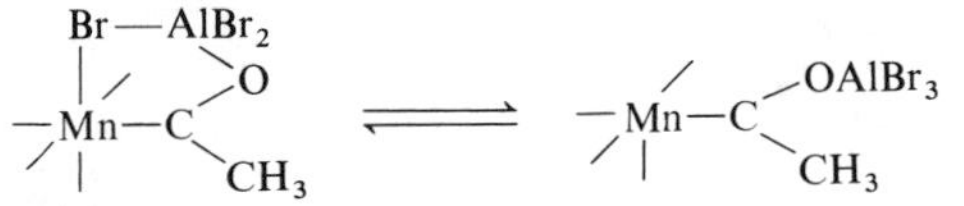

$$\tag{40a}$$

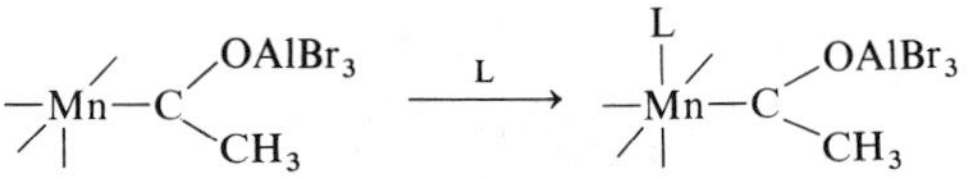

$$\tag{40b}$$

A detailed study of the stereochemistry of the BF_3-promoted carbonylation of $CpFe(CO)(L)(CH_3)$ has demonstrated retention of stereochemistry when the reaction is performed at low temperature [Eq. (41)] (209). Superficially this might be interpreted to indicate that the Lewis acid-promoted reaction

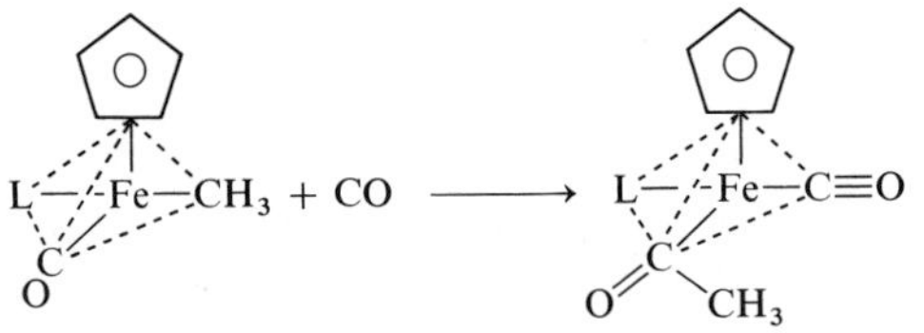

$$\tag{41}$$

occurs by alkyl migration. However the analysis of various mechanisms involving intermediates containing B—F—Fe bridges or an η^2-acetyl ligand indicates that the mode of migration cannot be deduced from the stereochemical data (209).

In addition to the kinetic enhancement, the formation of an acyl adduct stabilizes the insertion product. Thus by means of the introduction of Lewis acids it is possible to carry out insertion reactions which are thermodynamically unfavorable in the absence of the acid. A good example is given in Eq. (42) where the presence of $AlBr_3$ leads to the formation of the stable

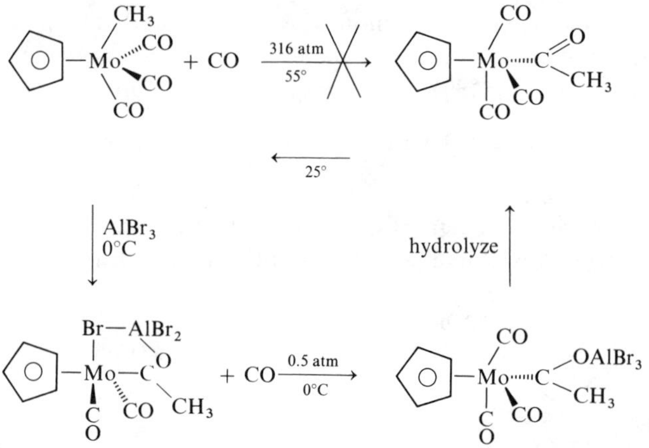

$$(42)$$

isolable acetyl adduct of $AlBr_3$, $CpMo(CO)_3(CCH_3OAlBr_3)$. The adduct undergoes hydrolysis to yield the simple acetyl, $CpMo(CO)_3(CCH_3O)$, which rapidly loses CO and reverts to the starting methyl complex (207), thus illustrating the thermodynamic instability of the simple metal acetyl.

The enhancement of migratory insertion is effected by a variety of Lewis acids, including triphenylaluminum (207), in addition to the aluminum halides and boron trifluoride discussed above. An alternative to these molecular Lewis acids is an alumina surface. The dehydration of γ-Al_2O_3 at elevated temperatures exposes both Lewis acid sites, Al^{3+}, and Lewis base sites, O^{2-} [Eq. (43a)]. When $Mn(CO)_5(CH_3)$ adsorbs on such a surface an

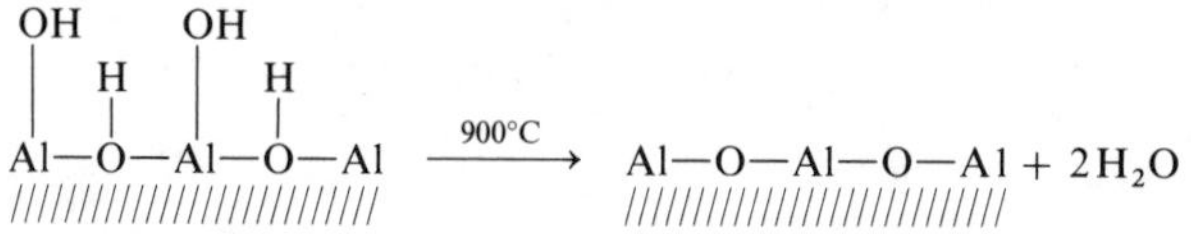

$$(43a)$$

immediate color change is observed. Both IR spectra and chemical evidence indicate that a surface-bound acetyl product is obtained with a cyclic structure analogous to that observed in the case of the molecular Lewis acids [Eq. (43b)] (210). On the basis of these results, it has been suggested that metal oxides, which are often used to support noble metal particles in heterogeneous catalysis, might promote CO insertion in some heterogeneous

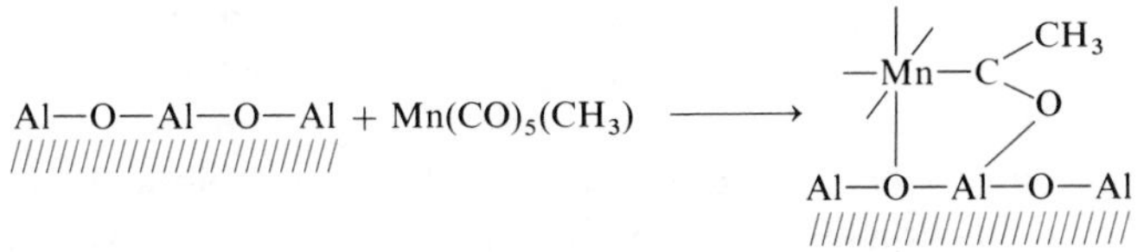

Al—O—Al—O—Al + Mn(CO)5(CH3)
—Mn—C
CH3
O
Al—O—Al—O—Al
(43b)

catalytic reactions of CO (*211*). As yet there appears to be no good test of this postulate.

A number of cyclic polyolefin complexes of metal carbonyls have been shown to react with aluminum halides to produce cyclic ketones [Eqs. (44)–(46)] (*212–214*). This type of reaction affords easy synthetic routes to

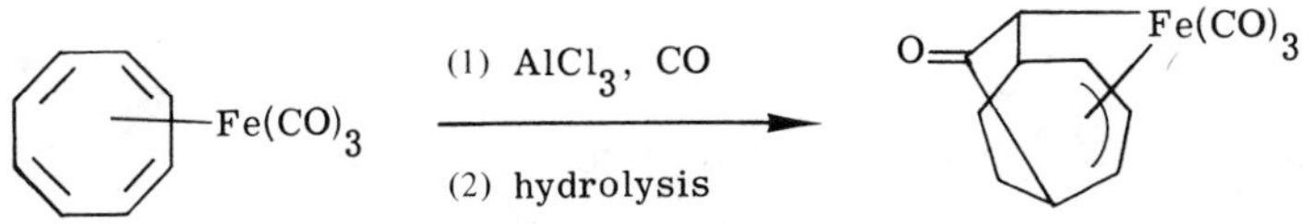

Fe(CO)3
(1) AlCl3, CO
(2) hydrolysis
O=
Fe(CO)3
(44)

Fe(CO)3
(1) AlCl3
(2) hydrolysis
O=
Fe(CO)3

Fe(CO)3
(1) AlCl3, CO
(2) hydrolysis
O
O
Fe
OC C C CO
O O
(46)

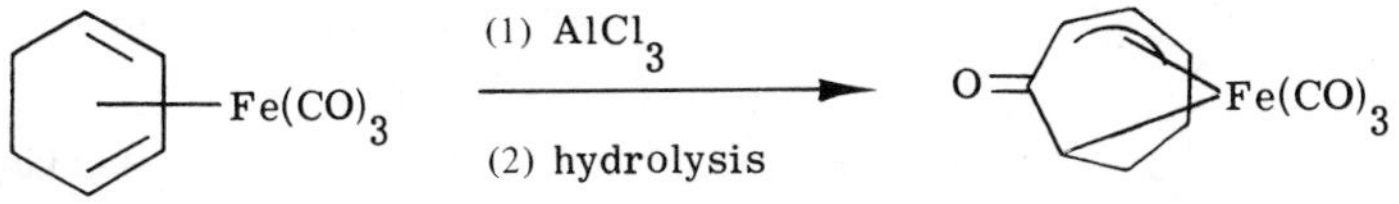

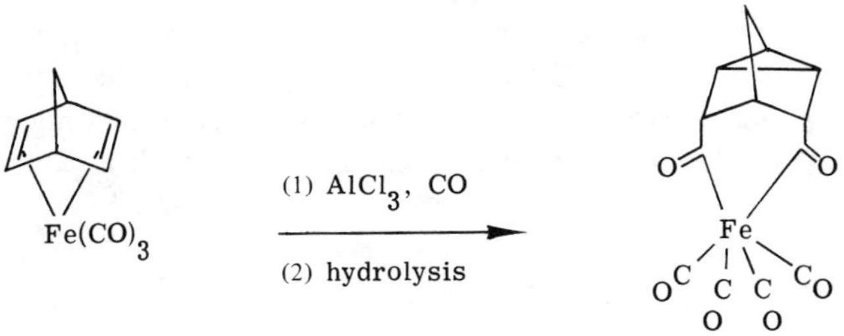

organic molecules that are otherwise difficult to obtain. The cyclic metal acyl product from the reaction of norbornadieneirontricarbonyl with aluminum chloride [Eq. (46)] suggests that at least some of these reactions are proceeding by the route outlined above for $Mn(CO)_5(CH_3)$, that is, insertion of coordinated CO into the metal–carbon bond is facilitated by Lewis acid attack on the oxygen of CO. In contrast to this suggestion, initial adduct formation has been observed between the Lewis acid and metal carbonyl moiety in one instance, and the product is thought to involve an Fe—Al donor–acceptor bond (*214*). If this species is germane to the formation of the cyclic ketone products, mechanisms other than the O-attack may be important.

Preparative-scale CO insertion reactions for transition metal hydrides to produce metal formyls are unknown in transition metal chemistry, but a reaction of this type is frequently invoked to explain the homogeneous hydrogenation of CO. Therefore it is of interest to explore the possibility that Lewis acids might facilitate this reaction and stabilize the formyl product. Studies in the interaction of Lewis acids with metal hydrides reveal no formyl formation. Instead, the observed reactions are (1) no reaction when the hydride is highly acidic, (2) metal–H-acid adduct formation for more basic hydrides, and (3) halide–hydride metathesis in the case of strongly basic hydrides (215, 216). Perfluoroalkyltransition metal carbonyls also do not undergo CO insertion. Again it was found that metathetical reactions occur such as the conversion of metal–CF_3 to metal–CCl_3 when BCl_3 is the acid (217).

4. *Amphoteric Molecule-Assisted Migratory Insertion*

The reaction of a molecule having both acidic and basic functionalities, Ph_2PN-t-$BuAlEt_2$, with $CpFe(CO)_2(CH_3)$ leads to the formation of a cyclic insertion product [Eq. (47)] (218). This product is related to those obtained

$$Et_2Al\text{-}t\text{-}BuNPPh_2 + CpFe(CO)_2(Me) \longrightarrow CpFe(CO)(\overline{C(Me)OAlEt_2N\text{-}t\text{-}BuPPh_2}) \quad (47)$$

from the reaction of aluminum halides with metal carbonyl alkyls, where the phosphine now occupies a coordination site on the iron center. The intermediate $Cp(CO)Fe(C(Me)OAlEt_2N\text{-}t\text{-}BuPPh_2)$ was isolated and crystallographically characterized. This compound (25) has an interesting η^2 orientation of the acetyl CO with respect to the metal center. It is possible that a

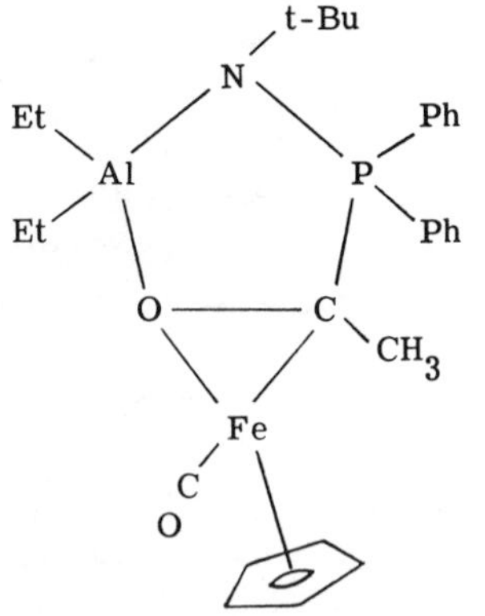

(25)

similar species is an intermediate in the AlX_3-assisted migratory insertion reaction but has so far eluded detection. Species of a related type also are seen in the transition metal-promoted insertion reaction discussed in the next section. The possibility of a CO migratory insertion in a transition metal carbonyl hydride was investigated with the above amphoteric ligand, and in the course of this study evidence for an interesting set of proton transfers from Mn to P to carbonyl C were found [Eq. (48)] (*219*).

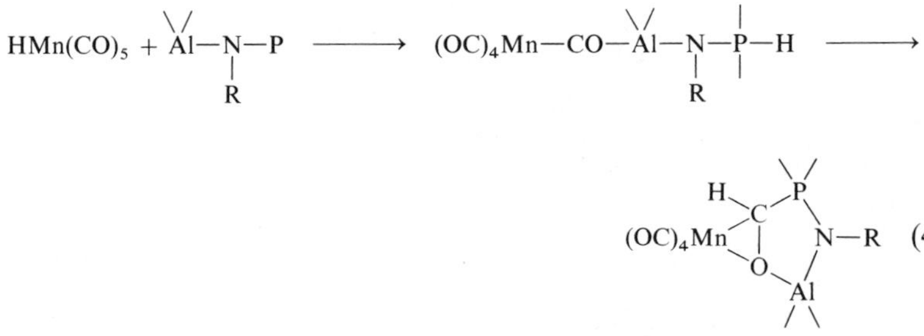

$$HMn(CO)_5 + \overset{\diagdown \diagup}{\underset{R}{Al}} - N - P \longrightarrow (OC)_4Mn - CO - \overset{\diagdown \diagup}{Al} - \underset{R}{N} - \overset{|}{P} - H \longrightarrow$$

$$(OC)_4Mn \underset{O}{\overset{H}{\diagdown}} \overset{\diagdown \diagup}{\underset{Al}{\diagdown}} \quad (48)$$

5. Transition Metal Lewis Acid-Assisted Migratory Insertion

Several cases have been reported in which organometallic Lewis acids promote the migratory insertion reaction, apparently by initial attack on the oxygen of coordinated CO (*220*). An example of this type of reaction is given in Eq. (49). Two types of products can be obtained in these reactions: those

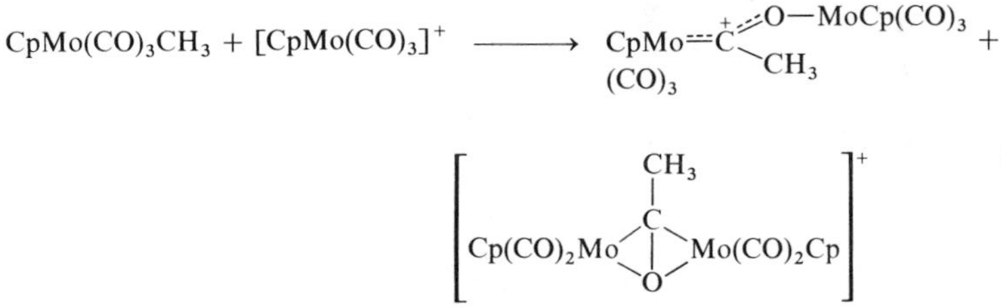

$$CpMo(CO)_3CH_3 + [CpMo(CO)_3]^+ \longrightarrow \underset{(CO)_3}{CpMo} \overset{+}{=} \overset{O-MoCp(CO)_3}{\underset{CH_3}{C}} +$$

$$\left[Cp(CO)_2Mo \underset{O}{\overset{\overset{CH_3}{|}}{\diagdown}} Mo(CO)_2Cp \right]^+ \quad (49)$$

containing a simple C-bonded acetyl and others in which the acetyl is η^2-bonded to both transition metal centers.

As an alternative to the mechanism above, it is possible that the reaction outlined in Eq. (49) proceeds by way of an oxidative-promoted insertion, in which a 17-electron alkyl carbonyl is formed which then rapidly undergoes insertion to produce an acyl complex. This redox-promoted reaction appears to occur in a rather large number of cases involving the action of metal or halogen oxidizing agents on metal alkyl carbonyls [Eq. (50)] (Ref. *221*, and

$$CpFe(CO)_2(CH_3) + Ce^{4+} + L \longrightarrow CpFe(CO)L[C(CH_3)O]^+ \quad (50)$$

references therein). Unless the initial redox process is mediated by the CO ligand, there appears to be no involvement of O-bonded CO in the oxidative-promoted reactions.

C- and O-bonded CO also is implicated in the partial reduction of CO in early transition metal and actinide organometallics (*222–225*). A characteristic of many of these reactions is the formation of metal–oxygen-bonded products such as acetyls and enediolates.

B. *CO Cleavage and Reduction*

The reduction and/or cleavage of the C—O bond in carbon monoxide is of fundamental importance in heterogeneous as well as homogeneous chemistry (*193, 226*). Scission of this bond is a requirement for the formation of high-molecular-weight hydrocarbons by the Fischer–Tropsch chemistry. The current preferred model for Fischer–Tropsch catalysis (*226–231*), which is close to the original proposal by Fischer and Tropsch (*232*), initially involves the cleavage of CO on the metal surface followed by hydrogenation of the surface carbide and oxide atoms, and catenation of surface hydrocarbon fragments. It is presumed that before cleavage the CO molecule is bound to the surface through both C and O (Scheme 6, lower line).

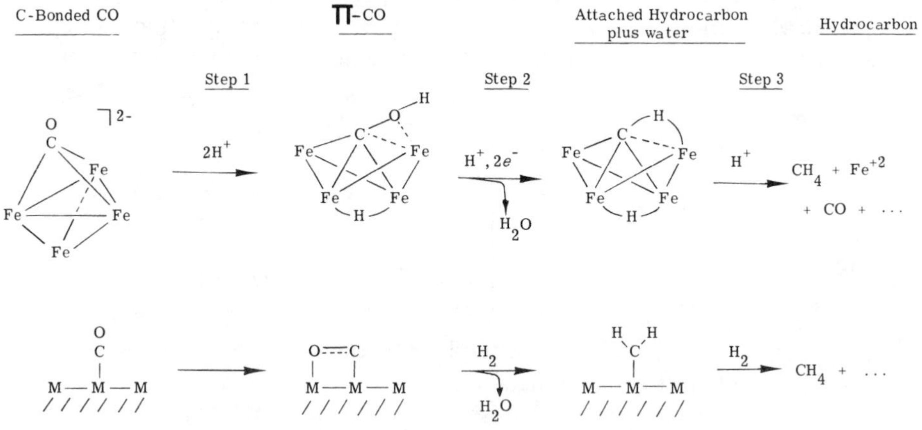

SCHEME 6.   Each Fe vertex possesses three CO ligands.

Evidence for the importance of surface methylene in hydrocarbon chain building is provided by the decomposition of CH_2N_2 on supported Fe, Co, and Ru catalyst surfaces to give the surface methylene species (*228*), which combine to yield a distribution of hydrocarbons consistent with that observed for a typical Fischer–Tropsch synthesis. Although there is a significant body of knowledge on the homogeneous and heterogeneous catenation of hydrocarbon fragments, information on the CO cleavage step is scarce.

As discussed in Sections II and III, the C—O bond is lengthened in both Σ —CO— and Π —CO environments. The lengthening suggests an activation of the C—O bond. The reduction of CO has been effected in molecular systems under conditions that are conducive to Σ —CO— or Π —CO formation (91, 141, 145, 190, 233, 234, 247).

The use of strong acid media has led to success in cleaving the CO bond and forming hydrocarbons (236–246). These acids include Brønsted acids, e.g., HSO_3CF_3 and HCl, as well as Lewis acids, BX_3 (X = H, F, Cl, or Br), AlH_3, and $AlCl_3$.

A number of parallels were found between the conversion of CO in $[Fe_4(CO)_{13}]^{2-}$ to CH_4 and the proposed mechanism for Fischer–Tropsch chemistry. The steps in the conversion of CO in the four-iron cluster to CH_4 have been elucidated by isotope tracer studies and identification of reactive intermediates, which are shown in simplified form in Scheme 6 (191, 236, 237). The overall reaction of $[Fe_4(CO)_{13}]^{2-}$ with neat HSO_3CF_3 results in degradation of the cluster and formation of approximately 0.6 mol CH_4/mol $[Fe_4(CO)_{13}]^{2-}$. The yield of CH_4 is increased to 1.0 CH_4/mol $[Fe_4(CO)_{13}]^{2-}$ when reducing agents are added to the reaction mixture. This added reducing agent appears to be necessary to effect the reductive cleavage of the protonated Π —CO in step 2 of Scheme 6. In the third step, $HFe_4(CO)_{12}(CH)$ is protonated to form a cation, $[H_2Fe_4(CO)_{12}(CH)]^+$, which slowly breaks down to form CH_4, Fe^{2+}, $Fe(CO)_5$, CO, and H_2 (237).

No details are available on the evolution of the four-iron butterfly cation to methane, but further protonation of the framework and reductive elimination of CH_4 seem likely. The four-metal butterfly framework appears to play a significant role in these reactions, particularly in activating carbon monoxide through Π —CO formation. Significantly, the proton-induced reduction has been observed with other four- and six-metal carbonyl clusters, but the reaction does not appear to occur with clusters with fewer nuclei (248). By analogy with the findings in the iron system, this minimum metal nucleus number requirement suggests that Π —CO may be involved in all of these reactions.

An example of CO hydrogenation in the presence of a strong Lewis acid is provided by the $Ir_4(CO)_{12}$ cluster in molten $AlCl_3 \cdot NaCl$ (243, 244, 246). This system has been investigated by two independent research groups, and the results obtained differ significantly. At this time there does not appear to be a good explanation for the observed differences. For example, the distribution of hydrocarbon products in one study (244) indicated the presence of hydrocarbons up to C_6 in significant quantities, while only CH_4 and C_2H_6

were detected in reasonable quantities in the other investigation (246). The $AlCl_3$ is directly implicated in the reduction of CO. A Σ—CO—$AlCl_3$ interaction is indicated by the presence of an absorption at 1630 cm^{-1} in the IR spectrum of a quenched melt (244), and CH_3Cl is isolated and shown to be an intermediate in the formation of CH_4 and C_2H_6 (246) in the presence of $Ir_4(CO)_{12}$. The related reaction of $Os_3(CO)_{12}$ with CO–H_2 in neat BBr_3 or BCl_3 (245) yields methane and a small amount of C_2H_6. In BBr_3 a substantial amount of the CO was reduced to CH_3Br (16%) as well as C_2H_5Br (3%).

The clusters $H_3M_3(CO)_{10}(COMe)$ (M = Ru or Os) (Section II,C,4) (91, 141, 142) have reduced bond order owing to the alkyl group attached to the carbonyl oxygen, and the COMe ligand is susceptible to reduction by H_2. When a hydrocarbon solution of $H_3Ru_3(CO)_{10}COMe$ is heated under a CO–H_2 atmosphere, dimethyl ether and other oxygen-containing organics are obtained. Similarly, the cluster $Co_3(CO)_9COMe$ undergoes hydrogenation to form a variety of oxygenated products [Eq. (51)] (233). The

$$Co_3(CO)_9COMe \xrightarrow[120°]{H_2-CO} H_3C\overset{\overset{O}{\|}}{C}CH_3 + H_3C-O-CH_2CH_2OH + H_3COCH_2CH_2\overset{\overset{O}{\|}}{C}OH \quad (51)$$

presence of oxygen in these products indicates that extensive CO cleavage does not occur under these conditions.

Cleavage of the C—O bond of an alkylated CO by external reducing agents also has been observed (247). In a two-step process involving initial attack with the hydride nucleophile, the COMe group in $HOs_3(CO)_{10}COMe$ is reduced to a methylidyne ligand with release of MeOH [Eq. (52)] (247).

$$HOs_3(CO)_{10}(COMe) + H^- \longrightarrow HOs_3(CO)_{10}CHOMe \xrightarrow{H^+}$$

$$\phantom{HOs_3(CO)_{10}(COMe) + H^- \longrightarrow} a \phantom{HOs_3(CO)_{10}(COMe) + H^- \longrightarrow} b$$

$$HOs_3(CO)_{10}CH \xrightarrow[2.\ H^+]{1.\ H^-} HOs_3(CO)_{10}CH_3 \quad (52)$$

$$\phantom{HOs_3(CO)_{10}CH} c \phantom{HOs_3(CO)_{10}CH} d$$

Addition of another equivalent of hydride to c followed by protonation leads to the cluster $HOs_3(CO)_{10}CH_3$ (d). Thus the conversion of a to d represents the hydrogenation of CO to CH_3.

Convenient high-yield syntheses of reactive carbide compounds have been devised which are based on the reductive cleavage of O-alkylated or O-acylated carbonyl ligands (145, 235). The acylation of a carbonyl oxygen in $[Fe_3(CO)_{11}]^{2-}$ leads to $[Fe_3(CO)_{10}(COC(O)CH_3)]^-$, the structure of which is analogous to $[Fe_3(CO)_{10}(COMe)]^-$ (Section II,B,3). Further reaction of the acylated cluster in THF with a reductant, sodium benzo-

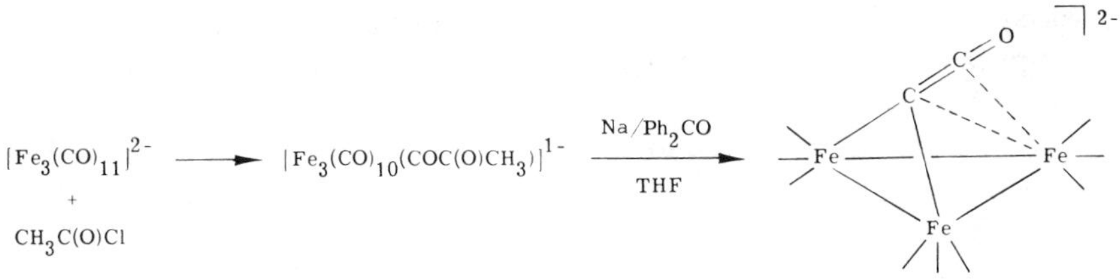

$$(53)$$

phenone ketyl, results in cleavage of the C—O bond and formation of a ketenylidene species $[Fe_3(CO)_9(CCO)]^{2-}$ [Eq. (53)] (235). Labeling of the parent compound with ^{13}CO conclusively shows that the carbon center of the ketenylidene functionality is derived from the C—OC(O)CH_3 group. Similar reactions with $[Fe_4(CO)_{13}]^{2-}$ result in formation of $[Fe_4(CO)_{12}C]^{2-}$ [Eq. (54)]. A variety of reducing agents has been investigated for this reaction, including some of the low-nuclearity iron carbonyl clusters (145). The capability of weak reducing agents to cleave the C—OR bond suggests that the alkyl substituent weakens the C—O bond substantially and serves as a good leaving group.

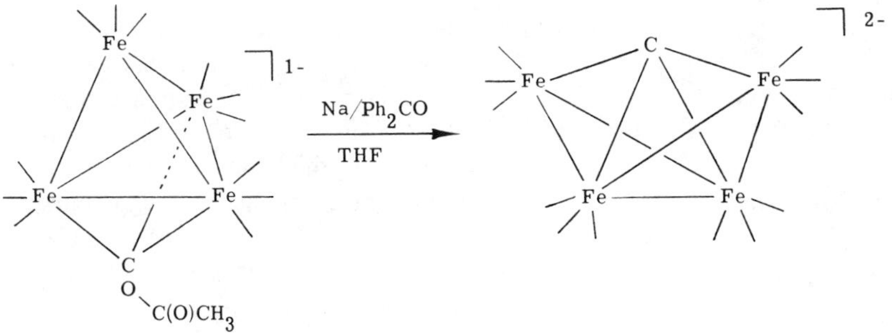

$$(54)$$

The Lewis acid-assisted migratory insertion reaction was discussed in Section IV,A. It was shown that the rate of methyl migration increased by a factor of at least 10^8 in the presence of a Lewis acid. The migratory insertion is a chain-lengthening process, and if the M—C(O)CH_3 group could be further reduced to a sigma-bound alkyl, M—CH_2CH_3, the alkyl chain will have been lengthened by one unit [Eq. (55)]. The extension to higher molecular weight hydrocarbons is obvious. Although CO insertion currently

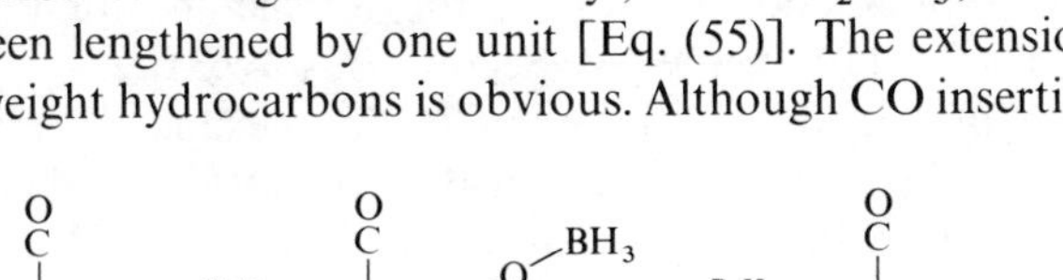

$$(55)$$

is rejected as the main route to hydrocarbons in Fischer–Tropsch catalysis, it is implicated as the chain terminating step which leads to oxygen-containing products (*249*). An apparent instance of hydrocarbon production by successive CO insertions and reductions is provided by the $CH_3Mn(CO)_5/B_2H_6$ system (*242*) [Eq. (56)]. Under the conditions of this experiment, Lewis

$$Mn(CO)_5(CH_3) + B_2H_6 \xrightarrow[\text{toluene}]{CO} \xrightarrow[C_2H_5OH]{HCl} \text{hydrocarbons} \qquad (56)$$

acid-promoted CO insertion is likely to occur. The hydrocarbons detectable in significant quantities by gas chromatography (GC) or GC–mass spectrometry range from C_2 to C_4. Isotopic labeling studies with ^{13}CO and B_2D_6 confirm that the hydrocarbon products are derived from reduced carbonyl ligands, and deuterium incorporation in the products establishes the borane as the reductant. The isotopic distribution of the products is consistent with the majority of hydrocarbons arising from the insertion pathway.

C. *CO-Mediated Electron Transfer*

Although examples of inner sphere redox processes are common in the chemistry of Werner complexes (*250*), this mechanism is invoked only rarely in organometallic chemistry (*251*). The primary requirement for an inner sphere mechanism is the formation of a bridge between the oxidizing and reducing centers by means of a bifunctional ligand, which acts as an electron mediator between the two sites. Migration of the bridging ligand often occurs following the redox reaction (*252*) but this is not always the case (*253*). Numerous examples of C and O bonding are known, and it is logical to ask whether CO might function as an effective mediator for electron transfer reactions.

A recent report on the disproportionation reaction of $V(CO)_6$ in weak bases (*53, 54*) provides the best available evidence for Σ —CO— as an electron mediator. Most oxygen and nitrogen donor ligands promote the disproportionation of $V(CO)_6$ to form $[V(B)_6][V(CO)_6]_2$ (*254–257*). However, in some instances complexes such as $V(B)_4[(OC)V(CO)_5]_2$, in which a V—O—C—V interaction is present, are isolable and for B = THF the structure has been determined 9 [see Section II,A,5 (*63*)].

The reaction of $V(CO)_6$ with base in CH_2Cl_2–hexane solvent was monitored by IR spectroscopy. In the case of B = py (pyridine), THF, or Et_2O, reaction according to Eq. (57) was observed. For B = THF and Et_2O no

$$3V(CO)_6 + 4B \longrightarrow V(B)_4[(OC)V(CO)_5]_2 + 6CO \qquad (57)$$

further changes occur, but for $B = py$ an additional reaction is observed [Eq. (58)]. The spectral changes indicate that in the case of $B = THF$ the reaction proceeds selectively to a single product with a Σ —CO— bridge.

$$V(py)_4[(OC)V(CO)_5]_2 + 2py \longrightarrow [V(py)_6][V(CO)_6]_2 \qquad (58)$$

For $B = py$ the species with a Σ —CO— bridge appears to be a transient kinetic product, which reacts with additional py according to Eq. (58). These observations suggest that the electron transfer process is inner sphere in nature and proceeds through the bridging carbonyl ligand.

The rate of the disproportionation reaction to form $V(B)_4[(OC)V(CO)_5]_2$ is dependent on both the concentration of $V(CO)_6$ as well as base, B. Furthermore, the activation parameters for $B = THF$, $\Delta H^{\ddagger} = 14.8$ kcal mol^{-1} and $\Delta S^{\ddagger} = -19.6$ cal mol^{-1} deg^{-1}, are appropriate for an associative pathway (258). The rate of reaction is also dependent on the basicity as well as steric bulk of the base, with the rate spanning a range of 10^4 between the fastest (py) and slowest (Et$_2$O) reactions.

A mechanism consistent with both the spectral data and the kinetic data is presented in Scheme 7. Step 2, formation of the CO bridge, provides the

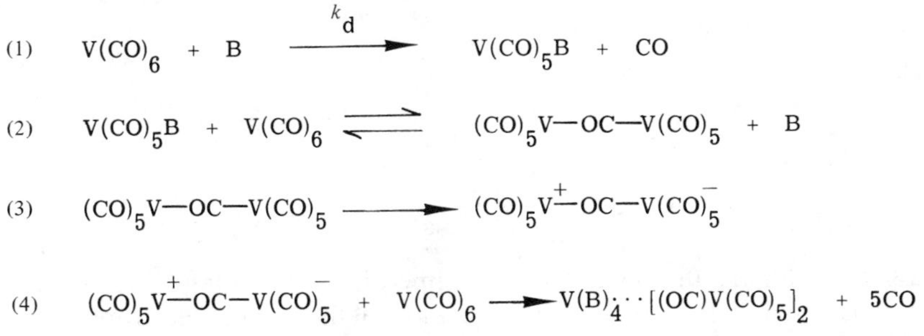

(1) $\quad V(CO)_6 + B \xrightarrow{\quad k_d \quad} V(CO)_5B + CO$

(2) $\quad V(CO)_5B + V(CO)_6 \underset{\longleftarrow}{\overset{\longrightarrow}{\rule{1cm}{0pt}}} (CO)_5V\text{—}OC\text{—}V(CO)_5 + B$

(3) $\quad (CO)_5V\text{—}OC\text{—}V(CO)_5 \longrightarrow (CO)_5\overset{+}{V}\text{—}OC\text{—}V(CO)_5^{-}$

(4) $\quad (CO)_5\overset{+}{V}\text{—}OC\text{—}V(CO)_5^{-} + V(CO)_6 \longrightarrow V(B)_4^{\cdots}[(OC)V(CO)_5]_2 + 5CO$

Scheme 7

configuration required for the inner sphere electron transfer. Step 4 combines the elements of the second electron transfer step with the loss of CO and coordination of the base molecules.

The reaction of $Cp_2Fe_2(CO)_{4-n}(CNMe)_n$ ($n = 0$–2) with electrophiles, such as SnX_2 (259, 260), SnX_4 (259, 261), BX_3 (261), and other strong Lewis acids, often leads to oxidative cleavage of the iron–iron bond. In some instances the electrophile is incorporated wholly or partially in the final products. It has been proposed that the bridging carbonyls (or isonitriles) act as electron mediators in the oxidation of the organoiron system and reduction of the acceptor molecule. A review on this chemistry has appeared recently (262); therefore, the discussion will be confined to issues involving electron transfer through CO.

Scheme 8 shows the coordination and reduction of the electrophile. Reactions which follow coordination at the oxygen appear to be rapid,

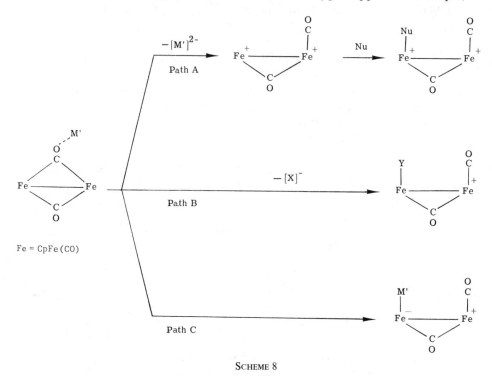

SCHEME 8

especially for the unsubstituted iron dimer, because no intermediates have been isolated. Path A involves complete loss of the reduced electrophile from the intermediate, and the open coordination site on the metal is occupied by a nucleophile (solvent or coordinating anion). In path B some component of the electrophile is incorporated into the intermediate, while in Path C all of the reduced acceptor becomes incorporated. Stability of the reduced acceptor will determine to a large extent which of these paths will be followed. The CO moiety in these proposed reactions appears unaltered in the final products.

Another example that may involve the CO-mediated oxidation of a metal carbonyl is the reaction of $trans$-ReCl(CO)(PMe$_2$Ph)$_4$ with TiCl$_4$, which was described in Section II,A,5 (57). A 2:1 ratio of rhenium complex to TiCl$_4$ results in the formation of [$trans$-ReCl(PMe$_2$Ph)$_4$(CO)]$_2$TiCl$_4$. A Ti—O—C interaction accounts for the appearance of a low-frequency CO stretch at 1610 cm^{-1}, but in the presence of an excess of TiCl$_4$ two additional CO bands appear at 1870 and 1920 cm^{-1} that are attributed to the oxidized species $ReCl_3(CO)(PR_3)_3$ and its $TiCl_4$ chloride-bridged adduct. It is likely that the electron transfer process is mediated by the Σ —CO—.

In contrast to the metal carbonyl oxidations described above is the reduction of $Cp_2Fe_2(CO)_4$ with $[Cp*_2ZrN_2]_2N_2$ (263). The product of this reaction is a species in which a C—C bond is formed between two carbonyl groups [Eq. (59)]. The C—C distance, 1.57 Å, is consistent with formation

$$[Cp*_2ZrN_2]_2N_2 + 2Cp_2Fe_2(CO)_4 \longrightarrow 3N_2 + 2Cp*_2Zr(CO)_4Fe_2Cp_2 \quad (59)$$

of a single bond (264) and suggests that Zr has been oxidized from +2 to +4. The reaction is reversed under an atmosphere of CO to regenerate $Cp_2Fe_2(CO)_4$ and $Cp*_2Zr(CO)_2$. In the proposed mechanism the zirconium center interacts simultaneously with two terminal CO ligands (Scheme 9).

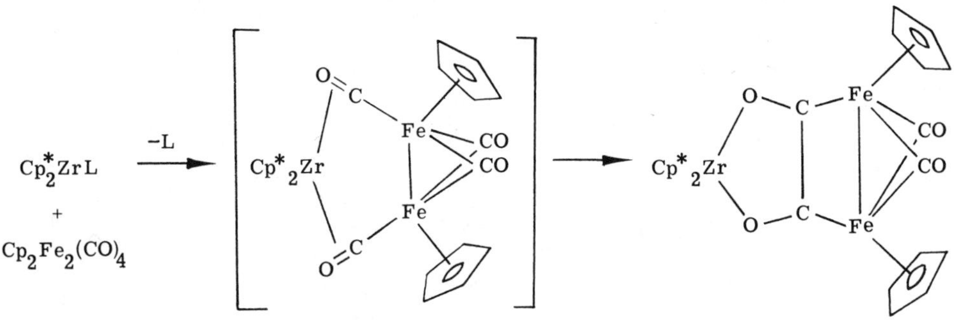

SCHEME 9

This "side-on" bonding is similar to the Π —CO bonding described in Section III. Assuming that this model is correct, the π system of the CO appears to be an effective mediator for electron transfer.

The reductive cleavage of M—M bonds can be achieved by a variety of reducing agents such as Na–Hg, Na benzophenone ketyl, and in some instances anionic metal carbonyls. These reagents would be expected to react by an outer sphere electron transfer mechanism. This does not preclude the possibility that the electrons are being transferred through the carbonyls to the metal centers.

In electrochemical redox reactions it is thought that the electron is transferred through the carbonyl ligand. Cyclic voltammetry experiments, in which the separation of the anodic and cathodic waves can be viewed as a measure of the ease of electron transfer, have been used to indirectly show that the carbonyl ligand can mediate electron transfer (265). For example, the electron transfer rate is inhibited by bulky ligands for $R'CCo_3(CO)_{9-n}L_n$ ($n = 1$–3 and L = PR_3) (266, 267). The interpretation is that steric bulk of the phosphines prevents the cluster carbonyls from approaching the platinum electrode with the proper orientation. In agreement with this interpretation, changing the apical substituent from H to Me_3Si has a minimal effect on the charge transfer kinetics, indicating that charge transfer does not occur through this group (268).

Recently a series of compounds has been prepared in which an organo-lanthanide, $Cp*_2Yb(OEt_2)$, is the reducing agent (55, 56, 110). Structures of the products were described in Sections II,A,4 and II,B,4. The compound $Cp*_2Yb(OEt_2)$ is a strong reducing agent, capable of being oxidized to a Yb(III) species (55). Accordingly, the reaction of $Cp*_2Yb(OEt_2)$ with $Co_2(CO)_8$ or $Fe_3(CO)_{12}$ leads to the formation of the compounds containing reduced Co or Fe moieties [see Figs. 12 and 15, Eqs. (5) and (16)]. A proposed mechanism for the reduction of $Co_2(CO)_8$ is given in Scheme 10. In a two-step mechanism, an electron transfer following initial adduct formation

$$(1) \quad (CO)_6Co_2(\mu\text{-}CO)_2 \;+\; Cp_2^*(Et_2O)Yb(II) \quad \xrightarrow{-Et_2O} \quad \left[(CO)_6Co_2^{\bar{}}(\mu\text{-}CO)(\mu\text{-}COYb^{III}Cp_2^*)\right]$$

$$(2) \quad \left[(CO)_6Co_2^{\bar{}}(\mu\text{-}CO)(\mu\text{-}COYb^{III}Cp_2^*)\right] \quad \xrightarrow[\text{THF}]{Cp_2^*(Et_2O)Yb(II)} \quad 2\;(CO)_3Co(CO)Yb(III)\dots$$

$$Yb(III) = Cp_2^*(THF)Yb$$

SCHEME 10

would make the second bridging carbonyl oxygen more susceptible to electrophilic attack. Superficially, the reduction of $Fe_3(CO)_{12}$ (109) may proceed in the same manner.

The reduction of the triple bond of $Cp_2Mo_2(CO)_4$ with $Cp_2Ti(CO)_2$ (59) appears to be similar to those given above with $Cp*_2Yb(OEt_2)$. In this instance the metal dimer is cleaved and a Ti(III) species is formed in which a Ti—O—C—Mo linkage occurs (Section II,A,5). Formulation of the product with a Ti(III) center is confirmed by magnetic susceptibility measurements. A postulated reaction pathway is shown in Eqs. (60a) and (60b). It is probable

$$Cp_2Ti(CO)_2 + Cp_2Mo_2(CO)_4 \longrightarrow \text{``}Cp_2Ti\text{''} + Cp_2Mo_2(CO)_6 \qquad (60a)$$

$$2\,\text{``}Cp_2Ti\text{''} + Cp_2Mo_2(CO)_6 \xrightarrow{THF} 2\,Cp_2Ti(thf)(CO)_3MoCp \qquad (60b)$$

that the "Cp_2—Ti" intermediate, a potentially strong Lewis acid, is attached to a carbonyl oxygen prior to reduction since aluminum alkyls are known to form adducts with the saturated molybdenum dimer (93).

This section has had to be speculative because there is little solid information on the role of CO as an electron mediator. However there are many indications that the CO ligand may be an effective mediator of redox reactions, and hopefully this aspect of CO chemistry will receive closer scrutiny.

V

CONCLUDING REMARKS

The variety of bonding arrangements in which CO is found ranges from the conventional C-bonded terminal and bridging groups, where CO is a formal two-electron donor, through the Π—CO interactions, where CO can be a two, four, or six-electron donor, and finally to the four-electron donor Σ—CO— ligand. This series of bonding patterns far exceeds that of any of the other common diatomic 10-valence electron ligands: C_2^{2-}, CN^-, N_2, and NO^+. Synthetic routes to Σ—CO— compounds are simple and well known, but the same is not true for the generation of Π—CO ligands, in which syntheses are particularly scarce for high-nuclearity clusters. Because of the importance of this type of bonding pattern in CO activation, systematic routes to Π—CO are an important goal.

Perhaps the greatest research opportunities in the area of C- and O-bonded carbon monoxide lie in the area of enhanced CO reactivity. Results which are presently available for Lewis acid-promoted alkyl migration onto CO and the reductive cleavage of CO indicate the huge influence which C and O bonding may have on reactivity. One area which is ripe for further development is the potential role of CO as a mediator for redox reactions, and another is the possibility that the carbonyl oxygen may be the site of initial attack by electrophiles on metal carbonyls. In view of the large effect of C and O bonding on the solution chemistry of metal carbonyls, an understanding of the role of this mode of bonding in the heterogeneous catalytic chemistry of CO is likely to pay large dividends.

ACKNOWLEDGMENTS

We thank the National Science Foundation for the support of our research on the solution chemistry of C- and O-bonded carbon monoxide, and the United States Department of Energy is likewise acknowledged for its support of related heterogeneous chemistry. The following individuals have kindly provided preprints or descriptions of unpublished research: R. A. Andersen, F. Basolo, J. M. Basset, R. Bau, J. M. Burlitch, C. H. Cheng, R. E. Cramer, J. P. Collman, M. D. Curtis, A. R. Cutler, M. Y. Darensbourg, W. F. Edgell, M. B. Hall, M. Ichikawa, H. D. Kaesz, A. R. Manning, and G. Schmid. Finally we appreciate the help of the following individuals in providing illustrations: K. G. Caulton, R. Colton, G. Fachinetti, W. A. Herrmann, M. McPartlin, J. D. Smith, and E. Weiss.

REFERENCES

1. W. F. Edgell, M. T. Yang, and N. Koizumi, *J. Am. Chem. Soc.* **87**, 2563 (1965).
2. N. J. Nelson, N. Kime, and D. F. Shriver, *J. Am. Chem. Soc.* **91**, 5173 (1969).
3. J. C. Kotz and C. D. Turnipseed, *J. Chem. Soc. D*, p. 41 (1970).

4. L. J. Todd and J. R. Wilkinson, *J. Organomet. Chem.* **77**, 1 (1974).
5. H. A. Hodali and D. F. Shriver, *Inorg. Chem.* **18**, 1236 (1979).
6. C. J. Commons and B. Hoskins, *Aust. J. Chem.* **28**, 1663 (1975); R. Colton and C. J. Commons, *ibid.* **28**, 1673 (1975).
7. R. J. Klinger, W. H. Butler, and M. D. Curtis, *J. Am. Chem. Soc.*, **100**, 5034 (1978).
8. B. V. Lokshin, E. B. Rusach, Z. P. Valueva, A. G. Ginzburg, and N. E. Kolobova, *J. Organomet. Chem.* **102**, 535 (1975).
9. S. Onaka and N. Furuichi, *J. Organomet. Chem.* **173**, 77 (1979).
10. F. Calderazzo, G. Fachinetti, F. Marchetti, and P. F. Zanazzi, *J. Chem. Soc. Chem. Commun.*, p. 181 (1981).
11. C. K. Chen and C. H. Cheng, *Inorg. Chem.* **22**, 3378 (1983).
12. W. F. Edgell, *in* "Ions and Ion Pairs in Organic Reactions," (M. Szwarc Ed.), Ch. 4, Vol. I. Wiley, New York, 1972.
13. M. Y. Darensbourg, *Prog. Inorg. Chem.* (1984), in press.
14. W. F. Edgell and T. Balch, *J. Am. Chem. Soc.* (1984), in press.
15. W. Edgell and S. Chanjamsri, *J. Am. Chem. Soc.* **102**, 147 (1980).
16. W. F. Edgell, S. Hegde, and A. Barbetta, *J. Am. Chem. Soc.* **100**, 1406 (1978).
17. W. C. Kaska, *J. Am. Chem. Soc.* **90**, 6340 (1968).
18. W. F. Edgell and N. Pauuwe, *J. Chem. Soc. D*, p. 284 (1969).
19. G. B. McVicker, *Inorg. Chem.* **14**, 2087 (1975).
20. K. H. Pannell and D. Jackson, *J. Am. Chem. Soc.* **98**, 4443 (1976).
21. M. Nitay and M. Rosenblum, *J. Organomet. Chem.* **136**, C23 (1977).
22a. H. B. Chin and R. Bau, *J. Am. Chem. Soc.* **98**, 2434 (1976).
22b. C. G. Pierpont, B. A. Sosinsky, and R. G. Strong, *Inorg. Chem.* **21**, 3247 (1982).
23. R. G. Teller, R. G. Finke, J. P. Collman, H. B. Chin, and R. Bau, *J. Am. Chem. Soc.* **99**, 1104 (1977).
24. J. L. Peterson, R. K. Brown, and J. M. Williams, *Inorg. Chem.* **20**, 158 (1981).
25. M. K. Cooper, P. A. Duckworth, K. Hendrick, and M. McPartlin, *J. Chem. Soc.*, *Dalton Trans.*, p. 2357 (1981).
26. G. Fachinetti, C. Floriani, P. F. Zanazzi, and A. R. Zanzari, *Inorg. Chem.* **17**, 3002 (1978).
27. S. W. Ulmer, P. M. Skarstad, J. M. Burlitch, and R. E. Hughes, *J. Am. Chem. Soc.* **95**, 4469 (1973).
28. G. B. McVicker and R. S. Matyas, *J. Chem. Soc.*, *Chem. Commun.*, p. 972 (1972).
29. J. M. Burlitch, S. W. Ulmer, J. J. Stezowski, R. C. Winterton, and R. E. Hughes, *Abstr. Sixth Int. Conf. Organomet. Chem.*, *Amherst, Mass. Aug. 1973*, p. 140.
30. D. F. Shriver, *Acc. Chem. Res.* **3**, 231 (1970).
31. J. M. Burlitch and R. B. Petersen, *J. Organomet. Chem.* **24**, C65 (1970).
32. R. B. Petersen, J. J. Stezowski, C. Wan, J. M. Burlitch, and R. E. Hughes, *J. Am. Chem. Soc.* **93**, 3532 (1971).
33. J. Chatt, R. H. Crabtree, and R. L. Richards, *J. Chem. Soc.*, *Chem. Commun.*, p. 534 (1972).
34. R. B. Petersen, Ph.D. Thesis, Cornell University (1973).
35. J. Chatt, R. H. Crabtree, E. A. Jeffrey, and R. L. Richards, *J. Chem. Soc.*, *Dalton Trans.*, p. 1167 (1973).
36. M. Aresta, *Gazz. Chim. Ital.* **102**, 781 (1972).
37. Y. Ben Taarit, J. L. Bilhou, M. Lecomte, and J. M. Basset, *J. Chem. Soc.*, *Chem. Commun.*, p. 38 (1978).
38. J. M. Burlitch, M. E. Leonowicz, R. B. Petersen, and R. E. Hughes, *Inorg. Chem.* **18**, 1097 (1979).

39. G. J. Gainsford, R. R. Schrieke, and J. D. Smith, *J. Chem. Soc., Chem. Commun.*, p. 650 (1972).
40. A. J. Conway, G. J. Gainsford, R. R. Schrieke, and J. D. Smith, *J. Chem. Soc., Dalton Trans.*, p. 2499 (1975).
41. D. F. Shriver and A. Alich, *Inorg. Chem.* **11**, 2984 (1972).
42. D. F. Shriver and A. Alich, *Coord. Chem. Rev.* **8**, 15 (1972).
43. J. S. Kristoff and D. F. Shriver, *Inorg. Chem.* **12**, 1788 (1973).
44. N. Calderon, J. P. Lawrence, and E. A. Ofstead, *Adv. Organomet. Chem.* **17**, 449 (1979).
45. R. H. Grubbs, *Prog. Inorg. Chem.* **24**, 1 (1978).
46. J. C. Mol and J. A. Moulijn, *Adv. Catal.* **24**, 131 (1975).
47. N. Calderon, *Acc. Chem. Res.* **5**, 127 (1972).
48. J. L. Bilhou, A. Theolier, A. K. Smith, and J. M. Basset, *J. Mol. Catal.* **3**, 245 (1977/78).
49. J. M. Basset, Y. Ben Taarit, J. L. Bilhou, R. Mutin, A. Theolier, and J. Bousquet, *Proc. Int. Congr. Catal. 6th*, 1976, p. 47 (1977).
50. J. L. Bilhou and J. M. Basset, *J. Organomet. Chem.* **132**, 395 (1977).
51. V. W. Motz and M. F. Farona, *Inorg. Chem.* **16**, 2545 (1977).
52. T. J. Katz, S. J. Lee, M. Nair, and E. B. Savage, *J. Am. Chem. Soc.* **102**, 7940 (1980).
53. A. E. Crease and P. Legzdins, *J. Chem. Soc., Chem. Commun.*, p. 268 (1972).
54. A. E. Crease and P. Legzdins, *J. Chem. Soc., Dalton Trans.*, p. 1501 (1973).
55. T. D. Tilley and R. A. Andersen, *J. Chem. Soc., Chem. Commun.*, p. 985 (1981).
56. J. M. Boncella and R. A. Andersen, *Inorg. Chem.* **23**, 432 (1984).
57. R. Robson, *Inorg. Chem.* **13**, 475 (1974).
58. D. M. Hamilton, Jr., W. S. Willis, and G. D. Stucky, *J. Am. Chem. Soc.* **103**, 4255 (1981).
59. J. S. Merola, R. A. Gentile, G. B. Ansell, M. A. Modrick, and S. Zentz, *Organometallics* **1**, 1731 (1982).
60. J. A. Marsella, J. C. Huffman, K. G. Caulton, B. Longato, and J. R. Norton, *J. Am. Chem. Soc.* **104**, 6360 (1982).
61. B. Longato, J. R. Norton, J. C. Huffman, J. A. Marsella, and K. G. Caulton, *J. Am. Chem. Soc.* **103**, 209 (1981).
62. J. Abys and W. M. Risen, Jr., *J. Organomet. Chem.* **204**, C5 (1981).
63. M. Schneider and E. Weiss, *J. Organomet. Chem.* **121**, 365 (1976).
64. T. G. Richmond, Q.-Z. Shi, W. C. Trogler, and F. Basolo, *J. Chem. Soc., Chem. Commun.*, p. 650 (1983).
65. T. G. Richmond, Q.-Z. Shi, W. C. Trogler, and F. Basolo, *J. Am. Chem. Soc.* (1984), in press.
66. T. Blackmore and J. M. Burlitch, *J. Chem. Soc., Chem. Commun.* p. 405 (1973).
67. E. L. Brown and D. B. Brown, *J. Chem. Soc. D*, p. 67 (1971).
68. J. E. Bulkowski, P. L. Burk, M. Widmann, and J. A. Osborn, *J. Chem. Soc., Chem. Commun.*, p. 498 (1977).
69a. J. Powell and C. J. May, *J. Am. Chem. Soc.* **104**, 2636 (1982).
69b. G. Doyle, K. A. Eriksen, and D. vanErgen, *Inorg. Chem.* **22**, 2892 (1983), and references therein.
70. J. St. Denis, W. Butler, M. D. Glick, and J. P. Oliver, *J. Am. Chem. Soc.* **96**, 5427 (1974).
71. J. E. Bercaw, R. H. Maravich, L. G. Bell, and H. H. Brintzinger, *J. Am. Chem. Soc.* **94**, 1219 (1972).
72. J. E. Bercaw, *J. Am. Chem. Soc.* **96**, 5087 (1974).
73. H. A. Hodali, D. F. Shriver, and C. A. Ammlung, *J. Am. Chem. Soc.* **100**, 5239 (1978).
74. J. B. Keister, *J. Organomet. Chem.* **190**, C36 (1980).
75. J. R. Wilkinson and L. J. Todd, *J. Organomet. Chem.* **118**, 199 (1976).
76. P. McArdle and A. R. Manning, *J. Chem. Soc. A*, p. 2133 (1970).

77. W. McFarlane and G. Wilkinson, *Inorg. Synth.* **8**, 181 (1966).
78. C. B. Dammann, P. Singh, and D. J. Hodgson, *J. Chem. Soc., Chem. Commun.*, p. 586 (1972).
79. L. Pauling, "Nature of the Chemical Bond." Cornell Univ. Press, Ithaca, New York, 1960.
80. L. F. Dahl, personal communication.
81. K. P. Schick, N. L. Jones, P. Sekula, N. M. Boag, J. A. Labinger, and H. D. Kaesz, *Inorg. Chem.* (1984), in press.
82. K. E. Inkrott and S. G. Shore, *J. Am. Chem. Soc.* **100**, 3954 (1978).
83. K. E. Inkrott and S. G. Shore, *Inorg. Chem.* **18**, 2817 (1979).
84. C. C. Nagel and S. G. Shore, *J. Chem. Soc., Chem. Commun.*, p. 530 (1980).
85. L. F. Dahl and J. F. Blount, *Inorg. Chem.* **4**, 1373 (1965).
86. J. L. Vidal, R. C. Schoening, and J. M. Troup, *Inorg. Chem.* **20**, 227 (1981).
87. H. N. Adams, G. Fachinetti, and J. Strahle, *Angew. Chem. Int. Ed. Engl.* **19**, 404 (1980).
88. D. F. Shriver, D. Lehman, and D. Strope, *J. Am. Chem. Soc.* **97**, 1594 (1975).
89. B. F. G. Johnson, J. Lewis, A. G. Orpen, P. R. Raithby, and G. Suess, *J. Organomet. Chem.* **173**, 187 (1979).
90. P. D. Gavens and M. J. Mays, *J. Organomet. Chem.* **162**, 389 (1978).
91. J. B. Keister, *J. Chem. Soc., Chem. Comm.*, p. 214 (1979).
92. M. Green, K. A. Mead, R. M. Mills, I. D. Salter, F. G. A. Stone, and P. Woodward, *J. Chem. Soc., Chem. Commun.*, p. 51 (1982).
93. A. Alich, N. J. Nelson, D. Strope, and D. F. Shriver, *Inorg. Chem.* **11**, 2976 (1972).
94. A. Alich, N. J. Nelson, and D. F. Shriver, *J. Chem. Soc. D*, p. 254 (1971).
95. J. S. Kristoff and D. F. Shriver, *Inorg. Chem.* **13**, 499 (1974).
96. G. Schmid and V. Batzel, *J. Organomet. Chem.* **81**, 321 (1974).
97. D. F. Shriver, S. Onaka, and D. Strope, *J. Organomet. Chem.* **117**, 277 (1976).
98a. S. Onaka, *Inorg. Chem.* **19**, 2132 (1980).
98b. T. J. Marks, J. S. Kristoff, A. Alich, and D. F. Shriver, *J. Organomet. Chem.* **33**, C35 (1971).
99. P. Chini and R. Ercoli, *Gazz. Chim. Ital.* **88**, 1170 (1958).
100. O. A. Gansow, A. R. Burke, and W. D. Vernon, *J. Am. Chem. Soc.* **94**, 2550 (1972).
101. R. D. Adams and F. A. Cotton, *J. Am. Chem. Soc.* **95**, 6589 (1973).
102. F. A. Cotton and G. Yagupsky, *Inorg. Chem.* **6**, 15 (1967).
103. P. McArdle and A. R. Manning, *J. Chem. Soc. A.*, p. 2128 (1970).
104. E. R. Corey and L. F. Dahl, *J. Am. Chem. Soc.* **83**, 2203 (1961).
105. R. Mason and A. I. McRae, *J. Chem. Soc. A.*, p. 778 (1968).
106. D. F. Shriver, *J. Organomet. Chem.* **94**, 259 (1975).
107. F. A. Cotton and J. M. Troup, *J. Chem. Soc., Dalton Trans.*, p. 800 (1974).
108. J. R. Moss and W. A. G. Graham, *J. Chem. Soc. D.*, p. 835 (1970).
109. T. D. Tilley and R. A. Andersen, *J. Am. Chem. Soc.* **104**, 1772 (1982).
110. J. W. Lauher, *J. Am. Chem. Soc.* **100**, 5305 (1978).
111. C. Tessier-Youngs, F. Correa, D. Pioch, R. L. Burwell, Jr., and D. F. Shriver, *Organometallics* **2**, 898 (1983).
112. F. Hughes, J. M. Basset, Y. Ben Taarit, A. Chaplin, M. Primet, D. Rojas, and A. K. Smith, *J. Am. Chem. Soc.* **104**, 7020 (1982).
113. M. Ichikawa, *Bull. Chem. Soc. Jpn.* **51**, 2268 (1978).
114. M. Iwamoto, H. Kusano, and S. Kagawa, *Inorg. Chem.* **22**, 3365 (1983).
115. T. Kruck, M. Hofler, and M. Noack, *Chem. Ber.* **99**, 1153 (1966).
116. L. Malatesta, G. Caglio, and M. Angoletta, *J. Chem. Soc.*, p. 6974 (1965).
117. R. L. Schneider, R. F. Howe, and K. L. Watters, *Inorg. Chem.* (1984), in press.

118. D. Ballivet-Tkatchenko and G. Courdurier, *Inorg. Chem.* **18**, 558 (1979).
119. M. Ichikawa, T. Fukushima, and K. Shikakura, *Proc. Int. Cong. Catal. 8th, 1984* (1985), in press.
120. G. Fachinetti, *J. Chem. Soc., Chem. Commun.*, p. 397 (1979).
121. G. Fachinetti, S. Pucci, P. F. Zanazzi, and U. Methong, *Angew. Chem. Int. Ed. Engl.* **18**, 619 (1979).
122. H. N. Adams, G. Fachinetti, and J. Strähle, *Angew. Chem. Int. Ed. Engl.* **20**, 125 (1981).
123. G. Fachinetti, L. Balocchi, F. Secco, and M. Venturini, *Angew. Chem. Int. Ed. Engl.* **20**, 204 (1981).
124. H. N. Adams, G. Fachinetti, and J. Strähle, *Angew. Chem. Int. Ed. Engl.* **19**, 404 (1980).
125. S. A. Fieldhouse, B. H. Freeland, C. D. M. Mann, and R. J. O'Brien, *J. Chem. Soc. D*, p. 181 (1970).
126. G. Fachinetti, *J. Chem. Soc., Chem. Commun.*, p. 396 (1979).
127. F. Klanberg, W. B. Askew, and L. J. Guggenbenger, *Inorg. Chem.* **7**, 2265 (1968).
128. C. D. M. Mann, A. J. Cleland, S. A. Fieldhouse, B. H. Freeland, and R. J. O'Brien, *J. Organomet. Chem.* **24**, C61 (1970).
129. G. Schmid and V. Bätzel, *J. Organomet. Chem.* **46**, 149 (1972).
130. G. Schmid, V. Bätzel, G. Etzrodt, and R. Pfeil, *J. Organomet. Chem.* **86**, 257 (1975).
131. V. Bätzel, U. Müller, and R. Allman, *J. Organomet. Chem.* **102**, 109 (1975).
132. V. Bätzel, *Z. Naturforsch. B: Anorg. Chem., Org. Chem.* **31B**, 342 (1976).
133. L. Bencze, V. Galamb, A. Guttmann, G. Palyi, *Atti. Accad. Nazl. Lincei, Cl. Sci. Fis., Mat., Nat. Rend.* **337**, 437 (1980).
134. D. Seyferth, *Adv. Organomet. Chem.* **14**, 97 (1976).
135. E. O. Fischer and C. Palm, *Chem. Ber.* **91**, 1725 (1958).
136. M. A. Neuman, Trinh-Toan, and L. F. Dahl, *J. Am. Chem. Soc.* **94**, 3383 (1972).
137. D. Seyferth, J. E. Hallgren, and P. L. K. Hung, *J. Organomet. Chem.* **50**, 265 (1973).
138. V. Bätzel and G. Schmid, *Chem. Ber.* **109**, 3339 (1976).
139. G. Mignani, H. Patin, and R. Dabard, *J. Organomet. Chem.* **169**, C19 (1979).
140. A. Bou, M. A. Pericas, and F. Serratosa, *Tetrahedron* **37**, 1441 (1981).
141. J. B. Keister and T. L. Horling, *Inorg. Chem.* **19**, 2304 (1980).
142. L. M. Bavaro, P. Montangero, and J. B. Keister, *J. Am. Chem. Soc.* **105**, 4977 (1983).
143. E. M. Holt, K. Whitmire, and D. F. Shriver, *J. Chem. Soc., Chem. Commun.*, p. 778 (1980).
144. P. A. Dawson, B. F. G. Johnson, J. Lewis, and P. R. Raithby, *J. Chem. Soc., Chem. Commun.*, p. 781 (1980).
145. A. Ceriotti, P. Chini, G. Longoni, and G. Piro, *Gazz. Chim. Ital.* **112**, 353 (1982).
146. W. Wong, K. W. Chiu, G. Wilkinson, A. M. R. Galas, M. Thornton-Pett, and M. B. Hursthouse, *J. Chem. Soc., Dalton Trans.*, p. 1557 (1983).
147. W. Wong, G. Wilkinson, A. M. Galas, M. B. Hursthouse, and M. Thornton-Pett, *J. Chem. Soc., Dalton Trans.*, p. 2496 (1981).
148. T. Pakkanen and R. C. Kerber, *Inorg. Chim. Acta* **49**, 47 (1981).
149. R. C. Kerber and T. Pakkanen, *Inorg. Chim. Acta* **37**, 61 (1979).
150. S. A. Fieldhouse, A. J. Cleland, B. H. Freeland, C. D. M. Mann, and R. J. O'Brien, *J. Chem. Soc. A*, p. 2536 (1971).
151. A. J. Chalk and J. F. Harrad, *J. Am. Chem. Soc.* **89**, 1640 (1967).
152. R. Ball, M. J. Bennett, E. H. Brooks, W. A. G. Graham, J. Hoyano, and S. M. Illingworth, *J. Chem. Soc. D*, p. 592 (1970).
153. R. J. Doedens and L. F. Dahl, *J. Am. Chem. Soc.* **88**, 4847 (1966).
154. A. J. Canty, B. F. G. Johnson, J. Lewis, and J. R. Norton, *J. Chem. Soc., Chem. Commun.*, p. 1331 (1972).
155. A. J. Deeming and M. Underhill, *J. Chem. Soc., Chem. Commun.*, p. 277 (1973).

156a. R. B. Calvert and J. R. Shapley, *J. Am. Chem. Soc.* **99**, 5225 (1977).
156b. K. S. Wong and T. P. Fehlner, *J. Am. Chem. Soc.* **103**, 966 (1981).
157. G. Schmid, V. Bätzel, and B. Stutte, *J. Organomet. Chem.* **113**, 67 (1976).
158. B. Stutte and G. Schmid, *J. Organomet. Chem.* **155**, 203 (1978).
159. G. Schmid, B. Stutte, and R. Boese, *Chem. Ber.* **111**, 1239 (1978).
160. B. Stutte, V. Bätzel, R. Boese, and G. Schmid, *Chem. Ber.* **111**, 1603 (1978).
161. M. Ishii, H. Ahsbahs, E. Hellner, and G. Schmid, *Ber. Bunsenges. Phys. Chem.* **83**, 1026 (1979).
162. E. L. Muetterties and J. Stein, *Chem. Rev.* **79**, 479 (1979).
163. N. D. Shinn, M. Trenary, M. R. McClellan, and F. R. McFeely, *J. Chem. Phys.* **75**, 3142 (1981).
164. R. Colton, C. J. Commons, and B. F. Hoskins, *J. Chem. Soc., Chem. Commun.*, p. 363 (1975).
165a. K. G. Caulton and P. Adair, *J. Organomet. Chem.* **114**, C11 (1976).
165b. J. A. Marsella and K. G. Caulton, *Organometallics* **1**, 274 (1982).
166. T. E. Wolff and L. P. Klemann, *Organometallics* **1**, 1667 (1982).
167. P. T. Barger and J. E. Bercaw, *J. Organomet. Chem.* **201**, C39 (1980).
168. G. M. Dawkins, M. Green, K. A. Mead, J. Y. Salaiin, F. G. A. Stone, and P. Woodward, *J. Chem. Soc., Dalton Trans.*, p. 527 (1983).
169. G. A. Carriedo, D. Hodgson, J. A. K. Howard, K. Marsden, F. G. A. Stone, M. J. Went, and P. Woodward, *J. Chem. Soc., Chem. Commun.*, p. 1006 (1982).
170. R. D. Barr, M. Green, J. A. K. Howard, T. B. Marder, I. Moore, and F. G. A. Stone, *J. Chem. Soc., Chem. Commun.*, p. 746 (1983).
171. A. A. Pasynskii, Yu. V. Skripkin, I. L. Eremenko, V. T. Kalinnikov, G. G. Aleksandrov, V. G. Andrianov, and Yu. T. Struchkov, *J. Organomet. Chem.* **165**, 49 (1979).
172. J. Potenza, P. Giordano, D. Mastropaolo, A. Efraty, and R. B. King, *J. Chem. Soc., Chem. Commun.*, p. 1333 (1972).
173. J. Potenza, P. Giordano, D. Mastropaolo, and A. Efraty, *Inorg. Chem.* **11**, 2540 (1974).
174. M. D. Curtis and W. M. Butler, *J. Organomet. Chem.* **155**, 131 (1978).
175. R. J. Klinger, W. Butler, and M. D. Curtis, *J. Am. Chem. Soc.* **97**, 3535 (1975).
176. R. H. Hooker, K. A. Mahmoud, and A. J. Rest, *J. Organomet. Chem.* **254**, C25 (1983).
177. G. L. Geoffroy and M. S. Wrighton, "Organometallic Photochemistry." Academic Press, New York, 1979.
178. D. E. Crotty, E. R. Corey, T. J. Anderson, and M. D. Glick, *Inorg. Chem.* **16**, 920 (1977).
179. F. A. Cotton, L. Kruczynski, and B. A. Frenz, *J. Am. Chem. Soc.* **95**, 951 (1973).
180. E. D. Jemmis, A. R. Pinhas, and R. Hoffmann, *J. Am. Chem. Soc.* **102**, 2576 (1980).
181. B. J. Morris-Sherwood, C. B. Powell, and M. B. Hall, personal communication from M. B. Hall, 1983.
182. W. A. Herrmann, M. L. Ziegler, K. Weidenhammer, and H. Biersack, *Angew. Chem. Int. Ed. Engl.* **18**, 960 (1979).
183. L. N. Lewis and K. G. Caulton, *Inorg. Chem.* **19**, 3201 (1980).
184. M. Manassero, M. Sansoni, and G. Longoni, *J. Chem. Soc., Chem. Commun.*, p. 919 (1976).
185. P. Brun, G. M. Dawkins, M. Green, A. D. Miles, A. G. Orpen, F. G. A. Stone, *J. Chem. Soc., Chem. Commun.*, p. 926 (1982).
186. W. Hieber and R. Werner, *Chem. Ber.* **90**, 286 (1957).
187. K. Farmery, M. Kilner, R. Greatrex, and N. N. Greenwood, *J. Chem. Soc. A*, p. 2339 (1969).
188. C. P. Horwitz and D. F. Shriver, *Organometallics*, **3**, 756 (1984).
189. C. P. Horwitz, E. M. Holt, and D. F. Shriver, *Inorg. Chem.*, (1984), in press.

190. K. H. Whitmire and D. F. Shriver, *J. Am. Chem. Soc.* **103**, 6754 (1981).

191. K. H. Whitmire, D. F. Shriver, and E. M. Holt, *J. Chem. Soc., Chem. Commun.*, p. 780 (1980).

192. E. L. Muetterties, T. N. Rhodin, E. Band, C. F. Brucker, and W. R. Pretzer, *Chem. Rev.* **79**, 91 (1979).

193a. D. F. Shriver, *Chem. Brit.* 482 (1983).

193b. G. W. Parshall, "Homogeneous Catalysis." Wiley, New York, 1980.

194. A. Wojcicki, *Adv. Organomet. Chem.* **11**, 87 (1973).

195. F. Calderazzo, *Angew. Chem. Int. Ed. Engl.* **16**, 299 (1977).

196. K. Nowack and F. Calderazzo, *J. Organomet. Chem.* **10**, 101 (1967).

197. T. C. Flood, J. E. Jensen, and J. A. Slater, *J. Am. Chem. Soc.* **103**, 4410 (1981).

198. T. C. Flood, K. D. Campbell, H. H. Downs, and S. Nakanishi, *Organometallics* **2**, 1590 (1983).

199. H. Berke and R. Hoffmann, *J. Am. Chem. Soc.* **100**, 7224 (1978).

200. S. B. Butts, T. G. Richmond, and D. F. Shriver, *Inorg. Chem.* **20**, 278 (1981).

201. J. P. Collman, J. N. Cawse, and J. I. Brauman, *J. Am. Chem. Soc.* **94**, 5905 (1972).

202. J. P. Collman, R. G. Finke, J. N. Cawse, and J. I. Brauman, *J. Am. Chem. Soc.* **100**, 4766 (1978).

203. R. E. Stimson and D. F. Shriver, *Inorg. Chem.* **19**, 1141 (1980).

204. J. Powell, A. Kuksis, C. J. May, S. C. Nyburg, and S. J. Smith, *J. Am. Chem. Soc.* **103**, 5941 (1981).

205. S. R. Winter, G. W. Cornett, and E. A. Thompson, *J.Organomet. Chem.* **133**, 339 (1977).

206. S. C. Kao, M. Y. Darensbourg, and W. Schenk, private communication from M. Y. Darensbourg, 1983.

207. S. B. Butts, S. H. Strauss, E. M. Holt, R. E. Stimson, N. W. Alcock, and D. F. Shriver, *J. Am. Chem. Soc.* **102**, 5093 (1980).

208. T. G. Richmond, F. Basolo, and D. F. Shriver, *Inorg. Chem.* **21**, 1272 (1982).

209. H. Brunner, B. Hammer, I. Bernal, and M. Draux, *Organometallics* **2**, 1595 (1983).

210. F. Correa, R. Nakamura, R. E. Stimson, R. L. Burwell, and D. F. Shriver, *J. Am. Chem. Soc.* **102**, 5112 (1980).

211. D. F. Shriver, *ACS Symp. Ser.* **152**, 1 (1981).

212. B. F. G. Johnson, J. Lewis, D. J. Thompson, and B. Heil, *J. Chem. Soc., Dalton Trans.* p. 567 (1975).

213. B. F. G. Johnson, J. Lewis, and D. J. Thompson, *Tetrahedron Lett.*, p. 3789 (1974).

214. B. F. G. Johnson, K. D. Karlin, and J. Lewis, *J. Organomet. Chem.* **174**, C29 (1979).

215. T. G. Richmond, F. Basolo, and D. F. Shriver, *Organometallics* **1**, 1624 (1982).

216. S. G. Slater, R. Lusk, B. F. Schumann, and M. Darensbourg, *Organometallics* **1**, 1662 (1982).

217. T. G. Richmond and D. F. Shriver, *Organometallics* **2**, 1061 (1983).

218. J. A. Labinger, J. N. Bonfiglio, D. L. Grimmett, S. T. Masuo, E. Shearing, and J. S. Miller, *Organometallics* **2**, 733 (1983).

219. D. L. Grimmett, J. A. Labinger, J. N. Bonfiglio, S. T. Masuo, E. Shearing, and J. S. Miller, *Organometallics* **2**, 1325 (1983).

220. S. J. LaCroce and A. R. Cutler, *J. Am. Chem. Soc.* **104**, 2312 (1982).

221. R. H. Magnuson, R. Zulu, W.-M. T'sai, and W. P. Giering, *J. Am. Chem. Soc.* **102**, 6887 (1980).

222. J. M. Manriquez, D. M. McAlister, R. D. Saner, and J. Bercaw, *J. Am. Chem. Soc.* **100**, 2716 (1978).

223. G. Fachinetti, G. Fochi, and C. Floriani, *J. Chem. Soc., Dalton Trans.*, p. 1946 (1977).

224. P. J. Fagan, E. A. Maata, and T. J. Marks, *ACS Symp. Ser.* **152**, 53 (1981).

225. J. A. Marsella, J. C. Huffman, and K. G. Caulton, *ACS Symp. Ser.* **152**, 35 (1981).
226. P. Biloen and W. M. H. Sachtler, *Adv. Catal.* **30**, 165 (1981).
227. P. Biloen, J. N. Helle, and W. M. H. Sachtler, *J. Catal.* **58**, 95 (1979).
228. R. C. Brady, III and R. Pettit, *J. Am. Chem. Soc.* **102**, 6182 (1980).
229. R. C. Brady, III and R. Pettit, *J. Am. Chem. Soc.* **103**, 1287 (1981).
230. T. N. Rhodin and G. Ertl, "The Nature of the Surface Chemical Bond." North-Holland Publ., Amsterdam, 1979.
231. R. W. Joyner, *J. Catal.* **50**, 176 (1977).
232. F. Fischer and H. Tropsch, *Chem. Ber.*, **59**, 830 (1926).
233. G. Fachinetti, R. Lazzaroni, and S. Pucci, *Angew. Chem. Int. Ed. Engl.* **20**, 1603 (1981).
234. J. B. Keister, M. W. Payne, and M. J. Muscatella, *Organometallics* **2**, 219 (1983).
235. J. W. Kolis, E. M. Holt, M. Drezdzon, K. H. Whitmire, and D. F. Shriver, *J. Am. Chem. Soc.* **104**, 6134 (1982).
236. K. H. Whitmire and D. F. Shriver, *J. Am. Chem. Soc.* **102**, 1456 (1980).
237. M. A. Drezdzon and D. F. Shriver, *J. Mol. Catal.* **21**, 81 (1983).
238. E. M. Holt, K. H. Whitmire, and D. F. Shriver, *J. Organomet. Chem.* **213**, 125 (1981).
239. A. Wong, M. Harris, and J. D. Atwood, *J. Am. Chem. Soc.* **102**, 4529 (1980).
240. C. Masters, C. van der Woude, and J. A. van Doorn, *J. Am. Chem. Soc.* **101**, 1633 (1979).
241. L. I. Shoer and J. Schwartz, *J. Am. Chem. Soc.* **99**, 5831 (1977).
242. R. E. Stimson and D. F. Shriver, *Organometallics* **1**, 787 (1982).
243. G. C. Demitras and E. L. Muetterties, *J. Am. Chem. Soc.* **99**, 2796 (1977).
244. H. K. Wang, H. W. Choi, and E. L. Muetterties, *Inorg. Chem.* **20**, 2661 (1981).
245. H. W. Choi and E. L. Muetterties, *Inorg. Chem.* **20**, 2664 (1981).
246. J. P. Collman, J. I. Brauman, G. Tustin, and G. S. Wann, III, *J. Am. Chem. Soc.* **105**, 3913 (1983).
247. J. R. Shapley, M. E. Cree-Uchiyama, and G. M. St. George, *J. Am. Chem. Soc.* **105**, 140 (1983).
248. M. A. Drezdzon, K. H. Whitmire, A. A. Bhattacharyya, W. L. Hsu, C. C. Nagel, S. G. Shore, and D. F. Shriver, *J. Am. Chem. Soc.* **104**, 5630 (1982).
249. W. M. H. Sachtler and L. Bostelaar, *Stud. Surf. Sci. Catal.* **17**, 207 (1983).
250. H. Taube, "Electron Transfer Reactions of Complex Ions in Solution." Academic Press, New York, 1970.
251. J. K. Kochi, "Organometallic Mechanisms and Catalysis." Academic Press, New York, 1978.
252. H. Taube, H. Myers, and R. L. Rich, *J. Am. Chem. Soc.* **75**, 4118 (1953).
253. B. Grossman and A. Haim, *J. Am. Chem. Soc.* **92**, 4853 (1970).
254. W. Hieber, J. Peterhaus, and E. Winter, *Chem. Ber.* **94**, 2572 (1961).
255. W. Hieber, E. Winter, and E. Schubert, *Chem. Ber.* **95**, 3070 (1962).
256. R. Ercoli, F. Calderazzo, and A. Alberola, *J. Am. Chem. Soc.* **82**, 2966 (1960).
257. J. E. Ellis, R. A. Faltynek, G. L. Rochfort, R. E. Stevens, and G. A. Zank, *Inorg. Chem.* **19**, 1082 (1980).
258. F. Basolo and R. G. Pearson, "Mechanisms of Inorganic Reactions," 2nd Ed. Wiley, New York, 1967.
259. P. Hackett and A. R. Manning, *J. Chem. Soc., Dalton Trans.*, p. 1487 (1972).
260. B. O'Dywer and A. R. Manning, *Inorg. Chim. Acta* **38**, 103 (1980).
261. A. R. Manning, R. Kumar, S. Willis, and F. S. Stephens, *Inorg. Chim. Acta* **61**, 141 (1982).
262. A. R. Manning, *Coord. Chem. Rev.* **51**, 41 (1983).
263. D. H. Berry, J. E. Bercaw, A. J. Jircitano, and K. B. Mertes, *J. Am. Chem. Soc.* **104**, 4712 (1982).
264. J. B. Hendrickson, D. J. Cram, and G. S. Hammond, "Organic Chemistry." McGraw-Hill, New York, 1970.

265. H. H. Bauer, *J. Electroanalyt. Chem. Interfacial. Electrochem.* **16**, 149 (1968).
266. T. W. Matheson, B. H. Robinson, and W. S. Tam, *J. Chem. Soc.*, p. 1457 (1971).
267. A. M. Bond, P. A. Dawson, B. M. Peake, P. H. Rieger, B. H. Robinson, and J. Simpson, *Inorg. Chem.* **18**, 1413 (1979).
268. A. M. Bond, B. M. Peake, B. H. Robinson, J. Simpson, and D. J. Watson, *Inorg. Chem.* **16**, 410 (1977).
269. J. S. Merola, K. S. Campo, R. A. Gentile, M. A. Modrick, and S. Zentz, *Organometallics* **3**, 334 (1984).
270. L. W. Bateman, M. Green, K. A. Mead, R. M. Mills, I. D. Salter, F. G. A. Stone, and P. Woodward, *J. Chem. Soc. Dalton Trans.*, p. 2599 (1983).
271. A. F. Hepp and M. S. Wrighton, *J. Am. Chem. Soc.* **105**, 5934 (1983).
272. S. P. Church, H. Hermann, F.-W. Grevels, and K. Schaffner, *J. Chem. Soc., Chem. Commun.*, in press (1984).
273. P. T. Barger and J. E. Bercaw, *Organometallics* **3**, 278 (1984).
274. U. Seip, M.-C. Tsai, K. Christmann, J. Küppers, and G. Ertl, *Surf. Sci.* **139**, 29 (1984).
275. M. Benard, A. Dedieu, and S. Nakamura, *Nouv. J. Chim.* **8**, 149 (1984).
276. T. P. Fehlner and C. E. Housecroft, *Organometallics* **3**, 764 (1984).

NOTE ADDED IN PROOF[1]

In connection with Section II,A a new Π —CO— complex, $[Cp_2Ti(OC)Mo(CO)_2Cp_2]_2$, has been reported (*269*) and several new complexes containing the —$COCH_3$ ligand have been prepared by the interaction of $HM_3(CO)_{10}(COCH_3)$ with $Au(CH_3)(PPh_3)$ [M = Fe, Ru (*270*)].

The following material pertains to Section III,A. The photolysis of $Mn_2(CO)_{10}$ in an inert matrix or in solution produces $Mn_2(CO)_9$ which shows a low frequency CO stretch at 1760 cm^{-1}. In one recent publication this frequency is assigned to a simple C-bridging CO (*271*) but in a second report, a Π —CO is postulated (*272*) by analogy with earlier work on photochemically produced $Cp_2Mo_2(CO)_5$ (*176*). Inspection of the published data for π-CO and C-bonded CO lends credence to the existence of a Π —CO in $Mn_2(CO)_9$, but the matter cannot be considered to be settled. Two new Π —CO compounds, $CpRh(CO)_2ZrCp_2^*$ and $CpHRh(CO)_2ZrCp_2^*$ have been reported (*273*). Electron energy loss spectroscopic evidence has been obtained for Π —CO formation on an Fe(III) surface (*274*). Finally, two theoretical discussions of Π —CO compounds have appeared. LCAOMO-SCF calculations on $Mn_2(CO)_5(PH_3)_4$, as a model for $Mn_2(CO)_5(dppm)_2$ indicate that little or no electron density is transferred from O to the attached Mn center; thus the authors conclude that CO is not a four-electron donor (*275*). It should be noted however that the calculations show that the CO π cloud is polarized toward Mn. Fenske–Hall calculations on $[HFe_4(CO)_{13}]^-$ indicate that O is an electron donor to an iron atom (*276*).

[1] This section summarizes recent findings published through June, 1984.

Index

Cumulative List of Contributors

Abel, E. W., **5**, 1; **8**, 117
Aguilo, A., **5**, 321
Albano, V. G., **14**, 285
Alper, H., **19**, 183
Anderson, G. K., **20**, 39
Armitage, D. A., **5**, 1
Armor, J. N., **19**, 1
Atwell, W. H., **4**, 1
Behrens, H., **18**, 1
Bennett, M. A., **4**, 353
Birmingham, J., **2**, 365
Blinka, T. A., **23**, 193
Bogdanović, B., **17**, 105
Bradley, J. S., **22**, 1
Brinckman, F. E., **20**, 313
Brook, A. G., **7**, 95
Brown, H. C., **11**, 1
Brown, T. L., **3**, 365
Bruce, M. I., **6**, 273; **10**, 273; **11**, 447; **12**, 379; **22**, 59
Brunner, H., **18**, 151
Cais, M., **8**, 211
Calderon, N., **17**, 449
Callahan, K. P., **14**, 145
Cartledge, F. K., **4**, 1
Chalk, A. J., **6**, 119
Chatt, J., **12**, 1
Chini, P., **14**, 285
Chiusoli, G. P., **17**, 195
Churchill, M. R., **5**, 93
Coates, G. E., **9**, 195
Collman, J. P., **7**, 53
Connelly, N. G., **23**, 1
Connolly, J. W., **19**, 123
Corey, J. Y., **13**, 139
Corriu, R. J. P., **20**, 265
Courtney, A., **16**, 241
Coutts, R. S. P., **9**, 135
Coyle, T. D., **10**, 237
Craig, P. J., **11**, 331
Cullen, W. R., **4**, 145
Cundy, C. S., **11**, 253
Curtis, M. D., **19**, 213
Darensbourg, D. J., **21**, 113; **22**, 129
de Boer, E., **2**, 115
Dessy, R. E., **4**, 267
Dickson, R. S., **12**, 323

Eisch, J. J., **16**, 67
Emerson, G. F., **1**, 1
Epstein, P. S., **19**, 213
Ernst, C. R., **10**, 79
Evans, J., **16**, 319
Faller, J. W., **16**, 211
Fehlner, T. P., **21**, 57
Fessenden, J. S., **18**, 275
Fessenden, R. J., **18**, 275
Fischer, E. O., **14**, 1
Forster, D., **17**, 255
Fraser, P. J., **12**, 323
Fritz, H. P., **1**, 239
Furukawa, J., **12**, 83
Fuson, R. C., **1**, 221
Garrou, P. E., **23**, 95
Geiger, W. E., **23**, 1
Geoffroy, G. L., **18**, 207
Gilman, H., **1**, 89; **4**, 1; **7**, 1
Gladfelter, W. L., **18**, 207
Gladysz, J. A., **20**, 1
Green, M. L. H., **2**, 325
Griffith, W. P., **7**, 211
Grovenstein, Jr., E., **16**, 167
Gubin, S. P., **10**, 347
Guerin, C., **20**, 265
Gysling, H., **9**, 361
Haiduc, I., **15**, 113
Halasa, A. F., **18**, 55
Harrod, J. F., **6**, 119
Hart, W. P., **21**, 1
Hartley, F. H., **15**, 189
Hawthorne, M. F., **14**, 145
Heck, R. F., **4**, 243
Heimbach, P., **8**, 29
Helmer, B. J., **23**, 193
Henry, P. M., **13**, 363
Herrmann, W. A., **20**, 159
Hieber, W., **8**, 1
Hill, E. A., **16**, 131
Hoff, C., **19**, 123
Horwitz, C. P., **23**, 219
Housecroft, C. E., **21**, 57
Huang, Yaozeng (Huang, Y. Z.), **20**, 115
Ibers, J. A., **14**, 33
Ishikawa, M., **19**, 51
Ittel, S. D., **14**, 33

317

Cumulative List of Titles